从入门到实战·微课视频

微信小程序开发从入门到实战

微课视频版

陈云贵 高旭 编著

清華大学出版社
北京

内容简介

本书以微信小程序开发入门到实战为定位，内容共 12 章：第 1 章带领读者认识微信小程序；第 2 章整体性地讨论小程序的框架问题；第 3 章介绍常见的组件；第 4、5 章对视图层的样式布局和逻辑层 JavaScript 进行细化讲解；第 6 章介绍小程序数据库操作；第 7～9 章介绍各类常用的网络、媒体、设备、交互和开放接口；第 10 章对云开发进行介绍；第 11、12 章编写两个综合项目，可有效提升学习者的商业开发实战能力。考虑到大部分高校开设过 Java 程序设计课程，本书中高级接口均采用 Java 作为后端开发语言，案例中的后端代码附有注释，没有 Java 基础的读者可以根据注释修改成自己需要的后端程序。全书的 110 个知识点案例代码、9 个小型实训项目代码和 2 个大型实训项目代码，均在微信开发者工具和真机中调试通过，并提供全套代码和视频讲解。

本书可作为计算机相关专业学生学习微信小程序的教材，也可供对小程序开发感兴趣的开发人员、广大科技工作者和研究人员参考。

图书在版编目（CIP）数据

微信小程序开发从入门到实战：微课视频版/陈云贵，高旭编著. —北京：清华大学出版社，2020.3
(2024.12重印)
(从入门到实战·微课视频)
ISBN 978-7-302-54818-8

Ⅰ. ①微…　Ⅱ. ①陈…　②高…　Ⅲ. ①移动终端－应用程序－程序设计　Ⅳ. ①TN929.53

中国版本图书馆 CIP 数据核字(2020)第 000513 号

策划编辑：魏江江
责任编辑：王冰飞
封面设计：刘　键
责任校对：李建庄
责任印制：杨　艳

出版发行：清华大学出版社
网　　址：https://www.tup.com.cn，https://www.wqxuetang.com
地　　址：北京清华大学学研大厦 A 座　　**邮　　编**：100084
社 总 机：010-83470000　　**邮　　购**：010-62786544
投稿与读者服务：010-62776969，c-service@tup.tsinghua.edu.cn
质量反馈：010-62772015，zhiliang@tup.tsinghua.edu.cn
课件下载：https://www.tup.com.cn，010-83470236
印 装 者：艺通印刷（天津）有限公司
经　　销：全国新华书店
开　　本：185mm×260mm　　**印　　张**：30.25　　**字　　数**：735 千字
版　　次：2020 年 3 月第 1 版　　**印　　次**：2024 年 12 月第 9 次印刷
印　　数：15501～17000
定　　价：79.80 元

产品编号：083516-01

微信小程序从2017年1月9日上线以来，取得了巨大的成功。2019年1月9日微信官方公布的数据显示，小程序已覆盖超过200个细分行业，2018年小程序服务超过1000亿人次用户，年交易增长超过600%，创造了超过5000亿的商业价值。微信小程序的蓬勃发展也产生了巨大的人才需求缺口，微信小程序进大学课堂已成必然趋势。

小程序(Mini Program)是运行于“大程序”之上的无须下载、无须安装、触手可及和用完即走的轻量级应用。微信团队为小程序提供的框架命名为MINA应用框架。MINA框架通过封装微信客户端提供的文件系统、网络通信、任务管理和数据安全等基础功能，对上层提供一整套JavaScript API，让开发者能够非常方便地使用微信客户端提供的各种基础功能与能力，快速构建一个应用。

本书以微信小程序开发入门到实战为定位，内容共12章：第1章带领读者认识微信小程序；第2章整体性地讨论小程序的框架问题；第3章介绍常见的组件；第4章介绍视图层的样式布局；第5章介绍逻辑层JavaScript；第6章介绍小程序数据库操作；第7章介绍常见的网络接口；第8章介绍常见的媒体和设备接口；第9章介绍常见的交互和开放接口；第10章介绍微信小程序最新技术——云开发；第11、12章是两个综合性项目。

考虑到大部分高校开设过“Java程序设计”课程，本书中的高级接口均采用Java作为后端开发语言，案例中的后端代码附有注释，没有Java基础的读者可以根据注释修改成自己需要的后端程序。全书的110个知识点案例代码、9个小型实训项目代码和2个大型实训项目代码，均在微信开发者工具和真机中调试通过。

本书提供850分钟的视频讲解，扫描书中相应位置的二维码可以在线观看；本书还提供教学大纲、教学课件、电子教案、程序源码和教学进度表，扫描封底的课件二维码可以下载。

本书可作为高等院校计算机相关专业学生学习微信小程序的教材，也可供对小程序开发感兴趣的开发人员、广大科技工作者和研究人员参考。

本书由陈云贵和高旭编著，其中，陈云贵负责编写第1章、第6～10章和第12章，高旭负责编写第2～5章和第11章。全书由陈云贵审阅定稿。

最后，感谢清华大学出版社的魏江江分社长、王冰飞编辑以及其他工作人员为本书出版付出的辛勤劳动；感谢澳门科技大学赵庆林教授、冯丽教授和广东科技学院计算机学院田立伟院长在本书写作过程中给予的指导和帮助；同时感谢我的家人的默默付出与支持。

愿本书对读者学习微信小程序起到抛砖引玉的作用，并真诚地欢迎读者批评指正。

编　者
2020 年 1 月

目录

源码下载

第9章 交互接口和开放接口

第10章 小程序云开发

认识小程序

微信团队在 2019 年 1 月 9 日的微信公开课上发布的《2018 微信数据报告》显示：微信与 WeChat 合并后活跃账户数高达 10.82 亿；消息日发送次数 450 亿，较 2017 年增长 18%；音/视频通话次数达 4.1 亿次，较 2017 年增长 100%；在出行、零售、餐饮、公共服务领域，2018 年每月支付人数都较去年有了较大幅度的增长；微信小程序开发者增加了 80%；从社交到商业，微信已经深入到了大众生活的方方面面。

本章主要目标

- 了解微信小程序产生的背景和应用前景；
- 了解小程序的定义、特征和“大程序”的含义；
- 熟练掌握微信小程序开发者工具和开发者管理账户的操作；
- 开发第一个微信小程序。

1.1 微信小程序介绍

微信小程序的推出并非一蹴而就，早在 2016 年 1 月的微信公开课上，微信之父张小龙就透露微信即将推出应用号，而彼时的应用号就是现在的微信小程序。

1.1.1 微信小程序产生的背景

在小程序发布之前，微信公众平台已经发布了服务号、订阅号和企业号。

- 服务号：连接人和商品，很多电商企业，以及在微信端提供产品和服务的企业都使用服务号。
- 订阅号：微信官方的定位是阅读，连接人和资讯。订阅号以媒体、政府等机构使用居多。
- 企业微信(原企业号)：企业微信其实是内部 OA 的集成，把 OA 搬到了微信端。

企业微信可以理解为具有办公管理功能的普通个人微信，企业微信与个人微信的关系就相当于企业QQ与个人QQ的关系。而订阅号与服务号的区别在于，订阅号侧重阅读，服务号侧重服务，再通俗点可以简单理解为订阅号相当于文章类网站，而服务号则定位为类App和类功能性Web系统，也可以理解为服务号想替代App，减少微信用户安装和使用其他App的概率，让流量在微信大生态中形成闭环。例如笔者自己的微信上就关注了“58同城”“家乐福中国”“唯品会”等服务号，而这些服务号各自对应了一个App，如果用户不想安装“唯品会”的App，使用“唯品会”服务号也可以实现用户想要的功能服务。

但从很多商家和企业的反馈来看，服务号依然没有达到微信团队预期的效果。因为服务号是从订阅号中拆分出来的，服务号保留了订阅号太多的基因，对于大部分的商家来说，二者只是出现的位置不一样，发送消息的次数限制不一样。在这样的背景下，2017年1月9日微信团队发布了微信小程序，可以说小程序是微信在服务号的基础上提高企业服务能力的一次尝试。

1.1.2 什么是微信小程序

根据腾讯官方微信小程序接入指南的定义，微信小程序是一种全新的连接用户与服务的方式，它可以在微信内被便捷地获取和传播，同时具有出色的使用体验。张小龙在微信朋友圈对小程序给出的定义（如图1.1所示）是：小程序是一种不需要下载安装即可使用的应用，它实现了应用“触手可及”的梦想，用户扫一扫或者搜一下即可打开应用。也体现了“用完即走”的理念，用户不用关心是否安装太多应用的问题。应用将无处不在，随时可用，但又无须安装卸载。

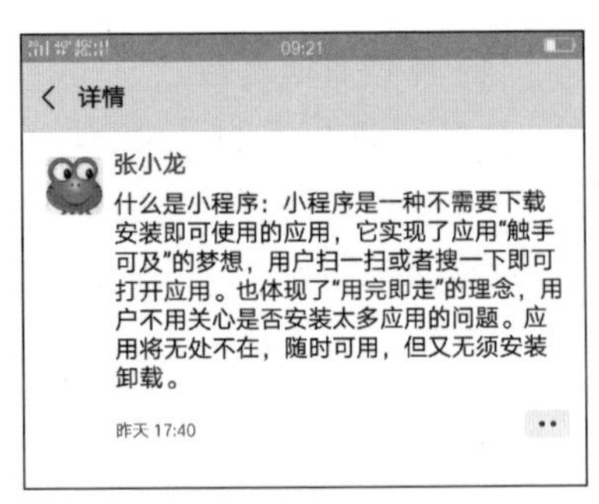

图1.1 小程序的定义

1.1.3 小程序之“大程序”

生物学中有寄生生物和宿主的概念，即两种生物在一起生活，一方受益，另一方受害，后者给前者提供营养物质和居住场所，这种生物的关系被称为寄生。小程序之所以称为小程序，也有类似的含义。微信小程序是运行在微信之上的，如支付宝小程序是运行在支付宝之上的，百度智能小程序是运行在百度App之上的。微信小程序、支付宝小程序和百度智能小程序不能离开微信、支付宝和百度App，后者为前者提供运行环境和底层接口，它们之间犹如寄生生物和宿主的关系一般。

同时，微信小程序、支付宝小程序和百度智能小程序与微信、支付宝和百度App的关系又不是简单的前者依赖后者，前者还有效地扩展了后者的功能。微信本来是不能打车的，“滴滴出行”小程序接入微信之后，用户可以不离开微信而实现打车功能；支付宝本来是没有查询快递的功能的，“菜鸟裹裹”小程序接入支付宝之后，用户在支付宝上就可以查询快递信息。可以说微信小程序、支付宝小程序和百度智能小程序让微信、支付宝和百度App变得更强大。

笔者从这层依赖和扩展关系出发，把微信、支付宝和百度App称为微信小程序、支付宝小程序和百度智能小程序的“大程序”。

1.5.2 节将会介绍百度智能小程序，百度发起了一个称之为“智能小程序开源联盟”的组织，该组织除了百度系产品外，还有“爱奇艺”“WiFi 万能钥匙”“快手”“猎豹移动”“携程”“bilibili”“58 同城”“汽车之家”“万年历”“宝宝巴士”等。该组织旨在共同制定标准、共建生态和共享流量，从“大程序”概念来看，“智能小程序开源联盟”成立的目的就是要让智能小程序的“大程序”更大。

有了“大程序”的概念之后，笔者把小程序重新定义为：小程序是运行在“大程序”之上的应用，它无须下载、安装和卸载，同时丰富和扩展了“大程序”的功能。

1.1.4　微信小程序应用前景

根据阿拉丁《微信小程序 2019 年 1 月份 TOP100 榜单》发布的数据显示，跻身 TOP100 活跃度的小程序指数从高到低依次分布在游戏、网络购物、生活服务、视频、内容资讯、工具、旅游、餐饮、图片摄影、大健康、线下零售、教育、音频、商业服务和社交等领域。人们经常在微信群看到各式各样的小程序转发信息，可以说小程序的应用领域是很广的。那什么场景合适开发小程序应用呢？笔者总结如下。

1. 快进快出的使用场景

无须安装，即扫即用，小程序非常适合一些快进快出的场景。所谓快进快出，就是说平时用不到，临时用一下，用完立即退出，以后可能永远不会再用。譬如餐厅点餐，商家在餐桌上放置一个小程序的二维码，用户使用微信扫描功能扫描一下二维码，即进入了点餐小程序。用户可以在小程序上实现浏览菜品、选择菜品、调整已选菜品、下单、结账付款和点评等功能。因为用户只是偶尔到某一餐厅吃顿饭而已，没有必要关注他们的公众号、服务号，更不愿意下载他们的 App，用小程序点完餐、结完账，与这家餐厅就没有关系了，这就是快进快出。试想一下如果一个餐厅强行地让客户下载他们的 App，哪怕是客户因为想拿奖券而下载了，客户又会保留这个 App 在自己的手机上多久呢？恐怕出门后就卸载了。

同类的场景还有很多，譬如医院挂号、各种会员卡、火车票订票抢票和小游戏等。很多失败的创业案例都已经证明过，让用户为低频应用下载一个 App 或关注公众号，都是非常难的事，营销成本很高。其实很多时候商家并不是很需要用户成为自己的“粉丝”，至少吸收“粉丝”不是商家的核心需求，商家的核心需求是让用户方便快捷地完成交易过程。小程序足够满足核心需求，双方都省心、省时、省力，何乐而不为呢？

在线教育平台“三节课”创始人认为大部分的服务和几乎所有的初创业务都是可以接入小程序的。他按照刚需/非刚需，高频/低频，将互联网产品分别放入四个象限，如图 1.2 所示。

如果服务是高频的，而且对于交互和界面体验的要求很高，还是要用原生 App(Native App)来做。但如果服务是低频/中频且重要的服务，建议商家加入微信小程序。图 1.3 显示的是相关象限中应用的优选方案。

象限 1：行业巨头、高频应用不应该选择小程序。这个象限基本上都是行业巨头，例如“360”“百度”“阿里巴巴”“高德地图”“滴滴出行”“支付宝”“招商银行”“新东方”等，这些应用因为用户经常需要打开，交互频次很高，用户对应用的体验要求也很高，可以说象限 1 的服务只合适应用 App，从某种程度上讲小程序在该象限是不合适的。

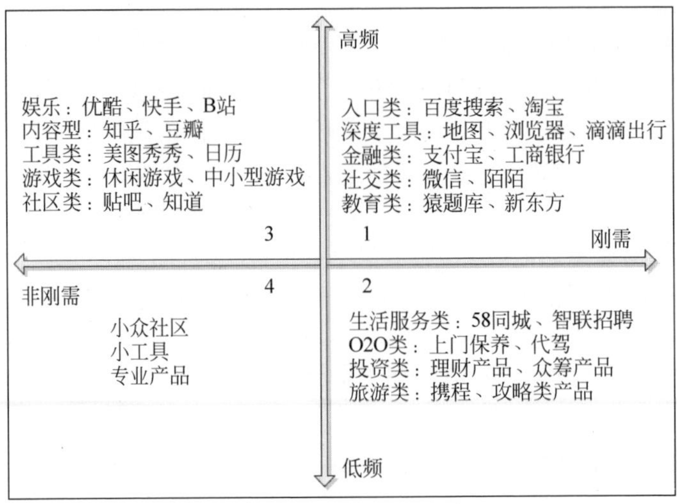

图 1.2　互联网产品象限分类

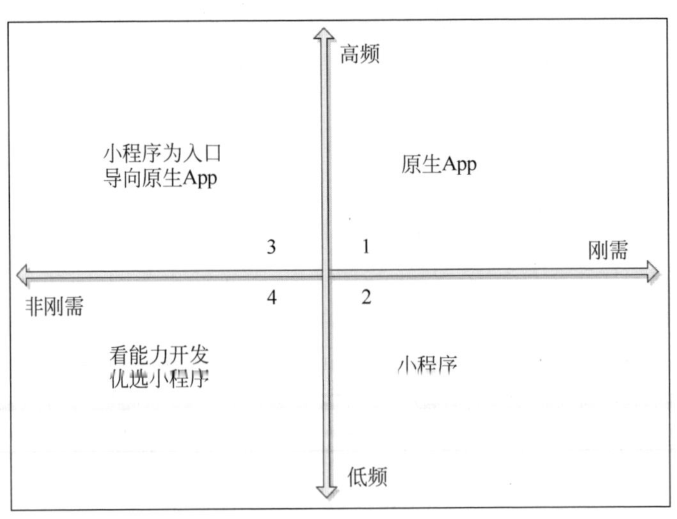

图 1.3　小程序在服务领域的应用匹配

象限 2：应该毫不犹豫地拥抱小程序。这个象限包含大量的服务类产品——教育、医疗、家政、求职招聘、二手买卖、旅游、票务和金融理财等，当然理论上还有“12306”。总之，但凡用户一年用一两次之后就再也想不起来的，是不应该用一个原生 App 应用的方式让用户下载的，而应该使用微信小程序来解决。初创型企业也应该通过小程序来试探 MVP(开发团队通过提供最小化可行产品获取用户反馈，并在这个最小化可行产品上持续快速迭代，直到产品达到一个相对稳定的状态)，因为微信拥有天然的传播能力和客户拓展能力，而原生 App 应用除了开发比较复杂外，推广成本极高，获客成本极高，这些都阻碍了 MVP 的产品试探。从这个角度来说，小程序能让初创的互联网公司减少试错成本，提升成功概率。

象限 3：利用微信的开放性，吸引用户到自有产品中。MVP 后，尽快引导到自有产品，因为自有产品能提供更好的服务，并且能留住用户，典型代表如“知乎”和“网易云音乐”。内容型的产品，通过微信获得新用户，然后转移到自有平台，也是一个很好的策略。

象限4：视情况接入，主要视开发能力而定。该象限很多都是个人兴趣产品、工具产品，可以从MVP的角度来做，或者以兴趣的角度来做，不考虑太多的商业产出，只考虑情怀，但这些开发者存在明显的问题，就是产品设计能力和开发能力有限。所以，如果公司拥有App开发能力，那就保留App；如果公司App方面的开发能力有限，则可以优选微信小程序。

2. 传统触屏机器的屏幕转移

传统触屏机器，包括自动柜员机、自动售货机，以及电影票、火车票等自动出票机等。这类机器的主要特征就是通过屏幕来显示信息，通过触屏或按键来进行交互。缺点主要是触屏不灵敏、输入信息很麻烦，如果涉及支付的话，那就更麻烦了。如果把这些屏幕去掉，而改用微信扫描机器上的二维码，在小程序中显示机器的交互界面，不但可以克服这些缺点，还可以提供更好的用户体验。这不仅是去掉了屏幕，节省了成本，更重要的是设备的形态，甚至商业模式都可能会发生改变。

以自动售货机为例，去掉机器上的屏幕，换为二维码，手机扫描后即可进入一个类似于电商App的小程序界面。传统自动售货机因为要通过透明玻璃门来展示货品，所以多是销售标准化的、常见的货品，如饮料等。使用小程序后，展示的信息更丰富，就可以售卖非标准化的、非常见的货品。顾客也不再需要通过货品的外观来选择货品，只需要在手机上浏览。这样，货品的摆放方式就可以更加有效利用空间。类似于电商App的界面设计，交互体验也会丰富得多，不仅可以查看货品的详情，还可以进行货品的组合促销，在小程序端直接完成支付，甚至还可以有积分等。这样的自动售货机可以有效降低硬件成本，提升购物体验。随着小程序的诞生和推广，很多领域已经不可以用旧的思维方式做经验管理，小程序甚至可能催生出新的商业模式。

3. 微信自身的深度结合

小程序可以分享给好友和微信群，这就相当于在不离开微信这个大生态的环境下，实现应用和微信的信息互通。这对于虚拟社区类产品、企业内部服务类产品来说，可能会是个福音。

例如对企业内部工作流程进行管理的小程序，就可以随时把工作流程通知给相关的参与者，参与者也能迅速进入工作流程中进行处理。同样还有社区类产品，一旦接入微信的通信能力，就能极大弥补社区类产品在即时通信方面的不足。即便仅把微信当成一个信息推送工具，也会非常有价值。传统App开发商都会遇到推送消息失败，或者被屏蔽的问题，当统一由微信来接管时，这种问题自然而然就消失了，没人会怀疑微信的消息处理能力。

小程序的应用前景是很光明的，腾讯和高校教育机构正在积极推广微信小程序进大学课堂，相信越来越多的IT人将从事小程序开发工作，越来越多的小程序会进入我们的生活中。

1.2 小程序特征

从张小龙对微信小程序的定义来看，小程序的特征是：无须安装、触手可及、用完即走、无须卸载，除此之外它还具有唯一性、新零售、入口丰富和传播能力强的特点。

(1) 唯一性——小程序名称具有唯一性,这一点和域名类似,某一名字被注册之后别人就不能注册同名小程序了,这也滋生了一些抢注行为,现在很多行业和地域性词汇已经被人抢注了。

(2) 新零售——小程序是新零售的最好落地工具,相信大家可以经常在小区附近的门店看到微信和支付宝小程序的二维码。

(3) 入口丰富——小程序目前入口有近 40 种,其中使用较为普遍的是手机桌面(暂只支持 Android)、微信搜索、附近的小程序、线下扫码、微信菜单"发现"、小程序识别码、好友分享、公众号关联和菜单直达等。

(4) 传播能力强——因为有近 40 种流量入口,加上小程序背靠微信这一大生态,使得微信小程序的传播能力极强,这也是初创企业选择小程序的一个重要原因。

小程序是微信公众平台上和公众号平行的产品,同时微信小程序从功能上又是 App 和 H5 的竞争对手。下面分别对公众号与小程序、小程序与 App、小程序与 H5 的特征区别做出总结,如表 1.1～表 1.3 所示。

表 1.1　公众号与小程序的特征区别

对　比　项	公　众　号	小　程　序
定位	服务于营销与信息传递	面向产品与服务
实现技术	基于 H5	基于微信自身开发环境与开发语言
用户体验	操作延时较大	体验接近原生 App
接口数量	较好	丰富的接口,详细请参考后续章节
入口	较多	入口丰富,新增一些入口,例如附件的小程序

表 1.1 显示了公众号与小程序的特征区别。公众号的定位是用于信息的传递,实现人与信息的连接,它借助 H5 技术实现,而小程序的定位则是面向产品和服务的类 App 应用,开发者需始终记住微信小程序是腾讯为了替代 App 形成微信流量闭环而打造的产品。公众号有粉丝关注的概念,而类 App 的小程序没有这个概念。因为有粉丝的概念,公众号可以发送消息,而小程序只能发送特殊的模板消息(见 9.5 节)。因为底层技术的不同,公众号页面的打开速度往往没有小程序快,而小程序可以达到接近 App 的速度体验。

表 1.2　小程序与 App 的特征区别

对　比　项	小　程　序	App
下载安装	无须下载、无须安装	需要下载安装
开发版本	开发一个版本	安卓和苹果两个版本
开发成本	较低	较高
推广成本	拥抱微信流量,较低	原生流量,较高
用户体验度	一般	较好

表 1.2 显示的是小程序与 App 之间的特征区别。很多人把小程序和 App 看成对手关系,也有人把它们看成补充关系。App 面向所有的智能手机用户,微信小程序面向微信用户。App 需要开发安卓和苹果两个版本,而小程序则不需要,另外小程序的开发效率也明显要比 App 高,当然小程序还做不到 App 那样良好的用户体验。

表 1.3 小程序与 H5 应用的特征区别

对 比 项	小 程 序	H5 应用
运行环境	运行在微信上	运行在浏览器上
用户体验	较流畅	实际上是打开一个网页,流畅度差点
开发成本	很多功能被封装成接口,开发成本低	比较依赖个人开始经验
系统权限	受益于微信,较多	较少
PC 支持问题	不支持 PC 打开	支持

表 1.3 显示了小程序与 H5 之间的特征区别。很多人把小程序和 H5 作对比,其实是不适合的,因为它们从概念上就不在一个层级,小程序是一种应用,而 H5 是超文本标记语言 HTML 的第五次修订版本,也就是说 H5 是一种技术,拿一种应用和一种技术作对比肯定是不合适的,应该对比的是小程序和 H5 应用。

小程序的运行环境是微信开发团队基于浏览器内核完全重构的一个内置解析器,有针对性地做了优化,配合自定义的开发语言标准,提升了小程序的性能,从流畅度上看小程序有对 H5 应用的优势。另外,因为微信团队把许多功能封装成了小程序接口,这使得小程序开发效率提高了很多,自然也就降低了开发成本。因为不同的开发者使用不同的工具,不同的开发者的开发经验也不一样,导致 H5 开发的效率非常依赖个人情况。当然小程序只能寄生在微信这个环境也是它的弱点,而 H5 应用可以计算机和手机自适应则是它的优点。

相信通过此节中公众号与小程序、小程序与 App、小程序与 H5 应用的对比,读者对公众号开发、小程序开发、App 开发和 H5 开发有了较深刻的理解,选择哪种应用开发需要具体问题具体分析。

1.3 微信小程序开发准备

学习者以前学习 C 语言或者 Java 语言之类编程语言的时候,下载开发工具并做一些环境配置就可以进行开发了。但是微信小程序的开发在下载开发者工具之前,开发者需要先在微信公众平台上申请好一个小程序管理账号,因为需要使用绑定了小程序管理账号的微信账号扫描二维码才可以进入开发者工具。下面介绍小程序管理账号的申请和配置过程。

1.3.1 申请小程序账号

开发者应该在微信公众平台网址 https://mp.weixin.qq.com/cgi-bin/wx 注册小程序管理账号,目前微信小程序注册面向个人、企业、政府、媒体和其他组织开放,而支付宝小程序目前个人开发者还处于公测阶段,支付宝小程序只正式对企业开发者开放。下面对注册流程做简单介绍。每一个学习者都需要按照如下的流程注册一个自己的小程序管理账号,因为如果没有注册这样一个账号,开发者就不可以登录到开发者工具上,也就谈不上开发、发布和运营小程序了。

图 1.4 显示了微信小程序的接入流程，依次是注册小程序账号、小程序信息完善、开发小程序及提交审核和发布。

接入流程

1 注册
在微信公众平台注册小程序，完成注册后可以同步进行信息完善和开发。

2 小程序信息完善
填写小程序基本信息，包括名称、头像、介绍及服务范围等。

3 开发小程序
完成小程序开发者绑定、开发信息配置后，开发者可下载开发者工具、参考开发文档进行小程序的开发和调试。

4 提交审核和发布
完成小程序开发后，提交代码至微信团队审核，审核通过后即可发布（公测期间不能发布）。

前往注册

图 1.4　微信小程序接入流程

下面对申请小程序管理账号进行讲解。

在填写注册信息的时候需要注意的是，填写的这个邮箱必须没有绑定过个人微信，也没有注册过微信公众平台下的订阅号或者服务号。假如一些自媒体朋友的邮箱已经注册过订阅号或者服务号，则需要更换新邮箱来注册小程序，如图 1.5 所示。

每个邮箱仅能申请一个小程序

邮箱　770059
作为登录帐号，请填写未被微信公众平台注册，未被微信开放平台注册，未被个人微信号绑定的邮箱

密码　•••••••••••••
字母、数字或者英文符号，最短8位，区分大小写

确认密码　•••••••••••••
请再次输入密码

验证码　mnlb　换一张

你已阅读并同意《微信公众平台服务协议》及《微信小程序平台服务条款》

注册

图 1.5　填写注册信息

小程序系统会发送一份确认邮件到注册时填写的邮箱中，如图 1.6 所示。单击收到的激活邮件的“确认”链接即可进入图 1.7 的界面。

图 1.6　邮箱确认注册

主体信息登记

身份证姓名　陈云贵
信息审核成功后身份证姓名不可修改；如果名字包含分隔号“·”，请勿省略。

身份证号码　360722
请输入您的身份证号码。一个身份证号码只能注册5个小程序。

管理员手机号码　188188　16秒后可重发
请输入您的手机号码，一个手机号码只能注册5个小程序。

短信验证码　787879　无法接收验证码？
请输入手机短信收到的6位验证码

管理员身份验证
请用管理员本人微信扫描二维码。本验证方式不扣除任何费用。注册后，扫码的微信号将成为该帐号的管理员微信号。

继续

图 1.7　“主体信息登记”界面

图 1.7 是比较关键的一步，这一步需要开发者填写真实的姓名、身份证号码和手机号码，并且需要用个人微信号扫描图中的二维码来绑定该小程序管理账户，这一点和公众号的注册类似，但是需要注意的是填写的姓名、身份证号码需要和用来绑定的这个微信保持身份一致。例如张三注册小程序管理账户，则需要用张三的微信来绑定，而且需要是张三已经实名认证过的微信来绑定，用李四的微信或者用一个没有实名认证的微信来扫描图中的二维

码都将出错。正确完成主体信息登记后，单击界面右下角的“继续”按钮，进入“确认信息”界面，如图 1.8 所示。

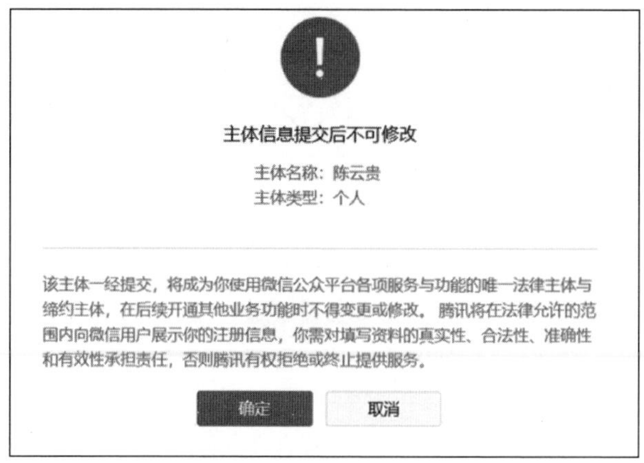

图 1.8 “确认信息”界面

单击图 1.8 中的“确定”按钮完成小程序管理账号的注册过程。

1.3.2 完善信息

完成上面的注册流程后，开发者可以使用注册好的账号进入小程序后台管理界面，需要对小程序进行信息完善，还需要设置该小程序的开发者信息和管理员，如图 1.9 所示。

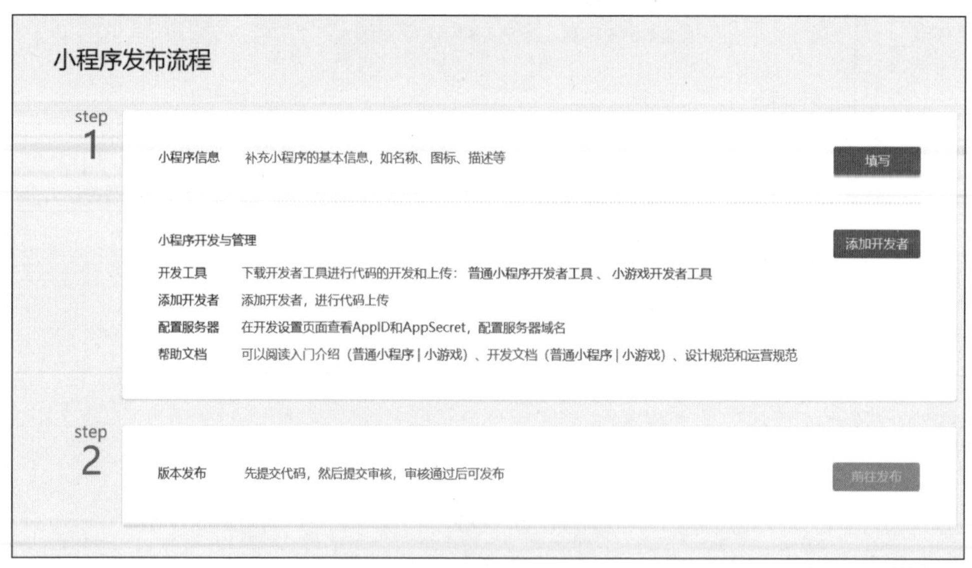

图 1.9 小程序信息完善

单击图 1.9 中的“填写”按钮，即可对小程序的信息进行完善。完善的内容有小程序名称、小程序简称、小程序头像、介绍、服务类目，需要说明的是小程序名称一年只可以修改两次，而小程序头像和介绍一个月可以修改 5 次。

单击图 1.9 中的“添加开发者”按钮，可以添加和修改小程序的管理员、开发者、运营者、数据分析者和体验成员，效果如图 1.10 所示，小程序后台成员权限如表 1.4 所示。

图 1.10　“成员管理”界面

表 1.4　小程序后台成员权限

权　　限	开　发　者	体　验　者	管　理　员
开发调试	√	√	√
模拟器功能	√	√	√
使用体验版	√	√	√
真机预览	√	×	√
上传上架	×	×	√
后台管理功能	×	×	√

从表 1.4 可以看出，管理员的权限大于开发者，而开发者的权限又大于体验者。这里需要说明的是体验版是指小程序开发完成后，开发者将小程序上传到小程序管理后台会形成开发版本，而这个开发版本在提交审核之前是不可以被外界访问的。为了让非开发身份的体验者可以先一步体验一下小程序的整体效果，管理员可以先发布一个体验版，这时候体验者可以通过扫描二维码等方式在微信上使用体验版，而外界的其他用户是找不到该小程序的。当管理员把该版本提交审核，而又被小程序官方审核通过，才是正式的发布成功，这时普通用户才可以访问该小程序。具体的体验版在管理后台的操作界面如图 1.11 所示。可

以简单地理解为体验版是一个内测版本，只有内部的体验者可以访问。

图 1.11　管理后台的体验版本

1.3.3　后台介绍

小程序后台提供了丰富的管理功能，具体的有版本管理、成员管理、反馈管理、统计、附近的小程序、物流助手、客服、模板消息、开发、推广和设置。后台操作界面如图 1.12 所示。

- 版本管理：展示线上版本、审核版本和开发版本（体验版本在这一范畴）。
- 成员管理：对管理员、开发者、运营者和体验者进行管理。
- 反馈管理：对反馈信息进行管理。
- 统计：小程序流量统计及分析功能，包括历史和实时的访问数据、访问分析、来源分析、人群画像分析和自定义分析等功能。
- 附近的小程序：开通和关闭附近的小程序功能。
- 物流助手：帮助有物流需求的开发者，快速高效地对接多家物流公司，对接后用户可通过微信服务查看实时物流状态，提升用户体验。需先通过微信认证才可以开通此项服务。
- 客服：可添加 100 个微信账号作为在线客服。
- 模板消息：模板库管理。

图 1.12 管理后台界面

• 开发：运维中心、开发设置、开发者工具和接口设置，首先单击“开发”菜单，然后单击“开发设置”按钮可以进入图 1.13 所示界面，图 1.13 中的 AppID 和 AppSecret 两个参数是后续需要经常用到的，新建项目的时候需要用到 AppID，而小程序支付功能需要用到 AppSecret。

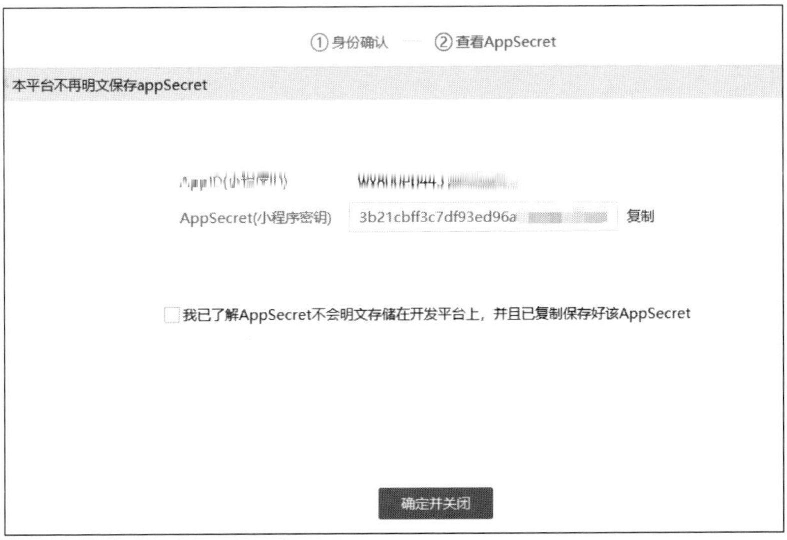

①身份确认 ②查看AppSecret
本平台不再明文保存appSecret
AppSecret(小程序密钥) 3b21cbff3c7df93ed96a 复制
我已了解AppSecret不会明文存储在开发平台上，并且已复制保存好该AppSecret
确定并关闭

图 1.13 AppID 和 AppSecret

- 推广：流量主和广告主功能，这一功能在公众号中也有。
- 设置：基本信息设置、第三方设置、关联设置、关注公众号和违规记录，其中关联设置指的是把小程序和公众号绑定起来。

1.4 第一个微信小程序

有了小程序管理账号之后，开发者就可以下载微信小程序开发者工具来开发第一个小程序了。本节将在开发者工具中新建一个 HelloWorld 项目。

1.4.1 开发者工具的安装

腾讯官方给出的开发者工具下载地址是 https://developers.weixin.qq.com/miniprogram/dev/devtools/download.html?t=19030621，开发者也可以百度搜索微信小程序开发者工具，然后在下载网站下载，根据自己计算机的情况选择合适的版本安装即可。因为篇幅所限，这里就不对安装过程做说明了。

1.4.2 第一个小程序

视频讲解

例 1-1 第一个微信小程序 HelloWorld 项目创建。

如图 1.14 所示，进入小程序开发者工具的时候需要扫码，而这个时候应该使用绑定了小程序管理账号的管理员或者开发者的微信账号来扫描。因为需要扫码进入，也就意味着计算机没有连接网络的话是进不去开发者工具的。

图 1.14 扫码进入开发者工具

进入开发者工具可以开发小程序、小游戏和公众号网页，如图 1.15 所示，选择“小程序”，然后单击界面中的“＋”按钮来新建小程序项目。

图 1.15　开发者工具新建项目之前界面

如图 1.16 所示，新建项目的时候合理的做法是先在“F:\miniprogram\ch1\”目录下新建 HelloWord 文件夹，然后单击图中虚线框中的向下箭头，在弹出的窗口中选择“F:\miniprogram\ch1\HelloWorld”文件夹，这时项目名称 HelloWorld 会被自动添加上。AppID 是小程序管理账号中取得的，如果项目不需要上传和发布也不需要使用云开发功能，可以单击图中的“测试号”，系统会自动生成一个临时的 AppID。如果项目不使用云开发功能，则选择“不使用云服务”，“小程序云开发”将在第 10 章中讲述。单击“新建”按钮即可成功创建项目 HelloWorld。

开发者工具会默认自动生成一个 HelloWorld 的 demo，所以在没有写代码的情况下项目已经编辑好了，程序效果如图 1.17 中左侧的模拟器所示。关于开发者工具的使用将在第 2 章中为读者详细介绍。

1.4.3　项目发布和提交审核

1.4.2 节创建了一个 HelloWorld 项目，本节介绍如何把自己开发的项目上传并发布以便供用户访问。如图 1.18 所示，单击开发者工具左上角的“上传”菜单，即可把项目上传到小程序云端，需要说明的是开发者不需要购买网络空间，腾讯免费为开发者提供了云端空间。

然后开发者需要登录到小程序管理账号中，上传的项目已经可以在小程序管理账号中

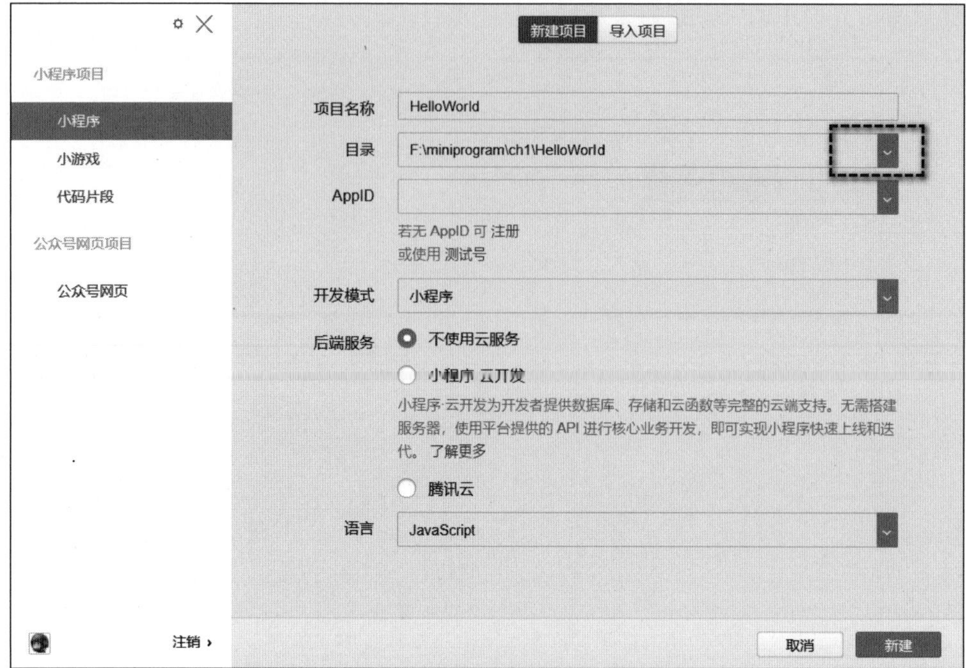

图 1.16　新建小程序项目

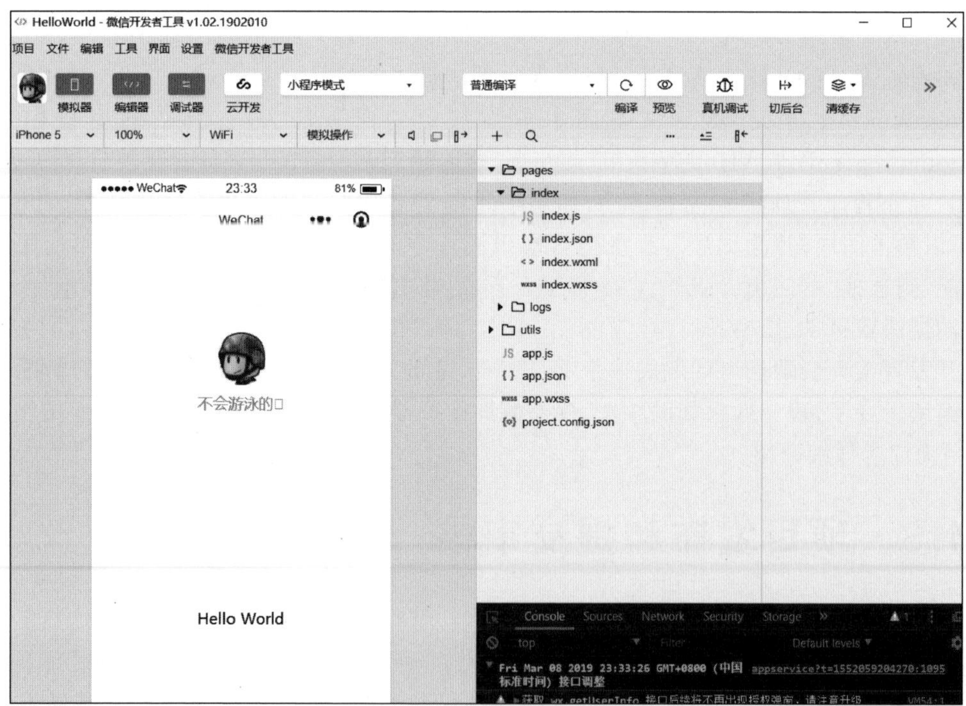

图 1.17　第一个小程序 HelloWorld

图 1.18　项目上传

查看，如图 1.19 所示，单击开发版本中的“提交审核”按钮，即可进入如图 1.20 所示界面。

图 1.19　小程序管理账号中的版本信息

在图 1.20 中，选中“已阅读并了解平台审核规则”，然后单击“下一步”按钮，即可进入如图 1.21 所示界面。开发者不得提交不合规的项目，否则无法通过微信团队的审核。

在提交审核的最后一步，开发者需要配置功能页面信息，如图 1.21 所示。开发者需要至少配置一个页面的信息，按照提示信息依次配置“功能页面”“标题”“所在服务类目”“标签”信息，然后单击“提交审核”按钮，完成提交审核操作。

在提交审核操作之后，审核版本中会出现开发者最新提交的审核版本，待微信团队审核通过，项目就会转化成线上版本，用户即可正常访问项目了。审核时间一般为两个工作日。

提交审核的相关须知

确认提交审核?

提交给微信团队审核前，请确保:

- 提交的小程序功能完整，可正常打开和运行，而不是测试版或 Demo
  小程序的调试和预览可在开发者工具进行。多次提交测试内容或 Demo，将受到相应处罚。
- 已仔细阅读《微信小程序平台运营规范》和《微信小程序平台审核常见被拒绝情形》

已阅读并了解平台审核规则

下一步　取消

图 1.20　确认提交审核

图 1.21　配置功能页面信息

1.5 支付宝和百度小程序

微信小程序是2017年1月9日正式发布的，腾讯公司是BAT中最早布局小程序的；支付宝小程序于2017年8月18日开始公测（对企业开发者开放），从支付宝小程序官方开发者社区来看，第一篇帖子发布在2017年8月17日；百度智能小程序则是2018年4月12日开始公测的，它是BAT小程序最迟上线的，但是它强调了“智能”二字。所以，有初学者问小程序有没有发展前景，能不能火起来，笔者的回答是观察BAT争相布局小程序就大概知道它的前景了。

1.5.1 支付宝小程序

支付宝小程序官方网址 https://open.alipay.com/channel/miniIndex.htm，蚂蚁金服开放平台官方给出的定义是：支付宝小程序是一种全新的开放模式，它运行在支付宝客户端，可以被便捷地获取和传播，为终端用户提供更优的用户体验。支付宝小程序开放给开发者更多的JSAPI（原生API）和OpenAPI（开发能力API）能力，通过小程序可以为用户提供多样化便捷服务。

需要说明的是支付宝小程序目前只正式对企业开发者开放，而个人开发者还处于公测阶段。考虑到大部分读者没有企业支付宝，在蚂蚁金服开放平台还没有开放个人开发者正式注册之前，大家可以申请公测注册。考虑到注册流程比较占篇幅，而且支付宝小程序和微信小程序注册的流程类似，所以这里就不对注册流程做介绍了。支付宝小程序管理后台如图1.22所示。

图1.22 支付宝小程序开发管理

支付宝小程序登录的时候也是通过扫描二维码进入的,进入管理后台之后笔者发现支付宝小程序管理后台和微信小程序管理后台雷同,具体功能就不做介绍了。

有了小程序账号之后,开发者就可以到官方网址 https://docs.alipay.com/mini/ide/download 下载开发工具了,选择适合自己计算机的版本下载安装即可。

支付宝开发者工具对比微信开发者工具最大的区别是提供了大量的模板,使得开发者可以快速建立小程序,这也许是淘宝应用市场思想的继承,如图 1.23 所示。支付宝小程序开发工具中的模拟器在窗口的右侧,如图 1.24 所示,而微信小程序开发工具的模拟器在窗口的左侧。

图 1.23　支付宝小程序新建项目

如图 1.25 所示,从项目结构上看,支付宝小程序沿用了微信小程序的项目目录结构,同样由全局文件和普通页面文件组成,但在文件扩展名上有所改变。微信小程序中的 .wxss 扩展名对应支付宝小程序的 .acss,微信小程序中 .wxml 扩展名对应支付宝小程序的 .axml,此外,.js 和 .json 扩展名没有变化。由此可见,开发者学习了微信小程序开发之后可以很快地迁移到支付宝小程序的开发上来。

1.5.2　百度智能小程序

"春晚红包抢的好,瓜分九亿不嫌少",百度作为央视 2019 年春节联欢晚会独家网络互动平台参与了 2019 年的春晚红包互动。和前几年的微信和支付宝在春节联欢晚会的抢红

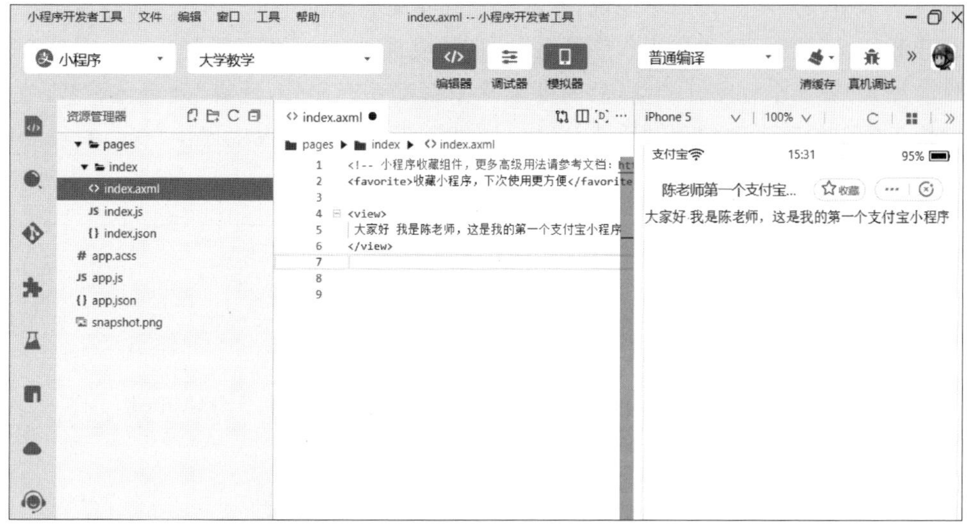

图 1.24 支付宝小程序项目窗口

图 1.25 支付宝小程序项目目录结构

包活动一样，百度此次的抢红包活动吸引了无数的眼球，而此次抢红包活动的承担者就是百度的智能小程序“春晚摇红包”，在元宵期间“春晚摇红包”智能小程序又变身成了“元宵摇红包”再次撒钱 2 亿元。

百度智能小程序目前只对企业和组织开放注册。相比微信和支付宝小程序，百度小程序的特点是“更自然、更开放、更智能”。所谓开放，是指不像微信小程序只能在微信上运行，支付宝小程序只能在支付宝上运行，百度小程序除了可以在百度 App 上打开之外，将来还可以在其他一些合作的 App 甚至浏览器上打开。说到这个，就要说到百度发起的“智能小程序开源联盟”。

“智能小程序开源联盟”的核心思想是生态共建、流量共享和商业共赢。生态共建指的是生态伙伴共建小程序技术标准和生态。流量共享是指生态伙伴之间共享海量分发资源，多种分发模式共享早期流量红利。商业共赢是指有效提升用户的使用时长及频次，创新商业变现模式带动收入增长。简单地说，微信和支付宝是自家的小程序平台，而百度的小程序平台是开源的，开放给了“智能小程序开源联盟”的成员，这些成员开发的小程序是互通互联的。目前加入“智能小程序开源联盟”的商家除了百度系产品外，还有“爱奇艺”“WiFi 万能钥匙”“快手”“猎豹移动”“携程”“bilibili”“58 同城”“汽车之家”“宝宝巴士”等。百度小程序官方给出的“智能小程序开源联盟”部分成员名单如图 1.26 所示。

图 1.26 “智能小程序开源联盟”部分成员

百度智能小程序同微信和支付宝的小程序存在很大的不同，而这些不同也就是百度智能小程序的优势，现简单总结如下。

(1) 百度作为搜索引擎为小程序提供搜索入口。BAT 是中国互联网最大的流量入口，微信和支付宝的流量是旗下网站和 App 的自有流量，但是百度的流量是搜索流量。搜索流量对于小程序从业者而言，是增量，不同于腾讯的社交流量和阿里的电商流量。在流量价格变得昂贵的今天，大量搜索流量对小程序从业者来说是极大的红利，这也促使智能小程序生态吸引更多的开发者。如图 1.27 所示，当用户在百度 App 搜索框中输入“申通快递”，搜索结果中出现了“申通快递智能小程序”。

但存在的问题是，如果用户使用的是百度手机 App 来搜索，出现了百度智能小程序，自然可以打开该小程序，但如果用户的手机浏览器不是百度浏览器，是不是搜索结果中不应该出现百度智能小程序呢？笔者在 OPPO 手机浏览器搜索“申通快递”，发现搜索结果中还是有“申通快递智能小程序”，但是点击该小程序之后出现了图 1.28 的界面，页面提示请先下载百度 App。也就是说，非“智能小程序开源联盟”成员的 App 是不支持百度智能小程序的，但相信后续会有更多的浏览器加入到“智能小程序开源联盟”的队伍中，这样才会有更多的浏览器能打开智能小程序。

图 1.27 搜索结果中的百度小程序

图 1.28 在非百度 App 中打开智能小程序

（2）智能小程序可以在百度 App 上运行，理论上也可以在“智能小程序开源联盟”成员的浏览器和 App 上运行。这是理想的状态，笔者尝试在“58 同城”和“快手”两个 App 上寻找小程序入口，结果发现上述 App 还没有提供小程序的入口，同样在“百度地图”App 上也没有找到小程序的入口，这说明百度智能小程序还有很长的路要走。不过可以试想一下，当用户使用“百度地图”的“发现周边”功能的时候，除了可以看到周边的学校酒店之外，还可以找到附近的小程序；当用户点开“58 同城”上一条房产信息的时候，在发布者信息中有一个按钮可以直达该房产中介公司的小程序；“快手”App 上的网红大 V 也可以使用小程序构建自己的个性空间，这一切正是百度智能小程序的遐想空间。

（3）除此之外，智能小程序为开发者提供了很多的特殊接口，例如来自百度地图的 LBS 接口、智能语音识别的接口、百度大脑 AI 核心技术接口等。智能小程序比起其他平台的小程序来，可以说更加智能了，更加 AI 化了，这种体验不仅比 H5 要好很多，甚至超出一般的原生 App。

第 2 章

小程序开发基础

本章将介绍小程序开发基础中涉及的众多概念，共 6 节内容，依次是开发者工具的介绍、小程序项目结构、生命周期函数、逻辑层、视图层和实训项目。通过本章的学习，读者能初步了解和掌握小程序开发的基础知识。

本章主要目标

- 熟练掌握微信 web 开发者工具的使用；
- 理解与掌握小程序的目录结构、全局文件的作用和定义、页面文件的定义、应用级和页面级生命周期函数的执行过程；
- 掌握小程序逻辑层数据的定义和修改、页面处理函数、自定义事件函数、页面的跳转、页面的参数传递和模板的概念和应用；
- 掌握小程序视图层中数据绑定、条件渲染、列表渲染、模板的概念和应用。

2.1 开发者工具介绍

当开发者创建项目之后，弹出如图 2.1 所示的微信 web 开发者工具的界面，图中的①～⑥虚线区域分别是菜单栏、工具栏、模拟器、目录树、编辑器和调试器。

2.1.1 菜单栏

菜单栏中分别有项目、文件、编辑、工具、界面、设置和微信开发者工具。现对这些菜单做简单介绍。

(1) 项目。

- 新建项目：快速新建项目；
- 打开最近：可以查看最近打开的项目列表，并选择是否进入对应项目；

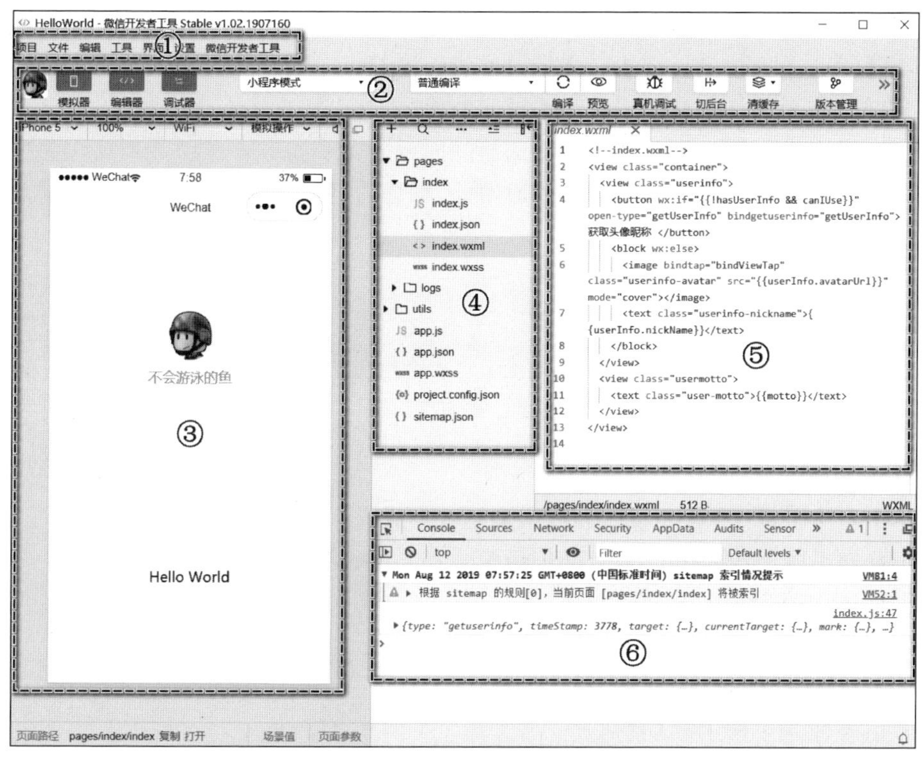

图 2.1 微信 web 开发者工具的界面

- 查看所有项目：新窗口打开启动页的项目列表页；
- 关闭当前项目：关闭当前项目，回到启动页的项目列表页。

(2) 文件：文件的新建、保存和关闭操作。在编辑修改一个文件之后需使用文件菜单下的“保存”功能来保存修改。

(3) 编辑：提供代码格式上的操作。

(4) 工具：项目的刷新、编译和预览、前后台切换、数据缓存的清理和项目的上传部署。

(5) 界面：工具栏、编辑器、模拟器、调试器和目录树的显示控制。

(6) 设置。

- 外观设置：控制编辑器的配色主题、字体、字号、行距；
- 编辑设置：控制文件保存的行为，编辑器的表现；
- 代理设置：选择直连网络、系统代理和手动设置代理；
- 通知设置：设置是否接受某种类型的通知。

(7) 微信开发者工具：账户的切换，开发者工具版本的更新。

2.1.2 工具栏

(1) 左侧区域：包含个人中心、模拟器、编辑器和调试器等，如图 2.2 所示。

图 2.2 左侧工具栏

具体说明如下。

- 个人中心：用于账户切换和消息提醒。
- 模拟器：单击切换显示或隐藏模拟器的面板。
- 编辑器：单击切换显示或隐藏编辑器的面板。
- 调试器：单击切换显示或隐藏调试器的面板。

（2）中间区域：包含小程序模式、普通编译、编译、预览、真机调试、切后台和清缓存，如图 2.3 所示。

图 2.3　中间工具栏

具体说明如下。

（1）小程序模式：小程序模式和搜索动态页模式。

（2）普通编译：普通模式、自定义编译模式和二维码编译模式。

（3）编译：重新编译小程序项目。

（4）预览：生成二维码进行真机预览。

（5）真机调试：生成二维码进行真机远程调试。

（6）切后台：用于切换场景值。

（7）清缓存：单独或同时清除数据缓存、文件缓存、授权数据、网络缓存、登录状态。

2.1.3　模拟器

模拟器用于在计算机中模拟小程序在手机上的执行效果，如图 2.4 所示。图 2.4 中的 3 张图分别为选择手机型号、选择显示比例和选择网络连接状态。

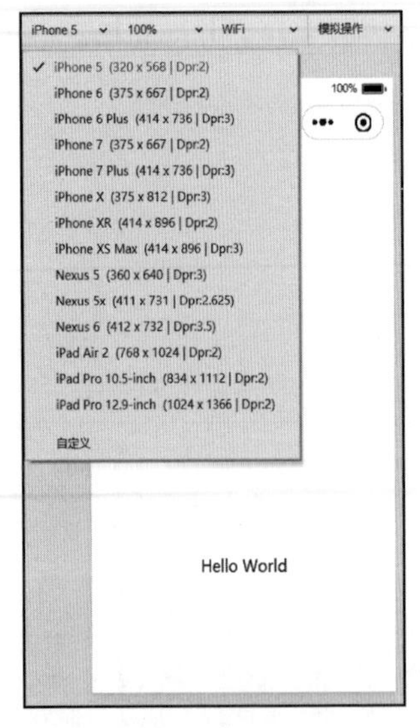

(a) 选择手机型号

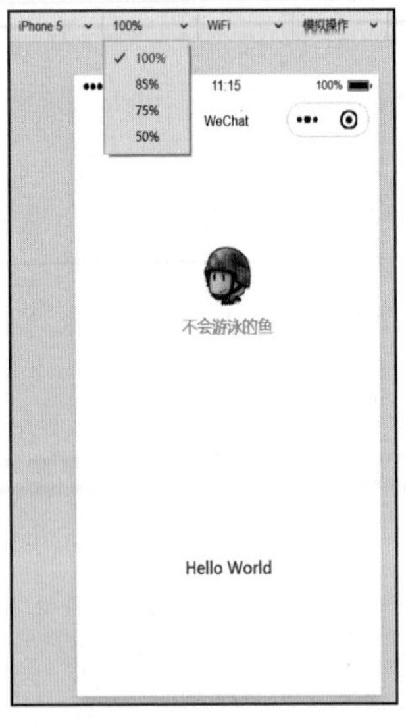

(b) 选择显示比例

(c) 选择网络连接状态

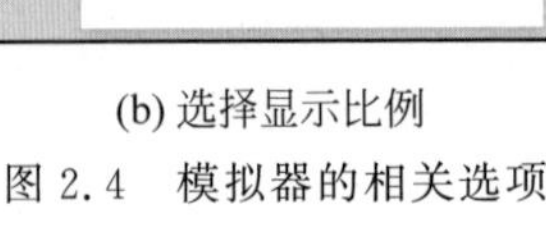

图 2.4　模拟器的相关选项

小程序的代码通过编译后可以在模拟器上直接运行。开发者可以选择不同的设备，也可以添加自定义设备来调试小程序在不同尺寸机型上的适配情况。

2.1.4　目录树

目录树如图 2.5 所示，在目录结构区中单击左上角的“＋”号可以添加新文件，文件类型包括 WXML、JS、WXSS 和 JSON。在 pages 文件下可以创建新的文件，其子文件夹下包含同名的 WXML、JS、WXSS 和 JSON 文件。关于微信小程序的目录结构详见本章 2.2 节。

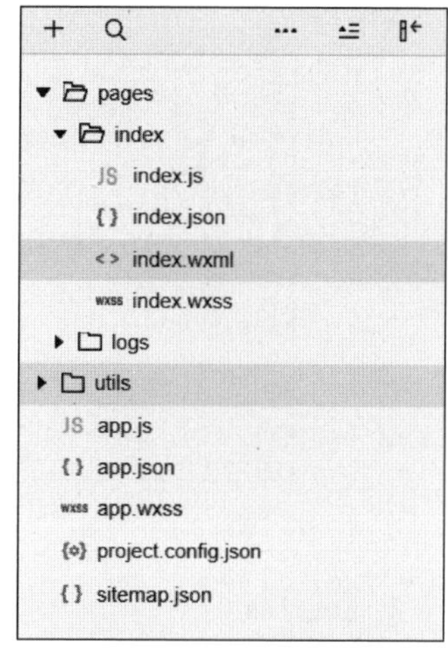

图 2.5　目录树

2.1.5　代码编辑区

在代码编辑区中可以打开多个页面切换查看，单击代码右上角的“×”号可以关闭当前的代码页面。在代码编辑区域，小程序提供自动联想功能，以输入一个< view >标签为例，代码联想功能如图 2.6 所示。

```
index.wxml
<!--index.wxml-->
<v>
<v  video                                   Tag
    view
    cover-view
称  movable-view
    picker-view
    scroll-view
    web-view
    </block>
  </view>
  <view class="usermotto">
    <text class="user-motto">{{motto}}</text>
  </view>
</view>
```

图 2.6　代码联想功能

2.1.6　调试器

小程序调试器主要包含 Console、Sources、Network、Security、Storage、AppData、WXML、Sensor 和 Trace 功能模块。

1. Console

Console 用于显示开发过程中的提示信息，例如当小程序编译或运行有误时，会在控制台显示 warning 和 error 等信息，这是小程序调试的基础功能。在 WXML 文件中输入错误格式代码时，Console 提示报错信息，如图 2.7 所示。

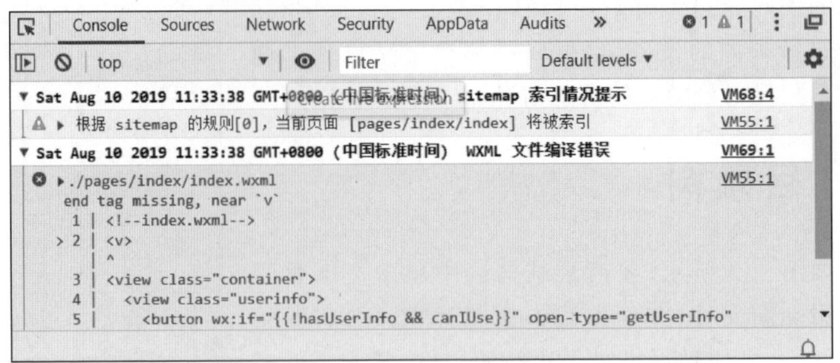

图 2.7　Console 控制台报错信息提示

2. Sources

Sources 面板是小程序的资源面板，用于显示本地和云端的相关资源文件，如图 2.8 所示。

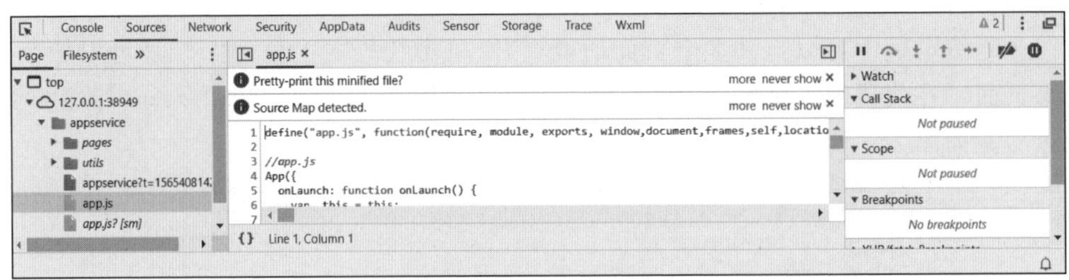

图 2.8　Sources 面板

3. Network

Network 用于观察和显示网络请求和响应的情况，如图 2.9 所示。

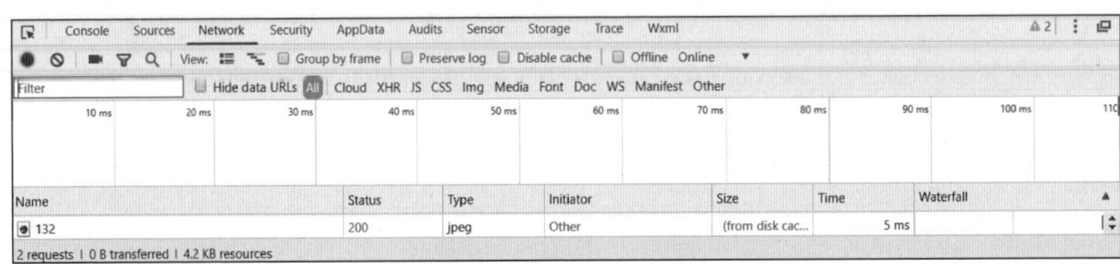

图 2.9　Network 面板

4. Security

Security 面板是小程序的安全面板，用于检测当发生网络请求时，记录所使用的域名来源是否安全，如图 2.10 所示。

5. Storage

Storage 面板用于显示当前小程序的缓存数据，如图 2.11 所示。

6. AppData

AppData 面板用于实时查看小程序页面 JS 文件中 data 数据的变化，允许开发者修改数据，修改后的效果会实时地在模拟器上发生变化，如图 2.12 所示。

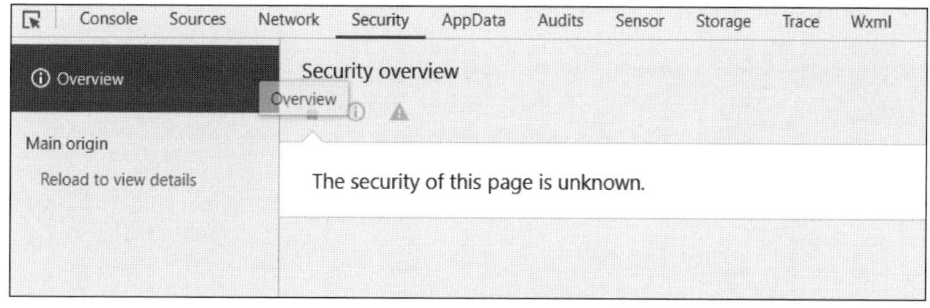

图 2.10 Security 面板

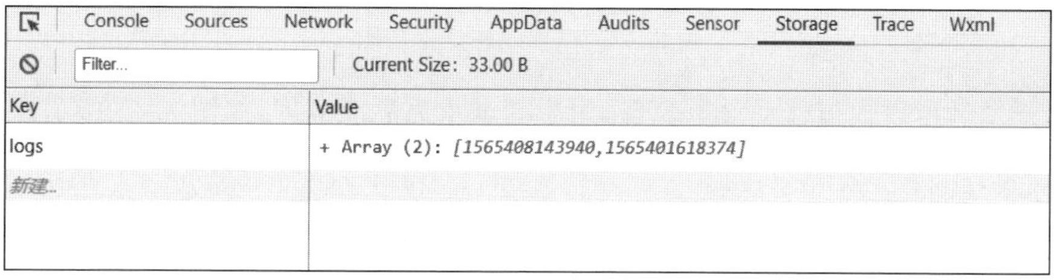

图 2.11 Storage 面板

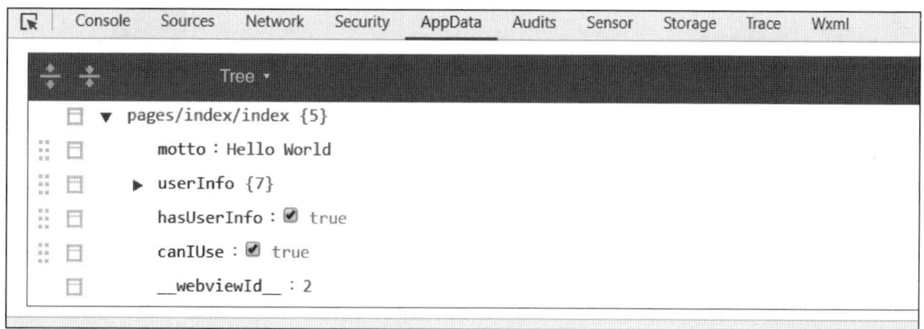

图 2.12 AppData 面板

7. WXML

WXML 面板提供开发者查看当前页面的 WXML 代码以及对应的渲染样式，此功能方便开发者查看页面的结构与对应的样式，如图 2.13 所示。

图 2.13 WXML 面板

8. Sensor

Sensor 面板有两大功能。

(1) 为开发者提供选择模拟地理位置的功能,如图 2.14 所示。

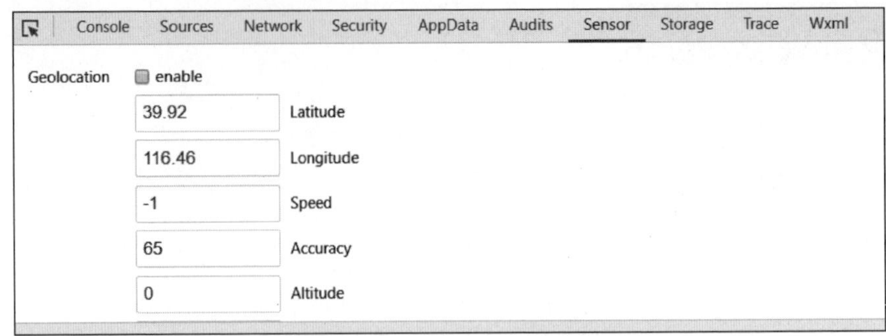

图 2.14 模拟地理位置功能

(2) 开发者可以模拟移动设备表现,用于调试重力感应 API,如图 2.15 所示。

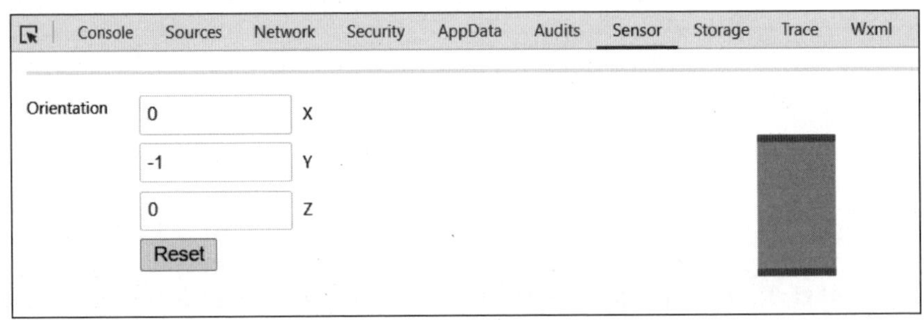

图 2.15 调试重力感应 API

2.2 小程序项目结构

小程序项目结构有自己的特点,项目结构分为全局文件和页面文件两部分。

2.2.1 项目文件结构

当开发者新建一个工程,如图 2.16 所示,项目文件包括根目录下的 pages 文件夹和 utils 文件夹,以及全局文件 app.js、app.json、app.wxss、project.config.json 和 sitemap.json。

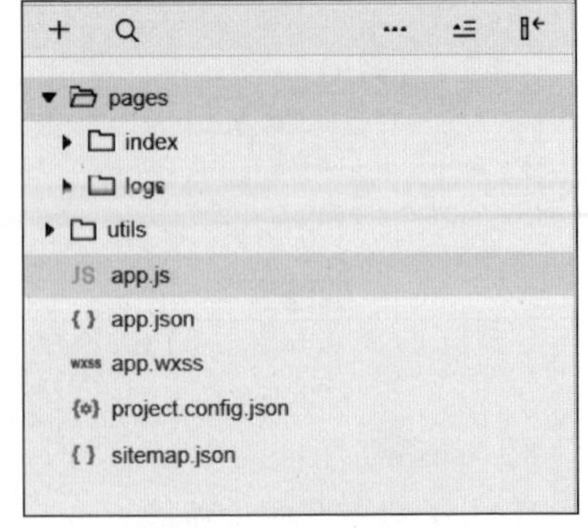

图 2.16 项目文件结构

全局文件是对整个小程序的全局属性的定义,其设置的属性优先级低于页面属性的优先级,即如果一个页面的某一属性在全部文件和页面文件中同时被设置的时候,页面属性设置将覆盖全局属性设置。

pages 文件夹是页面文件的所在,小程序中的一个页面对应一个文件夹,图 2.17 中 pages 文件夹下有

index 和 logs 两个文件夹即对应两个页面。

utils 文件夹下存放着 utils. js 文件，是工具类文件。

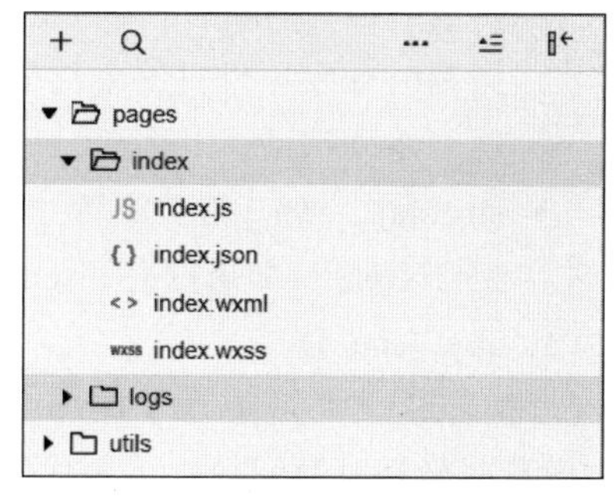

图 2.17 页面文件组成

2.2.2 页面文件

一个完整的小程序页面由四部分组成：

- WXML 文件：用于构建页面的结构；
- WXSS 文件：用于设置页面的样式，该文件定义的样式会覆盖 app. wxss 全局样式表中系统自定义的样式；
- JS 文件：用于设置当前页面的逻辑代码和用户交互；
- JSON 文件：用于重新设置 app. json 中 window 自定义的内容，新设置的选项只会显示在当前页面上，不会影响其他页面。

如图 2.17 所示是 index 页面对应的 index. js、index. json、index. wxml 和 index. wxss 4 个文件。其中 WXML 和 JS 文件是必不可少的，在不对页面文件进行相应设置或者不覆盖全局 JSON 和全局 WXSS 文件的时候，页面的 JSON 和 WXSS 文件可以没有。

当开发者新建一个小程序页面的时候，需要在全局文件 app. js 中注册，否则该页面将不能在项目中被执行。图 2.18 是 index 和 logs 两个页面在全局文件 app. js 中注册的代码，需写在 app. js 的 pages 属性中。

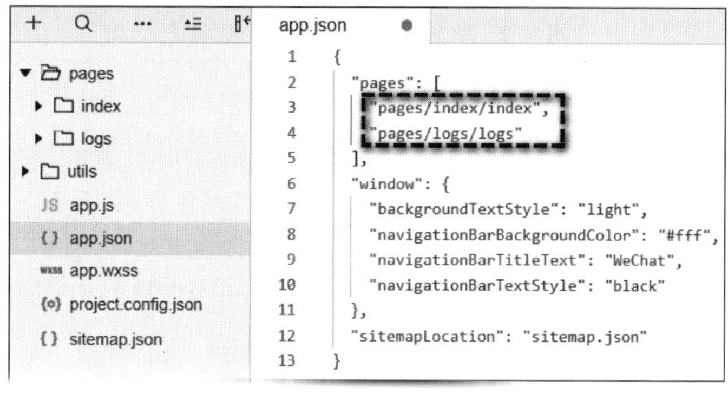

图 2.18 页面的注册

2.2.3 全局配置文件

全局配置文件包括 app. js、app. json、app. wxss、project. config. json 和 sitemap. json 5 个文件。其中前三个文件经常需要进行修改操作，而后两个文件则很少修改。

- app. js：必填文件，用于描述小程序的整体逻辑；
- app. json：必填文件，用于描述小程序的整体逻辑结构；
- app. wxss：可选文件，用于定义小程序的公共样式表；
- project. config. json：开发者工具上做的任何配置都会写入这个文件，当重新安装工具或者换计算机工作时，只要载入同一个项目的代码包，开发者工具就会自动恢复到当时开发项目时的个性化配置，其中会包括编辑器的颜色、代码上传时自动压缩等一系列选项；

- sitemap.json：用于配置小程序及其页面是否允许被微信索引，文件内容为一个JSON对象，如果没有sitemap.json，则默认为所有页面都允许被索引。

1. app.json

app.json文件是小程序中的全局配置文件，它决定页面的路径、窗口样式、tabBar样式、设置网络超时时间等。示例代码如下：

```
{
  "pages": [
    "pages/index/index",
    "pages/logs/logs"
  ],
  "window": {
    "backgroundTextStyle": "light",
    "navigationBarBackgroundColor": "#fff",
    "navigationBarTitleText": "WeChat",
    "navigationBarTextStyle": "black"
  },
  "sitemapLocation": "sitemap.json"
}
```

根据需要，app.json文件可以对17个属性进行设置，其属性如表2.1所示。

表2.1 全局配置文件app.json属性

属性	类型	必填	描述
pages	string Array	是	设置页面路径
window	Object	否	全局的默认窗口表现
tabBar	Object	否	底部tab栏的表现
networkTimeout	Object	否	网络超时
debug	boolean	否	是否开启debug模式，默认关闭
functionalPages	boolean	否	是否启用插件功能页，默认关闭
subpackages	Object[]	否	分包结构配置
workers	string	否	Worker代码放置的目录
requiredBackgroundModes	string[]	否	需要在后台使用的能力，如「音乐播放」
plugins	Object	否	使用到的插件
preloadRule	Object	否	分包预下载规则
resizable	boolean	否	iPad小程序是否支持屏幕旋转，默认关闭
navigateToMiniProgramAppIdList	string[]	否	需要跳转的小程序列表，详见wx.navigateToMiniProgram
usingComponents	Object	否	全局自定义组件配置
permission	Object	否	小程序接口权限相关设置
sitemapLocation	String	是	指明sitemap.json的位置
style	String	否	指定使用升级后的weui样式

下面对核心属性进行说明如下。

（1）pages 属性。

pages 为一个数组，每一项用字符串表示，字符串的格式为“路径＋文件名”，元素的个数为项目中页面的个数，第一项代表小程序的初始页面，即在编译后最先出现在模拟器中的页面，或者说是项目的首页，开发者可以通过调整页面在 pages 属性中的位置来查看不同页面在模拟器中的预览效果。当小程序中新增或减少页面时，都需要对 pages 属性进行修改。

（2）windows 属性。

windows 属性用于设置小程序的导航条、标题、窗口背景色等，其属性值如表 2.2 所示。

表 2.2　app.json 文件中的 windows 属性值

属　性	类　型	默认值	描　述
navigationBarBackgroundColor	HexColor	#000000	导航栏背景颜色，如 #000000
navigationBarTextStyle	string	white	导航栏标题颜色，仅支持 black/white
navigationBarTitleText	string		设置导航栏标题文字内容
navigationStyle	string	default	导航栏样式，仅支持以下值：default 默认样式、custom 自定义导航栏，只保留右上角胶囊按钮
backgroundColor	HexColor	#ffffff	窗口的背景色
backgroundTextStyle	string	dark	下拉 loading 的样式，仅支持 dark/light
backgroundColorTop	string	#ffffff	顶部窗口的背景色，仅 iOS 支持
backgroundColorBottom	string	#ffffff	底部窗口的背景色，仅 iOS 支持
enablePullDownRefresh	boolean	false	是否开启当前页面的下拉刷新
onReachBottomDistance	number	50	页面上拉触底事件触发时距页面底部的距离，单位：px

例 2-1　小程序 windows 属性设置案例，运行效果如图 2.19 所示。

视频讲解

图 2.19　设置 windows 属性后的预览图

app.json 文件代码如下：

```
{
  "pages": [
    "pages/index/index",
    "pages/logs/logs"
```

```
    ],
    "window": {
      "backgroundTextStyle": "light",
      "navigationBarBackgroundColor": "#004A80",
      "navigationBarTitleText": "设置 windows 属性",
      "navigationBarTextStyle": "white"
    },
    "sitemapLocation": "sitemap.json"
}
```

【代码讲解】 本例在 app.json 中通过 pages/index/index 语句，在 pages 属性中注册 index 页面。

- “"navigationBarBackgroundColor": "#004A80"”用于设置导航栏背景颜色为“天蓝色”。
- “"navigationBarTitleText": "设置 windows 属性"”用于设置导航栏标题。
- “"navigationBarTextStyle": "white"”用于设置导航栏标题文字为白色。

(3) tabBar 属性。

tabBar 属性用于设置 tab 栏包含的页面及显示样式，其配置属性值如表 2.3 所示。

表 2.3 app.json 文件中的 tabBar 属性值

属　性	类　型	默认值	描　述
color	HexColor		tab 上的文字默认颜色，仅支持十六进制颜色
selectedColor	HexColor	white	tab 上的文字选中时的颜色，仅支持十六进制颜色
backgroundColor	HexColor		tab 的背景色，仅支持十六进制颜色
borderStyle	string	black	tabBar 上边框的颜色，仅支持 black/white
list	Array		tab 的列表，详见表 2.4 list 属性
position	string	bottom	tabBar 的位置，仅支持 bottom/top

list 接收一个数组，配置最少为 2 个，最多为 5 个，按数组的顺序排序，每项都可以通过设置其属性而改变其样式，其属性值如表 2.4 所示。

表 2.4 list 属性值

属　性	类型	必填	描　述
pagePath	string	是	页面路径，必须在 pages 中先定义
text	string	是	tab 上按钮文字
iconPath	string	否	未选中时的图片路径，icon 大小限制为 40KB，建议尺寸为 81px×81px，不支持网络图片。当 position 为 top 时，不显示 icon
selectedIconPath	string	否	选中时的图片路径，icon 大小限制为 40KB，建议尺寸为 81px×81px，不支持网络图片。当 position 为 top 时，不显示 icon

当开发者需要自定义一个 tabBar 时，可以在 text 属性中设置标题，在 iconPath 中设置未选中时图片的来源路径，在 selectedIconPath 中设置选中图标后图标的来源路径，在 pagePath 中设置选中图标时对应的页面路径。当 position 属性值为 top 时，tabBar 会在页

面顶端，iconPath 和 selectedlconPath 属性无效，不显示图标。

视频讲解

例 2-2　设置 tabBar 属性小案例，程序运行效果如图 2.20 所示。

图 2.20　tabBar 效果

app.json 文件代码如下：

```
{
  "pages": [
    "pages/index/index",
    "pages/logs/logs"
  ],
  "window": {
    "backgroundTextStyle": "light",
    "navigationBarBackgroundColor": "#fff",
    "navigationBarTitleText": "WeChat",
    "navigationBarTextStyle": "black"
  },
  "tabBar": {
    "backgroundColor": "#ffffff",
    "selectedColor": "#E4393C",
    "list": [
      {
        "pagePath": "pages/index/index",
        "selectedIconPath": "images/icon1-live.png",
        "iconPath": "images/icon1.png",
        "text": "首页"
```

```
        },
        {
          "pagePath": "pages/logs/logs",
          "selectedIconPath": "images/icon2 - live.png",
          "iconPath": "images/icon2.png",
          "text": "我的"
        }
      ]
    },
    "sitemapLocation": "sitemap.json"
}
```

【代码讲解】 本例在 list 中配置两个 tabBar，分别为其设置未选择的图标、选择后的图标和字体的颜色 3 个属性，"首页"对应的页面路径为 index 页面，"我的"对应的页面路径为 logs 页面。

(4) networkTimeout 属性。

networkTimeout 属性用于设置各类网络请求的超时时间，单位均为毫秒(ms)，默认没有配置，开发者可以在 app.json 文件中自行增加，其属性值如表 2.5 所示。

表 2.5 app.json 文件中的 networkTimeout 属性值

属　性	类型	默认值	说　明
request	number	60000	wx.request 的超时时间，单位：ms
connectSocket	number	60000	wx.connectSocket 的超时时间，单位：ms
uploadFile	number	60000	wx.uploadFile 的超时时间，单位：ms
downloadFile	number	60000	wx.downloadFile 的超时时间，单位：ms

示例代码如下：

```
{
  "networkTimeout": {
    "downloadFile ": 3000
  }
}
```

上述代码中，设置下载文件 wx.downloadFile()方法的超时时间为 3s。

(5) debug 属性。

用户可以在开发者工具中开启 debug 模式，在开发者工具的控制台面板，调试信息以 info 的形式给出，其信息有页面的注册、页面路由、数据更新和事件触发等，可以帮助开发者快速定位一些常见的问题。

视频讲解

2. app.js

app.js 文件是小程序的全局逻辑文件，其生命周期函数详见 2.3.1 节。app.js 文件最常用的功能就是定义全局数据和全局函数。

例 2-3 本案例在全局文件 app.js 中定义变量和函数，index.js 引用 app.js 中定义的变量和函数。程序运行效果如图 2.21 所示。

图 2.21　app.js 向 index.js 传递数据

app.js 文件代码如下：

```
App({
  globalData: {
    Info: "加法"
  },
  add:function(a,b){
    return a + b
  }
})
```

pages/index/index.wxml 文件代码如下：

```
<view class = "usermotto">
    <text class = "user - motto">{{a}} + {{b}}{{Info}}结果是{{c}}</text>
  </view>
```

pages/index/index.js 文件代码如下：

```
const app  =  getApp()
Page({
  data: {
   a:5,
   b:5,
   c:0,
   Info:""
```

```
  },
  onLoad: function() {
    var c = app.add(this.data.a,this.data.b)
    var Info = app.globalData.Info
    this.setData({
      c:c,
      Info:Info
      })
  }
})
```

【代码讲解】 本例中的 index.js 引用了 app.js 中的变量 Info 和函数 add,在 index.js 中声明 const app = getApp()语句之后,app 对象即包含了 app.js 中所有的变量和函数,而后 app.Info 和 app.add()即可引用到 app.js 中的变量 Info 和函数 add。

3. app.wxss

app.wxss 文件是小程序的全局样式文件,代码如图 2.22 所示。

app.wxss 文件用于规定项目中所有页面的公共样式,关于样式的定义详见第 4 章样式与布局。

```
app.wxss  ×
/**app.wxss**/
.container {
  height: 100%;
  display: flex;
  flex-direction: column;
  align-items: center;
  justify-content: space-between;
  padding: 200rpx 0;
  box-sizing: border-box;
}
```

图 2.22　app.wxss 文件的代码

2.3　生命周期函数

小程序的生命周期分为两类:小程序应用级生命周期和页面级生命周期。

当打开小程序时,首先会触发应用级生命周期函数 onLaunch()进行程序的启动,完成后调用 onShow()准备显示页面,当被切换进入后台会调用 onHide(),直到下次程序在销毁前重新被唤起会再次调用 onShow()。

当小程序应用级生命周期调用完 onShow()以后,就会准备触发小程序页面级生命周期,在一个页面加载显示的过程中会分别触发 onLoad()、onShow()、onReady()函数。当页面被切换到后台,会调用页面 onHide(),从后台被唤醒会调用页面 onShow()。直到页面关闭会调用 onUnload(),当用户重新打开页面还会再依次触发 onLoad()、onShow()、onReady()函数。

在 tab 栏上页面之间互相切换以及在当前页面上跳转到一个新页面也会触发 onHide()函数,当用户再次切换到之前的页面会调用 onShow()函数。

2.3.1　应用级生命周期

小程序的应用级注册是通过重写 App()函数的各种回调事件来达到影响整个应用行为的目的,App()函数必须在 app.js 中注册,并且有且仅有一个,它接受一个 object 参数,指定小程序的生命周期。object 参数属性如表 2.6 所示。

表 2.6　object 参数属性

属　性	类型	必填	说　明
onLaunch	function	否	生命周期回调——监听小程序初始化
onShow	function	否	生命周期回调——监听小程序启动或切前台
onHide	function	否	生命周期回调——监听小程序切后台
onError	function	否	错误监听函数
onPageNotFound	function	否	页面不存在监听函数
其他	any	否	开发者可以添加任意函数或数据变量到 Object 参数中，用 this 可以访问

前台和后台：当用户单击左上角的“关闭”按钮时，或者设备中的界面离开微信，这时小程序进入后台，而并没有销毁；当小程序再次切换到当前屏幕时，小程序又会从后台进入前台，只有当小程序进入后台一定时间后，或者系统资源占用过高，小程序才会被销毁。

微信小程序提供了全局的 getApp()函数，可以获取小程序实例，在定义了 App 的函数后，this 即可获得实例。onLaunch()和 onShow()方法返回的参数如表 2.7 所示。

表 2.7　onLaunch()和 onShow()方法返回的参数

属　性	类　型	必　填
path	string	启动小程序的路径
scene	number	启动小程序的场景值
query	Object	启动小程序的 query 参数
shareTicket	string	小程序被转发时会生成一个 shareTicket，打开被转发的小程序页面可以获取该参数
referrerInfo	string	来源信息，当从另一个小程序、公众号或 App 进入小程序时返回，否则返回{}
referrerInfo. appId	string	来源小程序、公众号或 App 的 AppID
referrerInfo. extraData	Object	来源小程序传过来的数据，当 scene=1037 或 1038 时才支持

返回有效 referrerInfo 的场景值如表 2.8 所示。

表 2.8　返回 referrerInfo 的场景值

场　景　值	场　　景	AppID 含义
1020	公众号 profile 页相关小程序列表	来源公众号
1035	公众号自定义菜单	来源公众号
1036	App 分享消息卡片	来源 App
1037	小程序打开小程序	来源小程序
1038	从另一个小程序返回	来源小程序
1043	公众号模板消息	来源公众号

部分版本在无 referrerInfo 的时候会返回 undefined，建议使用 options. referrerInfo && options. referrerInfo. appId 进行判断。

2.3.2 页面级生命周期

注册小程序中的一个页面，可以在 JS 文件中使用 Page(Object)方法进行注册，接受一个 Object 类型参数后，就能够指定页面的初始数据、生命周期回调、事件处理函数等。页面级生命周期函数和应用级生命周期函数类似，读者可以参考表 2.6。

在 Page()方法中默认生成的 onLoad()、onShow()、onReady()、onHide()以及 onUnload()均是页面的生命周期回调函数，具体说明如下。

- onLoad(Object query)：页面加载时触发。一个页面只会调用一次，可以在 onLoad 的参数中获取打开当前页面路径的参数。
- onShow()：页面显示或切入前台时触发。
- onReady()：页面初次渲染完成时触发。一个页面只会调用一次，代表页面已经准备完毕，可以和视图层进行交互。
- onHide()：页面隐藏或切入后台时触发。如调用 wx.navigateTo()或底部 Tab 切换到其他页面，使小程序切入后台时。
- onUnload()：页面卸载时触发。如调用 wx.redirectTo()或 wx.navigateBack()到其他页面时。

开发者可以根据实际情况，删除不需要的函数或者保留该函数内部空白。现介绍 JS 文件中的代码。

视频讲解

例 2-4 本例通过 app.js 和 index.js 两个 JS 文件，来验证应用级生命周期函数和页面级生命周期函数的执行过程。项目启动时，生命级周期函数执行过程如图 2.23 所示；项目切换到后台时，生命级周期函数执行过程如图 2.24 所示。

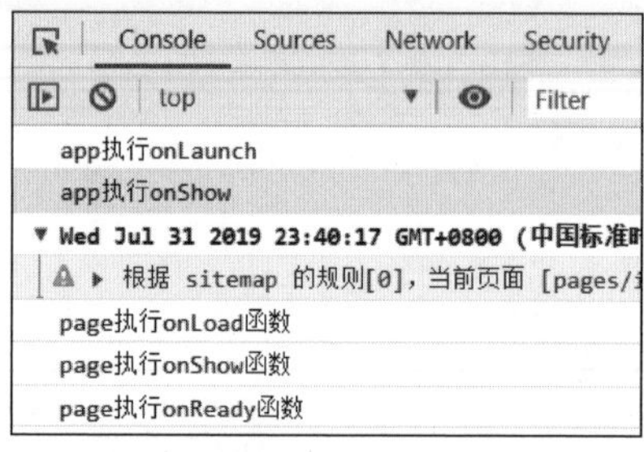

图 2.23 项目启动时生命级周期函数的执行过程

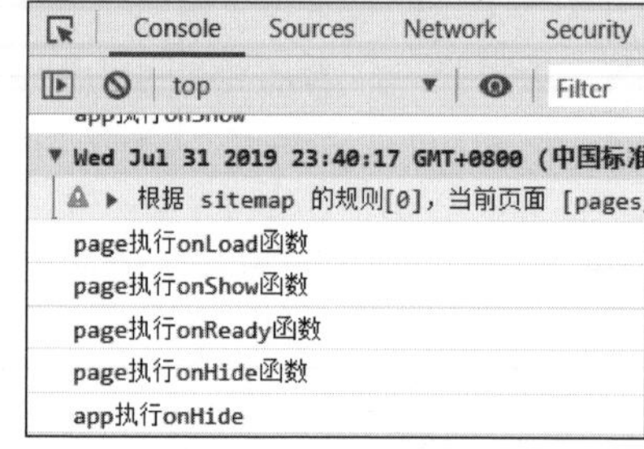

图 2.24 项目切换到后台时生命级周期函数的执行过程

app.js 文件代码如下：

```
App({
  /**
   * 当小程序初始化完成时，会触发 onLaunch(全局只触发一次)
   */
  onLaunch: function() {
    console.log("app 执行 onLaunch")
  },
```

```
  /**
   * 当小程序启动,或从后台进入前台显示,会触发 onShow
   */
  onShow: function(options) {
    console.log("app 执行 onShow")
  },
  /**
   * 当小程序从前台进入后台,会触发 onHide
   */
  onHide: function() {
    console.log("app 执行 onHide")
  },
  /**
   * 当小程序发生脚本错误,或者 API 调用失败时,会触发 onError 并带上错误信息
   */
  onError: function(msg) {
    console.log("app 执行 onError")
  }
})
```

pages/index/index.js 文件代码如下：

```
var app = getApp();
Page({
  data: {
    motto: 'Hello World',
    userInfo: {},
    hasUserInfo: false,
    canIUse: wx.canIUse('button.open-type.getUserInfo')
  },
  onLoad: function(options) {
    console.log("page 执行 onLoad 函数")
  },
  /**
   * 生命周期函数——监听页面初次渲染完成
   */
  onReady: function() {
    console.log("page 执行 onReady 函数")
  },
  /**
   * 生命周期函数——监听页面显示
   */
  onShow: function() {
    console.log("page 执行 onShow 函数")
  },
  /**
   * 生命周期函数——监听页面隐藏
   */
  onHide: function() {
    console.log("page 执行 onHide 函数")
  }
})
```

【代码讲解】 项目启动时先执行应用级生命周期函数，再执行页面级生命周期函数。项目切换到后台的时候，先执行页面级生命周期函数，再执行应用级生命周期函数。

2.4 逻辑层

逻辑层是事务逻辑处理的地方。对于小程序而言，逻辑层就是.js脚本文件的集合。逻辑层对数据进行处理后发送给视图层，同时接收视图层的事件反馈。

微信小程序开发框架的逻辑层是由JavaScript编写的。在JavaScript的基础上，微信团队做了一些适当的修改，以便提高小程序的开发效率。主要修改包括：

(1) 增加App和page函数，进行程序和页面的注册。

(2) 提供丰富的API，如扫一扫、支付等微信特有的功能。

(3) 每个页面有独立的作用域，并提供模块化功能。

逻辑层的实现就是编写各个页面的.js脚本文件。但由于小程序并非运行在浏览器中，所以JavaScript在Web中的一些功能无法使用，如document、window等。

小程序开发编写的所有代码最终会打包成一份JavaScript，并在小程序启动的时候运行，直到小程序销毁。

2.4.1 页面数据

1. 页面数据的定义

页面JS文件page函数中第一项为data属性，在data中定义本页面逻辑处理需要用到的数据，其中很大一部分数据将用于WXML文件的数据渲染。因为小程序JS文件是基于JavaScript编写的，所以在JS文件中可以定义字符串、数字、布尔值、对象和数组等类型的数据。

示例代码：

```
Page({
  data:{
     msg01:"Hello",
     msg02:2019
}
})
```

2. 使用setData()修改数据取值

除了使用数据的初始化，还可以使用page原型实例的setData()函数修改数据的取值，这种方法能够将相关数据异步更新到WXML页面上。setData()函数用于将数据从逻辑层发送到视图层(异步)，同时改变对应的this.data的值(同步)。Object以key：value的形式表示，将this.data中的key对应的值改变成value。其参数说明如表2.9所示。

表 2.9　setData()参数

属　性	类　型	说　明
data	object	这次要改变的数据
callback	function	setData 引起的界面更新渲染完毕后的回调函数

示例代码：

```
Page({
  data: {
    date: "2019 - 7 - 15"
  },
  changeData: function() {
    this.setData({
      date: "2019 - 8 - 15"
    })
  }
})
```

setData()函数修改数据的取值经常用于 WXML 文件数据绑定和用户的交互场景。

2.4.2　页面事件处理函数

在 page()函数中默认产生一系列页面事件处理函数，用于响应用户对页面执行某一动作而执行的处理。函数的说明如下。

- onPullDownRefresh()：监听用户下拉刷新事件。需要在 app.json 的 window 选项中或页面配置中开启 enablePullDownRefresh。可以通过 wx.startPullDownRefresh()触发下拉刷新，调用后触发下拉刷新动画，效果与用户手动下拉刷新一致。当处理完数据刷新后，wx.stopPullDownRefresh()可以停止当前页面的下拉刷新。
- onReachBottom()：监听用户上拉触底事件。可以在 app.json 的 window 选项中或页面配置中设置触发距离 onReachBottomDistance。在触发距离内滑动期间，本事件只会被触发一次。
- onPageScroll(Object)：监听用户滑动页面。其参数 Object 具有唯一属性 scrollTop，为 Number 类型，表示页面在垂直方向已滚动的距离(单位为 px)。
- onShareAppMessage(Object)：监听用户单击页面内转发按钮<button>组件(open-type="share")或右上角菜单“转发”按钮的行为，并自定义转发内容。需要注意的是只有定义了此事件处理函数，右上角菜单才会显示“转发”按钮。Object 的参数如表 2.10 所示。

表 2.10　onShareAppMessage()方法的 Object 参数

参　数	类型	说　明	最低版本
from	string	转发事件来源。button：页面内转发按钮；menu：右上角转发菜单	1.2.4
target	object	如果 from 值是 button，则 target 是触发这次转发事件的 button，否则为 undefined	1.2.4
webViewUrl	string	页面中包含<web-view>组件时，返回当前<web-view>的 URL	1.6.4

此事件处理函数需要返回一个 Object，用于自定义转发内容，返回内容如表 2.11 所示。

表 2.11　onShareAppMessage()方法返回的 Object 对象

字　段	说　　明
title	转发标题，当前小程序名称
path	转发路径，当前页面 path，必须是以"/"开头的完整路径
imageUrl	自定义图片路径，可以是本地文件路径、代码包文件路径或者网络图片路径。支持 PNG 及 JPG，显示图片长宽比是 5：4，另外使用默认截图

onShareAppMessage(Object)方法的示例代码如下：

```
Page({
  onShareAppMessage(res) {
    if (res.from == 'button') {           //页面内"转发"按钮
      console.log(res.target)
    }
    return {
      title: '自定义转发标题',
      path: '/page/user?id = 123'         //自定义转发页面路径
    }
  }
})
```

- onResize(Object object)：小程序屏幕旋转时触发。从基础库 2.4.0 开始支持，低版本需做兼容处理。
- onTabItemTap(Object object)：点击 Tab 时触发。从基础库 1.9.0 开始支持，低版本需做兼容处理。Object 的参数如表 2.12 所示。

表 2.12　onTabItemTap()方法的 Object 参数

参　数	类　型	说　　明	最低版本
index	string	被点击 tabItem 的序号，从 0 开始	1.9.0
pagePath	string	被点击 tabItem 的页面路径	1.9.0
text	string	被点击 tabItem 的按钮文字	1.9.0

onTabItemTap(Object)方法的示例代码如下：

```
Page({
  onTabItemTap(item) {
    console.log(item.index)
    console.log(item.pagePath)
    console.log(item.text)
  }
})
})
```

2.4.3　页面跳转

1. 跳转方式

页面跳转在小程序中被称为页面路由，所有页面的路由全部由框架进行管理。框架以栈的形式维护当前的所有页面。当发生路由切换的时候，页面栈的表现如表 2.13 所示。

表 2.13 页面栈

路 由 方 式	页面栈表现
初始化	新页面入栈
打开新页面	新页面入栈
页面重定向	当前页面出栈，新页面入栈
页面返回	页面不断出栈，直到返回目标页
Tab 切换	页面全部出栈，只留下新的 Tab 页面
重加载	页面全部出栈，只留下新的页面

开发者可以使用 getCurrentPages()函数获取当前页面栈，以数组形式按栈的顺序给出，修改页面栈会导致路由和页面状态发生错误。路由方式与生命周期函数的对应关系如表 2.14 所示。

表 2.14 路由方式与生命级周期函数的对应关系

路由方式	触 发 时 机	路由前页面调用函数	路由后页面调用函数
初始化	小程序打开的第一个页面		onLoad ()、onShow()
打开新页面	调用 API wx. navigateTo 或使用组件< navigator open-type="navigateTo"/>	onHide()	onLoad()、onShow()
页面重定向	调用 API wx. redirectTo 或使用组件< navigator open-type="redirectTo"/>	onUnload()	onLoad()、onShow()
页面返回	调用 API wx. navigateBack 或使用组件< navigator open-type=" navigateBack">用户按左上角返回按钮	onUnload()	onShow()
Tab 切换	调用 API wx. switchTab 或使用组件< navigator open-type="switchTab"/>或用户切换 Tab		见表 2.15
重启动	调用 API wx. reLaunch 或使用组件< navigator open-type="reLaunch"/>	onUnload()	onLoad()、onShow()

假设 A、B 页面为 tabBar 页面，C 页面是从 A 页面打开的页面，D 页面是从 C 页面打开的页面，Tab 切换对应的生命周期如表 2.15 所示。

表 2.15 Tab 与生命周期

当 前 页 面	路由后页面	触发的生命周期(按顺序)
A	A	无
A	B	A. onHide()、B. onLoad()、B. onShow()
A	B(再次打开)	A. onHide()、B. onShow()
C	A	C. onUnload()、A. onShow()
C	B	C. onUnload()、B. onLoad()、B. onShow()
D	B	D. onUnload()、C. onUnload()、B. onLoad()、B. onShow()
D(从转发进入)	A	D. onUnload()、A. onLoad()、A. onShow()
D(从转发进入)	B	D. onUnload()、B. onLoad()、B. onShow()

视频讲解

路由方式存在不同点：navigateTo 和 redirectTo 只能打开非 tabBar 页面；switchTab 只能打开 tabBar 页面；reLaunch 可以打开任意页面；调用页面路由传递的参数可以在目标页面的 onLoad 中获取。

例 2-5 页面跳转小案例，运行效果如图 2.25 所示。

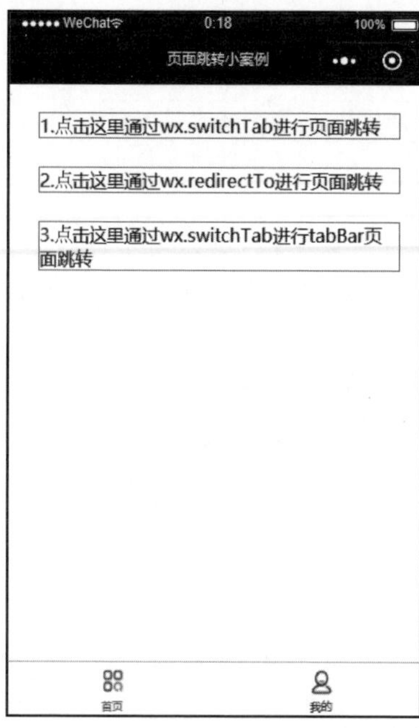

(a) 页面初始效果

(b) wx.navigateTo()方式

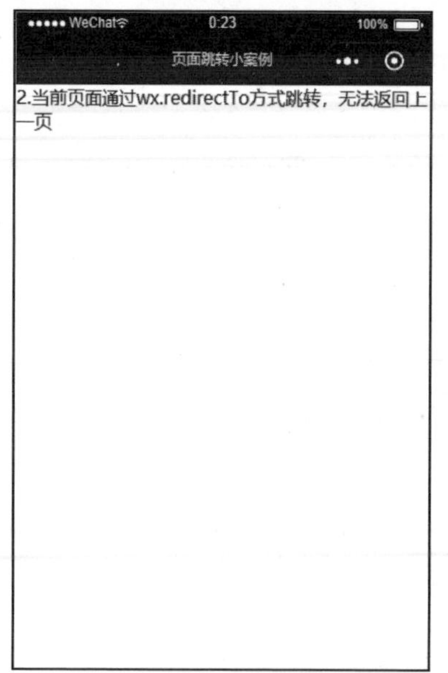

(c) wx.redirectTo()方式

(d) wx.switchTab()方式

图 2.25　3 种方式的页面跳转

pages/pageturn/pageturn. wxml 文件代码如下：

```
< view bindtap = "topageone"> 1.点击这里通过 wx.switchTab 进行页面跳转</view >
< view bindtap = "topagetwo"> 2.点击这里通过 wx.redirectTo 进行页面跳转</view >
< view bindtap = "totabpage"> 3.点击这里通过 wx.switchTab 进行 tabBar 页面跳转</view >
```

pages/pageturn/pageturn. js 文件代码如下：

```
Page({
  topageone: function() {
    wx.navigateTo({
      url: '/pages/pageone/pageone',
    })
  },
  topagetwo: function() {
    wx.redirectTo({
      url: '/pages/pagetwo/pagetwo',
    })
  },
  totabpage: function() {
    wx.switchTab({
      url: '/pages/mytab/mytab',
    })
  }
})
```

pages/pageturn/pageturn. wxss 文件代码如下：

```
view {
  margin: 50rpx;
  border: 1px solid gray;
}
```

【代码讲解】 本例创建 4 个页面，pageturn 页面是项目首页，pageone、pagetwo 和 mytab 页面是 pageturn 页面中 3 种不同的跳转方式的着陆页。在 pageturn. wxml 文件中放置 3 组< view >组件，并且分别绑定了点击事件 topageone()函数、topagetwo()函数和 switchTab()函数，而这 3 个点击事件使用了小程序的 3 种不同的跳转接口 wx. navigateTo()、wx. redirectTo()和 wx. switchTab()。

图 2.25(b)页面由 wx. navigateTo()方式跳转而来，着陆页左上角的返回图标可以返回上一页；图 2.25(c)页面由 wx. redirectTo()方式跳转而来，着陆页左上角没有返回图标；图 2.25(d)页面由 wx. switchTab()方式跳转而来，着陆页页面为 tabBar 页面。

2.4.4 页面间参数传递

在页面跳转中，当前页可将数据传递到着陆页。示例代码如下：

```
wx.navigateTo({
 url: 'page2/page2?a = ' + a + '&b = ' + b
})
```

视频讲解

在跳转页中，使用 onLoad()函数在加载页面时接收数据，在 options 中获取传递过来的数值。

例 2-6 跨页面传递参数小案例，运行效果如图 2.26 所示。

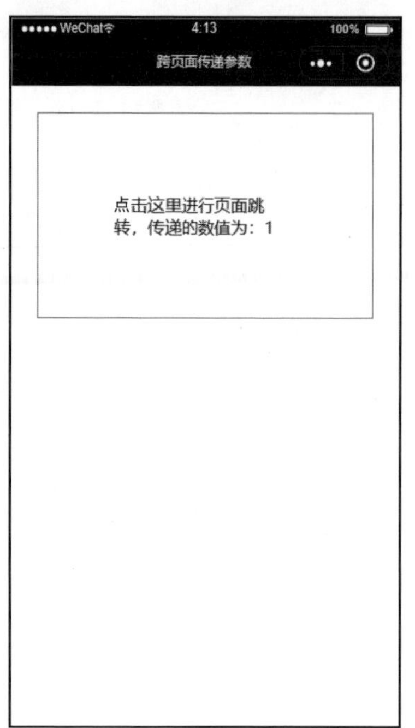

(a) 页面初始效果

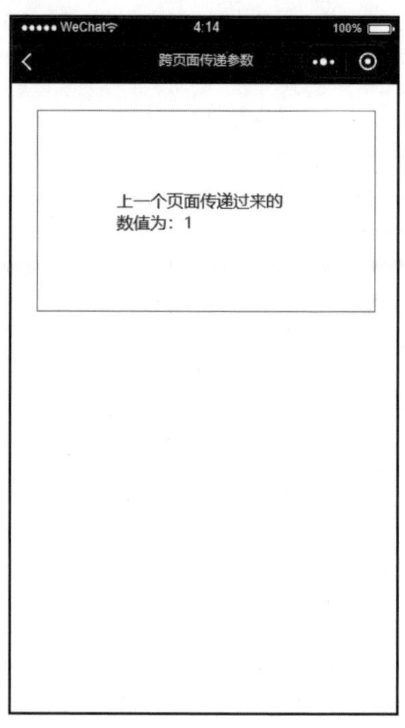

(b) 页面跳转接受参数

图 2.26 跨页面传递参数

pages/transfervalue/transfervalue.wxml 文件代码如下：

```
<view bindtap = "toreceive">点击这里进行页面跳转，传递的数值为：1 </view>
```

pages/transfervalue/transfervalue.js 文件代码如下：

```
Page({
  toreceive: function() {
    wx.navigateTo({
      url: '/pages/receive/receive?id = 1'      //页面跳转并传递参数
    })
  }
})
```

pages/receive/receive.wxml 文件代码如下：

```
<view>上一个页面传递过来的数值为：{{receiveData}}</view>
```

pages/receive/receive.js 文件代码如下：

```
Page({
  data: {
    receiveData: null                         //数据初始化为空值
```

```
    },
    onLoad: function(options) {
      var receiveData = options.id;              //接受参数并赋值
      console.log(options);
      this.setData({
        receiveData: receiveData                 //更新 receiveData 值
      })
    },
})
```

【代码讲解】 本例在 transfervalue. wxml 文件中为页面区域绑定点击事件；在 JS 文件中自定义函数 toreceive()；通过设置 url：'/pages/receive/receive?id=1'进行页面跳转并传递值为 1 的参数 id；在着陆页的 JS 文件中自定义 onLoad()函数；通过 options. id 接收到传递过来的参数并赋值给变量 receiveData；通过 setData()函数修改页面 data 属性 receiveData 的值，然后在 receive. wxml 文件中利用数据绑定显示该值。图 2. 26(a)为跳转页；图 2. 26(b)为着陆页。

2.4.5　模块化

模块化是软件工程中的一个重要概念。模块化程序设计是指在进行程序设计时将一个大程序按照功能划分为若干小程序模块，每个小程序模块完成一个确定的功能，并在这些模块之间建立必要的联系，通过模块的互相协作完成整个功能的程序设计方法。使用模块化思想可以有效提高软件开发和维护的效率，小程序开发也需要遵循这一原则。

1. 局部变量和全局变量

在页面文件中声明的变量和函数只在该文件中有效，在不同的文件中可以声明相同名字的变量和函数，互相不影响。当开发者需要让数据在不同页面中共享时，可以在 app. js 中定义全局变量，然后在其他 JS 文件中使用 getApp()获取全局变量。

app. js 中示例代码如下：

```
App({
  globalData: {
    msg: "hello xiaochengxu"
  }
})
```

其他的 JS 文件获取 app. js 中的全局变量的示例代码如下：

```
var app = getApp()
var msg = app.globalData.msg
```

2. 代码模块化

根据软件工程“耦合”和“内聚”的要求，小程序开发过程中可以把一些内聚性不强的功能代码写在外部 JS 文件中，小程序提供了 module. exports 和 exports 接口对外部 JS 代码进行定义和引用。exports 是 module. exports 的一个引用，因此在模块中随意更改 exports 的指向会造成未知的错误，所以更推荐开发者采用 module. exports 来定义模块接口。小程序目前不支持直接引入 node_modules,开发者需要使用 node_modules 的时候建议复制相关的代码到小程序的目录中,或者使用小程序支持的 npm 功能。

公共 JS 文件 common. js 中定义公用函数的示例代码如下:

```
function sayHello(name) {
  console.log("Hello ${name} !")
}
function sayGoodbye(name) {
  console.log("Goodbye ${name} !")
}
module.exports.sayHello = sayHello              //方式 1 推荐使用
exports.sayGoodbye = sayGoodbye
```

页面 JS 文件中调用公用函数的示例代码如下:

```
var common = require("../../utils/common.common.js")
Page({
  helloTom: function() {
    common.sayHello("Tom")
  },
  goodbyeTom: function() {
    common.sayGoodbye("Tom")
  }
})
```

2.4.6 页面自定义事件函数

1. 事件的定义和特点

事件是视图层到逻辑层的通信方式,特点如下:

- 事件可以将用户行为反馈到逻辑层进行处理;
- 事件可以绑定在组件上,当事件被触发,就会执行逻辑层中对应的事件处理函数;
- 事件对象可以携带额外信息,如 id、dataset、touches。

2. 事件的使用方式

事件的使用需要在 WXML 页面中某一个组件绑定事件函数名,在 JS 文件中的 Page 函数中自定义相应的事件处理函数。

WMXL 绑定事件的示例代码如下:

```
<button id = "buttonTest" bindtap = "tapName">我是按钮组件</button>
```

JS 文件中定义事件的示例代码如下:

```
Page({
  tapName: function() {
    console.log("你点击了按钮")
  }
})
```

3. 事件的分类

- 冒泡事件：当一个组件上的事件被触发后，该事件会向父节点传递。
- 非冒泡事件：当一个组件上的事件被触发后，该事件不会向父节点传递。

冒泡事件的类型及说明如表 2.16 所示。

表 2.16　冒泡事件的类型及说明

类　型	说　明
touchstart	手指触摸动作开始
touchmove	手指触摸后移动
touchcancel	手指触摸动作被打断，如来电提醒，弹窗
touchend	手指触摸动作结束
tap	手指触摸后马上离开
longpress	手指触摸后，超过 350ms 再离开，如果指定了事件回调函数并触发了这个事件，tap 事件将不被触发
longtap	手指触摸后，超过 350ms 再离开（推荐使用 longpress 事件代替）
transitionend	会在 wxss transition 或 wx.createAnimation 动画结束后触发
animationstart	会在一个 wxss animation 动画开始时触发
animationiteration	会在一个 wxss animation 一次迭代结束时触发
animationend	会在一个 wxss animation 动画完成时触发
touchforcechange	在支持 3D Touch 的 iPhone 设备，重按时会触发

除表 2.16 之外的其他组件自定义事件如无特殊声明都是非冒泡事件，如 form 的 submit 事件、input 的 input 事件、scroll-view 的 scroll 事件等。

4. 事件绑定和冒泡

事件绑定的写法同组件的属性，用 key="value"的形式。

- key 以 bind 或 catch 开头，跟上事件的类型，如 bindtap、catchtouchstart。从基础库版本 1.5.0 起，在非原生组件中，bind 和 catch 后可以紧接一个冒号，其含义不变，如 bind:tap、catch:touchstart。
- value 是一个字符串，需要在对应的 Page 中定义同名的函数，否则当触发事件的时候会报错。bind 事件绑定不会阻止冒泡事件向上冒泡；catch 事件绑定可以阻止冒泡事件向上冒泡。

例如有 3 个<view>组件，其中 A 包含 B，B 包含 C，示例代码如下：

```
<view id = "outer" bindtap = "tap1">
  外层容器 A
  <view id = "middle" catchtap = "tap2">
    中间容器 B
    <view id = "inner" bindtap = "tap3">
      内层容器 C
    </view>
  </view>
</view>
```

当点击容器 C 时会触发 tap3 事件，由于 tap3 是 bindtap 类型的冒泡事件，所以其父节点容器 B 的 tap2 事件会被触发,而容器 B 的 tap2 事件是 catchtap 类型,catchtap 类型阻止事件继续向上冒泡,即容器 A 的 tap1 事件不会被触发。当点击容器 B 和容器 A 时,只能分别触发自己的 tap2 和 tap1 事件。

5. 事件对象

如无特殊说明,当组件触发事件时,逻辑层绑定该事件的处理函数会收到一个事件对象。事件对象分为基础事件(BaseEvent)、自定义事件(CustomEvent)和触摸事件(TouchEvent)。基础事件(BaseEvent)对象属性如表 2.17 所示。

表 2.17 基础事件(BaseEvent)对象属性

属　　性	类型	说　　明
type	string	事件类型
timeStamp	integer	事件生成时的时间戳
target	object	触发事件的组件的一些属性值集合
currentTarget	object	当前组件的一些属性值集合

自定义事件(CustomEvent)对象属性列表(继承 BaseEvent)如表 2.18 所示。

表 2.18 自定义事件(CustomEvent)对象属性

属　　性	类型	说　　明
detail	object	额外的信息

触摸事件(TouchEvent)对象属性列表(继承 BaseEvent)如表 2.19 所示。

表 2.19 触摸事件(TouchEvent)对象属性

属　　性	类型	说　　明
touches	array	触摸事件,当前停留在屏幕中的触摸点信息的数组
changedTouches	array	触摸事件,当前变化的触摸点信息的数组

注意:canvas 画布组件中的触摸事件不可冒泡,因此没有 currentTarget。

各种事件属性的含义如下。

- type:代表事件的类型。
- timeStamp:页面打开到触发事件所经过的毫秒数。
- target:触发事件的源组件。
- currentTarget:事件绑定的当前组件。
- detail:自定义事件所携带的数据,如表单组件的提交事件会携带用户的输入,媒体的错误事件会携带错误信息。
- touches:是一个数组,每个元素为一个 Touch 对象,表示当前停留在屏幕上的触摸点。
- changedTouches:数据格式同 touches,表示有变化的触摸点,如从无变有(touchstart),位置变化(touchmove),从有变无(touchend、touchcancel)。

基础事件对象中的 target 和 currentTarget 属性包含的参数相同,如表 2.20 所示。

表 2.20 target 和 currentTarget 参数

属　性	类型	说　明
id	string	当前组件的 id
tagName	string	当前组件的类型
dataset	object	当前组件上由 data-开头的自定义属性组成的集合

- dataset：在组件节点中可以附加一些自定义数据。这样，在事件中可以获取这些自定义的节点数据，用于事件的逻辑处理。在 WXML 中，这些自定义数据以 data-开头，多个单词由连字符-连接。这种写法中，连字符写法会转换成驼峰写法，而大写字符会自动转成小写字符。例如：

  data-element-type 会转换为 event.currentTarget.dataset.elementType；

  data-elementType 会转换为 event.currentTarget.dataset.elementtype。

touches 对象属性如表 2.21 所示。

表 2.21 touches 对象属性

属　性	类型	说　明
identifier	number	触摸点的标识符
pageX, pageY	number	距离文档左上角的距离，文档的左上角为原点，横向为 X 轴，纵向为 Y 轴
clientX,clientY	number	距离页面可显示区域(屏幕除去导航条)左上角的距离，横向为 X 轴，纵向为 Y 轴

canvas 画布组件的触摸事件中携带的 touches 是 CanvasTouch 数组，属性如表 2.22 所示。

表 2.22 CanvasTouch 对象属性

属　性	类型	说　明
identifier	number	触摸点的标识符
x, y	number	距离 Canvas 左上角的距离，Canvas 的左上角为原点，横向为 X 轴，纵向为 Y 轴

2.5 视图层

微信小程序通过视图层与用户进行交互，WXML 文件决定页面的结构，WXSS 文件决定页面的样式。视图层的构成如下。

- WXML(WeiXin Markup Language)：用于定义页面的结构，类似于 HTML，是一种标记性语言；
- 组件(Component)：是构成页面的基本元素，也是小程序界面的基本组成单元；
- WXSS(WeiXin Style Sheet)：用于定义页面的样式，类似于 CSS 语法格式。

本节介绍视图层的基本概念,组件详见第 3 章微信小程序组件,WXML 和 WXSS 详见第 4 章样式与布局。

2.5.1 数据绑定

小程序数据绑定指的是在 WXML 页面中存在动态数据,这些动态数据在 JS 文件中定义和维护。根据场景的不同,数据绑定分为以下类别。

1. 简单绑定

在 WXML 文件中使用 Mustache 语法{{变量名}},用双大括号将变量包起来,用来绑定动态数据。示例代码如下:

WXML 文件中:

```
<view>{{message}}</view>
```

JS 文件中:

```
Page({
  data: {
    message: "Hello xiaochengxu"
  }
})
```

2. 组件属性绑定

组件属性绑定指被绑定的数据存在于组件的属性内,示例代码如下:

```
<view id="{{id}}"></view>
```

3. 控制属性绑定

控制属性绑定指被绑定的数据存在于组件的控制属性中,示例代码如下:

```
<view wx:if="{{condition}}"></view>
```

在 JS 文件中设置变量 condition 的值为 true 时,<view>组件会显示;当 condition 的值修改为 false 时,<view>组件不会显示。

4. 运算绑定

当需要对动态数据进行运算时,可以使用运算绑定符{{ }}把运算表达式包含起来,在页面被加载的时候,{{ }}内的运算表达式将会执行运算。运算绑定支持的运算类型有三元运算、算术运算、逻辑判断、字符串运算和数据路径运算。

三元运算示例代码如下:

```
<view hidden="{{result? true:false}}">隐藏该组件</view>
```

算术运算示例代码如下:

```
<view> {{a + b}} + {{c}} + d </view>
```

逻辑判断示例代码如下：

```
<view wx:if = "{{x > 10}}">显示该组件</view>
```

字符串运算示例代码如下：

```
<view> {{"hello" + name}} </view>
```

数据路径运算示例代码如下：

```
<view>{{object.key1}} {{array[1]}}</view>
```

5. 组合绑定

开发者使用绑定符号{{}}对变量、常量和符号进行组合，构成新的对象或者数组。数组组合的示例代码如下：

```
<view wx:for = "{{[1, x, 3, 4]}}"> {{item}} </view>
```

对象组合的示例代码如下：

```
<template is = "objectCombine" data = "{{for: a, bar: b}}"></template>
```

扩展运算符"..."将一个对象展开，示例代码如下：

```
<template is = "objectCombine" data = "{{...obj1, ...obj2, e: 5}}"></template>
```

2.5.2 条件渲染

1. <view wx:if>条件渲染

小程序<view>组件中使用 wx:if="{{condition}}"来控制代码块显示与否，示例代码如下。

WXML 文件中：

```
<view wx:if = "{{condition}}">组件</view>
```

上述代码中，如果变量 condition 取值为 false，组件不会显示。

开发者可以通过 wx:elif 和 wx:else 来增加一个 else 块，示例代码如下：

```
<view wx:if = "{{length > 5}}">这是组件 1 </view>
<view wx:elif = "{{length > 2}}">这是组件 2 </view>
<view wx:else>这是组件 3 </view>
```

上述代码中，当 length=3 时，length＞5 不成立，组件 1 不被显示；length＞2 成立，组件 2 被显示，wx:else 语句被忽略不执行。

2. <block wx:if>条件渲染

wx:if 不仅可以在<view>组件内使用，还可以在<block>标签内使用，示例代码如下：

```
<block wx:if = "{{false}}">
  <view>这是组件 1 </view>
  <view>这是组件 2 </view>
</block>
```

例 2-7　wx:if 实例,程序运行效果如图 2.27 所示。

视频讲解

图 2.27　wx:if 条件渲染

pages/wxif/wxif.wxml 文件的代码如下:

```
<view class="hello" wx:if="{{hours<=12}}">
goodmorning
</view>
 <view class="hello" wx:elif="{{hours<=20}}">
goodafternoon
</view>
 <view class="hello" wx:else>
goodevening
</view>
```

pages/wxif/wxif.js 文件的代码如下:

```
Page({
  data: {
    hours:0,
  },
  onLoad: function() {
    var myDate = new Date();
    var hours = myDate.getHours();
    this.setData({
      hours: hours
```

```
    })
  }
})
```

【代码讲解】 本例在 wxif.js 文件中获取当前时间的时钟 hours，然后 hours 被绑定到 wx:if 的条件属性中，根据 hours 值的不同提示不同的问候语。

2.5.3 列表渲染

1. <view wx:for>循环渲染

在小程序<view>组件上使用 wx:for 属性遍历数组元素，实现代码块的列表渲染。示例代码如下。

WXML 文件中：

```
<view wx:for="{{array}}">水果{{index}}: {{item}}</view>
```

JS 文件中：

```
Page({
  data: {
    array: ["苹果", "香蕉", "菠萝"]
  }
})
```

运行结果与下方代码一致：

```
<view>水果 0: 苹果</view>
<view>水果 1: 香蕉</view>
<view>水果 2: 菠萝</view>
```

上述代码中，index 是数组在循环遍历时当前项下标，item 代表数组在循环遍历时的当前元素。开发者可以使用 wx:for-item 和 wx:for-index 自定义当前元素和下标的变量名，示例代码如下：

```
wx:for-index="goods-index" wx:for-item="goods-item"
```

视频讲解

例 2-8 列表渲染小案例，程序运行效果如图 2.28 所示。

pages/wxfor/wxfor.wxml 文件代码如下：

```
<!-- 商品展示部分 -->
<view class="demo-box">
  <block wx:for="{{goodsdata}}" wx:for-item="item">
    <view class="goods-box" bindtap="btntodetail" data-id="{{item.id}}">
      <image class="goods-pic" src="{{item.image}}"></image>
      <view class="goods-title">{{item.title}}</view>
      <view class="goods-titleTwo">{{item.titleTwo}}</view>
      <view class="row">
        <view class="goods-price">¥{{item.price}}</view>
        <text class="goods-btn">看相似</text>
```

```
      </view>
    </view>
  </block>
</view>
```

图 2.28　wx:for 循环渲染

pages/wxfor/wxfor.js 文件代码如下：

```
Page({
  data:{
    goodsdata: [
      {id: 0,
      title: "(HUAWEI)P30",
      titleTwo: "超感光徕卡三摄|逆光智美自拍",
      price: "4288.00",
      image: "https://res.vmallres.com/pimages//product/6901443293513
        /800_800_1555464685019mp.png"
    },
    { id: 1,
      title: "(HUAWEI)荣耀 V20",
      titleTwo: "魅眼视屏 4800 万深感相机",
      price: "3699.00",
      image: "//img12.360buyimg.com/n7/jfs/t25954/134/1930444050/488286
        /31587d0d/5bbf1fc9N3ced3749.jpg"
    },
    { id: 2,
```

```
        title: "IE80S 入耳式监听耳机",
        titleTwo: "森海塞尔 HiFi 音乐耳机",
        price: "2399.00",
        image: "https://images.wincheers.net/UpLoad/Web/ProductImg
          /2018 - 06 - 15/NEW_XM/IE80S.jpg"
      },
      { id: 3,
        title: "IE60 入耳式 HiFi 耳机",
        titleTwo: "森海塞尔入耳式 HIFI 耳机",
        price: "799.00",
        image: "https://images.wincheers.net/UpLoad/Web/ProductImg
        /2017 - 08 - 14/NEW_XM/IE - 60 - 4.png"
    }
      ]
    }
})
```

pages/wxfor/wxfor.wxss 文件代码如下：

```
  flex - direction: row; flex - wrap: wrap; padding: 30rpx;
}
.goods - box {
  width: 49 %; margin - bottom: 20rpx;
}
/ * 商品内容样式 * /
.goods - pic {
  width: 220rpx; height: 250rpx;
  margin: 0 auto; display: block;
}
.row {
  display: flex; flex - direction: row;
  justify - content: space - around;
}
/ * 看相似样式 * /
.goods - btn {
  border: 1px solid #e3e3e3; border - radius: 20rpx;
  margin - top: 20rpx; width: 100rpx;
  height: 40rpx; line - height: 40rpx;
  font - size: 24rpx; color: #aaa;
  letter - spacing: 2rpx; text - align: center;
}
/ * 商品标题样式 * /
.goods - title {
  font - size: 30rpx; font - weight: 600; text - align: center;
}
.goods - titleTwo {
  font - size: 24rpx; margin - top: 10rpx; text - align: center;
}
/ * 商品价格样式 * /
.goods - price {
  font - size: 30rpx; margin - top: 20rpx; color: #ee3b3b;
}
```

【代码讲解】 wxfor.js 中的 goodsdata 数组存放了商品的信息；wxfor.wxml 使用 wx:for 条件循环渲染数组中多件商品的 image、title、titleTwo 和 price 等值。循环渲染经常用于商品和新闻的列表页。

2. <block wx：for>

类似于<block wx:if>,<block>标签也可以添加循环控制信息 wx:for,用来循环渲染代码块。示例代码如下：

```
<block wx:for = "{{['苹果', '香蕉', '菠萝']}}">
  <view> {{index}}: </view>
  <view> {{item}} </view>
</block>
```

效果等同于下面代码：

```
<view>水果: 0 </view>
<view>种类: 苹果</view>
<view>水果: 1 </view>
<view>种类: 香蕉</view>
<view>水果: 2 </view>
<view>种类: 菠萝</view>
```

3. 嵌套循环渲染

wx:for 可以嵌套,类似于循环结构中的多重循环,例如,小程序中九九乘法表的示例代码如下：

```
<view wx:for = "{{[1, 2, 3, 4, 5, 6, 7, 8, 9]}}" wx:for-item = "i">
  <view wx:for = "{{[1, 2, 3, 4, 5, 6, 7, 8, 9]}}" wx:for-item = "j">
    <view wx:if = "{{i <= j}}">
      {{i}} * {{j}} = {{i * j}}
    </view>
  </view>
</view>
```

2.5.4 模板

小程序框架允许开发者在 WXML 文件中使用模板(template),模板的使用分为模板的定义和模板的引用。

1. 定义模板

小程序使用<template>之前需要在 WXML 文件中把代码片段定义为模板,使用 name 属性自定义其名称。示例代码如下：

```
<template name = "myTemp">
  <view>
    <text> Name: {{name}}</text>
    <text> Age: {{age}}</text>
  </view>
</template>
```

上述代码中,自定义了一个名称为 myTemp 的模板。

2. 模板的引用

在需要引用模板的页面中使用< template >标签可以引用模板内容,使用 is 属性,引用需要引用的模板,示例代码如下。

WXML 文件中:

```
< template is = "myTemp" data = "{{...person}}"/>
```

使用< template >模板的时候,如果需要传递参数给模板代码,则参数通过 data 属性传递。

视频讲解

例 2-9 < template >模板小案例,设计两个模板,一个模板需要接收参数,另一个模板不需要接收参数,程序运行效果如图 2.29 所示。

●●●●● WeChat　17:23　100%
template
我是头部文件，我不需要参数
我是正文
我是尾部文件 传递过来的参数是2019

图 2.29　template 模板

pages/template/body. wxml 文件代码如下:

```
< import src = "header"/>
< template is = "header"/>
我是正文
< import src = "footer"/>
< template is = "footer" data = "{{Data}}" />
```

pages/templatc/body. js 文件代码如下:

```
Page({
  data: {
    Data:2019
  },
})
```

pages/template/header. wxml 文件代码如下:

```
< template name = "header">
< view >我是头部文件,我不需要参数</view >
</template >
```

pages/template/body. wxml 文件代码如下:

```
< template name = "footer">
< view >我是尾部文件 < text >传递过来的参数是{{Data}}</text >
```

```
</view>
</template>
```

【代码讲解】 本例包括3个页面：body、header和footer，header和footer页面定义了模板，其中header模板不需要引用方的参数传递，而footer模板需要引用方的参数传递。本例还提醒开发者在引用模板的时候，需要使用<import>引用标签把模板所在的页面引入进来。

2.5.5 引用

WXML提供两种文件引用方式：import和include。

1. import

import可以在文件中使用目标文件定义的<template>模板，示例代码如下：

```
<import src="item.wxml" />
<template is="item" data="{{text:'hello,xiaochengxu'}}" />
```

<import>具有作用域，假设有A、B和C 3个页面，其中B import A，且C import B，B页面可以使用A页面定义的<template>模板，C页面可以使用B页面定义的<template>模板，但是C页面不可以使用A页面定义的<template>模板。示例代码如下。

A页面a.wxml文件中：

```
<template name="A">
  <text>页面模板</text>
</template>
```

B页面b.wxml文件中：

```
<import src="a.wxml" />
<template name="B">
  <text>页面模板</text>
</template>
```

C页面c.wxml文件中：

```
<import src="b.wxml" />
<template is="A" /> <!--引入模板不成功,C页面需要自己 import A-->
<template is="B" /> <!--引入模板成功,C页面有 import B-->
```

开发者在引用中应注意页面之间的关系，正确引用。

2. include

使用<include>可以将目标文件中除了<template>模板以外的全部代码引入，相当于复制到<include>位置。从软件工程的角度看，一个文件的代码行尽量不要超过3个屏幕，当文件代码行过多的时候，可以将功能相关的代码独立成一个文件，然后再使用<include>标签将独立出去的代码引用回来，从而增强了页面的可读性和可维护性。

2.6　实训项目——商品列表页和详情页

视频讲解

本项目有商品列表页和商品详情页两个页面，当点击商品列表页中的某件商品时，页面跳转到商品详情页，项目的知识点涉及 app.js 全局数据和页面 JS 之间的数据交互、数据绑定、循环渲染、页面跳转和页面间参数传递等知识点。本项目是对本章所学知识点的有效检测。

项目的实现过程如下。

第一步：打开微信 Web 开发者工具，如图 2.30 所示，新建项目“第二章实训项目”，项目结构如图 2.31 所示，新建 list 和 detail 两个页面。

图 2.30　新建项目

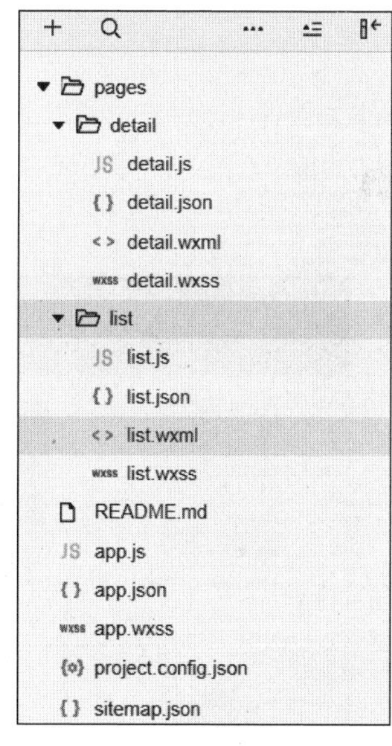

图 2.31　项目结构

第二步：在 app.js 文件中定义全局数据。

第三步：实现 list 和 detail 两个页面，它们的执行效果如图 2.32 和图 2.33 所示。

app.js 文件代码如下：

```
App({
  goodsdata: [
    { id: 0,
      title: "(HUAWEI)P30",
      titleTwo: "超感光徕卡三摄|逆光智美自拍",
      price: "4288.00",
      image: "https://res.vmallres.com/pimages//product/6901443293513
```

```
                /800_800_1555464685019mp.png"
      },
      { id: 1,
        title: "(HUAWEI)荣耀 V20",
        titleTwo: "魅眼视屏 4800 万深感相机",
        price: "3699.00",
        image: "//img12.360buyimg.com/n7/jfs/t25954/134/1930444050
                /488286/31587d0d/5bbf1fc9N3ced3749.jpg"
      },
      { id: 2,
        title: "IE80S 入耳式监听耳机",
        titleTwo: "森海塞尔 HiFi 音乐耳机",
        price: "2399.00",
        image: "https://images.wincheers.net/UpLoad/Web/ProductImg
                /2018-06-15/NEW_XM/IE80S.jpg"
      },
      { id: 3,
        title: "IE60 入耳式 HiFi 耳机",
        titleTwo: "森海塞尔入耳式 HIFI 耳机",
        price: "799.00",
        image: "https://images.wincheers.net/UpLoad/Web/ProductImg
                /2017-08-14/NEW_XM/IE-60-4.png"
      }
    ]
})
```

图 2.32 list 页

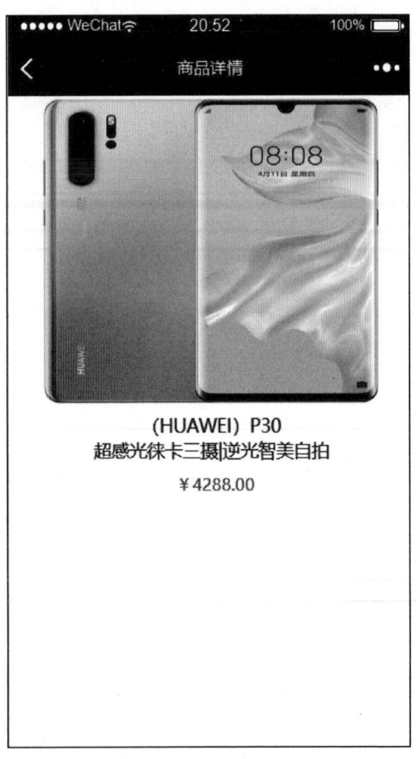

图 2.33 detail 页

pages/list/list.wxml 文件代码如下：

```
<view class = "demo - box">
  <block wx:for = "{{goodsdata}}" wx:for - item = "item">
    <view class = "goods - box" bindtap = "todetail" data - id = "{{item.id}}">
      <image class = "goods - pic" src = "{{item.image}}"></image>
      <view class = "goods - title">{{item.title}}</view>
      <view class = "goods - titleTwo">{{item.titleTwo}}</view>
      <view class = "row">
        <view class = "goods - price">¥{{item.price}}</view>
        <text class = "goods - btn">看相似</text>
      </view>
    </view>
  </block>
</view>
```

pages/list/list.js 文件代码如下：

```
var app = getApp()
Page({
  data:{
    goodsdata:null
  },
  onLoad(){
   this.setData({
     goodsdata:app.goodsdata
   })
  },
  todetail: function(e) {
    var listid = e.currentTarget.dataset.id
    console.log("你点击了第" + (listid + 1) + "个商品")
    wx.navigateTo({
      url: '../detail/detail?listid = ' + listid
    })
  }
})
```

pages/list/list.wxss 文件代码如下：

```
/* 商品标题样式 */
.goods{
  text - align: center;
}
.title {
  font - size: 34rpx;
}
/* 商品价格样式 */
.price {
  color: #f00; font - size: 32rpx;
  padding - top: 20rpx; padding - left: 15rpx;
}
/* 商品展示外部样式 */
```

```
.box - demo {
  display: flex; flex - direction: column;
  width: 100 %; height: 100rpx;
}
```

pages/detail/detail.wxml 文件代码如下：

```
< view class = "goods">
< view class = 'title'> < image src = "{{data.image}}"></image ></view >
< view class = 'title'>{{data.title}} </view >
< view class = 'title'> {{data.titleTwo}}</view >
 < view class = 'price'> ¥ {{data.price}}
</view >
</view >
```

pages/detail/detail.js 文件代码如下：

```
const app = getApp();
Page({
  data: {                                    //数据初始化
    data:null
  },
  //页面加载时获得 app.js 中的数据
  onLoad: function(option) {
    var id = option.listid
    this.setData({
      data: app.goodsdata[id]
    })
  }
})
```

pages/detail/detail.wxss 文件代码如下：

```
/ * 商品标题样式 * /
.goods{
  text - align: center;
}
.title {
  font - size: 34rpx;
}
/ * 商品价格样式 * /
.price {
  color: #f00; font - size: 32rpx;
  padding - top: 20rpx; padding - left: 15rpx;
}
/ * 商品展示外部样式 * /
.box - demo {
  display: flex; flex - direction: column;
  width: 100 %; height: 100rpx;
}
```

【代码讲解】 因为 list 和 detail 两个页面都需要用到同一商品数据，所以本实训项目中涉及的商品数据存放在 app. js 中，list. js 和 detail. js 文件的 onLoad()函数执行语句 main_key：goodsdata：app. goodsdata 获取 app. js 文件中的商品数据。

list. wxml 文件再使用< block wx：for >循环渲染把所有的商品在页面中显示出来。点击事件 todetail()实现商品列表页到商品详情页的跳转。

detail. wxml 页面使用数据绑定的方法把 detail. js 获取的数据在页面中显示出来。detail. js 获取 app. js 的全局数据和 list. js 获取全局数据有所不同，detail. js 使用语句 data：app. goodsdata[id]获取数据，语句中的 id 是由 list 页面传递而来。页面跳转时的 url：'../detail/detail?listid=' + listid 是两个跳转页面之间传递数据的有效方法，在着陆页 detail. js 中传递过来的数据被封装在 option 对象中。

第 3 章

微信小程序组件

微信小程序框架为开发者提供了一系列的基础组件，这些原生组件让小程序具有良好的用户体验，同时也方便开发者快速开发。

本章主要目标

- 了解小程序组件的含义；
- 熟练掌握常见的容器组件、内容组件、表单组件、导航组件、媒体组件和地图组件的属性以及用法；
- 综合运用小程序组件完成问卷调查项目的设计与开发。

3.1 组件概述

组件是视图层基本的组成单元，具备 UI 风格样式以及特定的功能效果。当打开某款小程序后，界面中的图片、文字等元素都需要使用组件，小程序组件使用灵活，组件之间通过相互嵌套进行界面设计，开发者可以通过组件的选择和样式属性设计出不同的界面效果。一个组件包括开始标签和结束标签，属性用来装饰这个组件的样式。

其语法格式如下：

```
<标签名称 属性 = "值">
内容
</标签名称>
```

示例代码如下：

```
< button class = "btn">我是按钮组件</button>
```

上述代码用< button ></button >表示一个按钮组件，在< button >标签中通过 class="btn"为< button >组件添加样式 btn。小程序目前提供的通用属性如表 3.1 所示。

表 3.1　小程序组件通用属性

属 性 名	类 型	说 明	备 注
id	string	组件的唯一标识	在当前界面中用 id 值标识唯一的组件，并且不能有两个及以上 id 值同名
class	string	组件的样式类	为一个或多个组件设置样式类
style	string	组件的内联样式	动态设置内联样式
hidden	boolean	组件的显示/隐藏	组件均默认为显示状态
data-*	any	自定义属性	当组件触发事件时会附带将该属性和值发送给对应的事件处理函数
bind*/catch*	eventHandler	组件的事件	为组件绑定/捕获事件

3.2　视图容器组件

视图容器(View Container)组件用于排版页面为其他组件提供载体。视图容器有 view、scroll-view 和 swiper 3 种。

3.2.1　view

view 容器是页面中最基本的容器组件，通过高度和宽度来定义容器大小。< view >相当于 HTML 中的< div >标签，是一个页面中最外层的容器，能够接受其他组件的嵌入，例如，多个 view 容器的嵌套。view 容器可以通过 flex 布局定义内部项目的排列方式(详见第 4 章 flex 布局)。其属性如表 3.2 所示。

表 3.2　<view>组件属性

属 性 名	类型	默认值	说 明
hover	boolean	false	是否启动点击态
hover-class	string	none	按住容器后的样式。当属性设置为 hover-class="none"时，没有点击效果
hover-stop-propagation	boolean	false	指定是否阻止本容器的祖先节点出现点击态
hover-start-time	number	50	按住容器后多久出现点击态，单位为 ms
hover-stay-time	number	400	手指离开后点击态的保留时长，单位为 ms

例 3-1　本例设计了两组父子 view 容器的点击态，第一组父子 view 容器中子 view 容器不阻止点击态向父容器传递，第二组父子 view 容器中子 view 容器阻止点击态向父容器传递，程序运行效果如图 3.1 所示。

视频讲解

pages/view/view. wxml 文件代码如下：

```
<view class = "demo - box">
  <view class = "title"> 1.view 小案例</view>
  <view class = "title">(1)不阻止父容器的 view - hover </view>
```

```
  <view class = "view - parent" hover - class = "view - hover">我是父类容器
    <view class = "view - son" hover - class = "view - hover">我是子类容器</view>
  </view>
  <view class = "title">(2)阻止父容器的 view - hover </view>
  <view class = "view - parent" hover - class = "view - hover">我是父类容器
    <view class = "view - son" hover - class = "view - hover" hover - stop - propagation hover -
start - time = "3000" hover - stay - time = "4000">我是子类容器</view>
  </view>
</view>
```

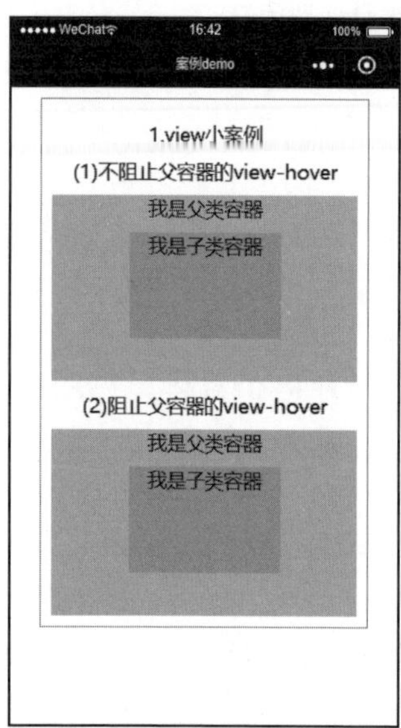

(a) 页面初始效果

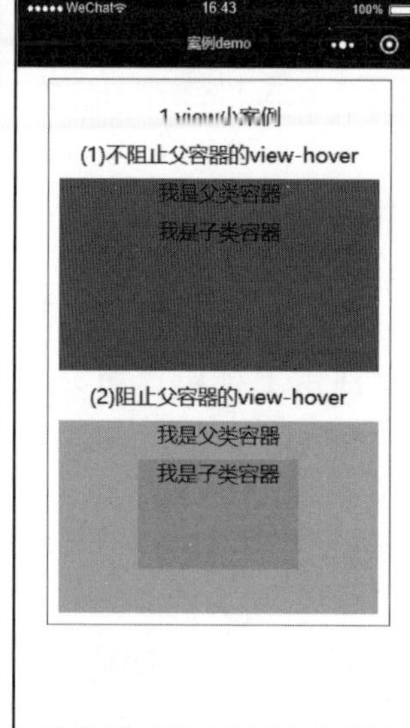

(b) 点击第1组子容器

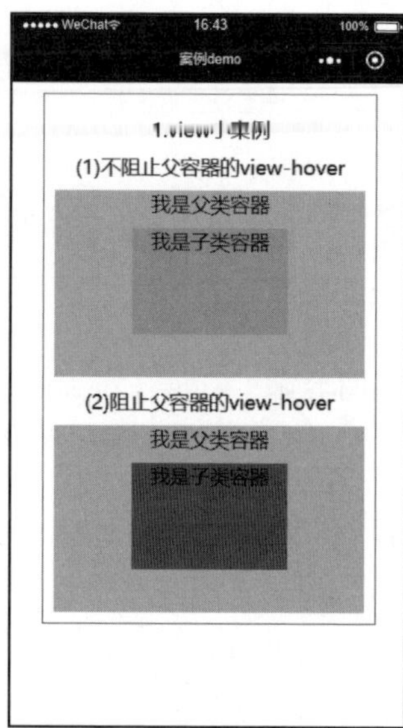

(c) 点击第2组子容器3s后

图 3.1　组件 view 小案例

pages/view/view.wxss 文件代码如下：

```
.view - parent {
  width: 100 %;
  height: 350rpx;
  background - color: pink;
  text - align: center;
}
.view - son {
  width: 50 %;
  height: 200rpx;
  background - color: skyblue;
  margin: 20rpx auto;
  text - align: center;
}
.view - hover {
  background - color: red;
}
```

app.wxss 文件代码如下：

```
.demo - box {
  padding: 20rpx; margin: 20rpx 60rpx; border: 1rpx solid gray;
}
.title {
  display: flex;
  flex - direction: row;
  margin: 20rpx;
  justify - content: center;
}
```

app. wxss 文件中代码为公共样式，用于设置页面的布局以及标题样式，在本章所有案例中均相同，在后面的案例中省略。

【代码讲解】 本例在 view. wxml 文件中放置两组< view >容器，在 app. wxss 文件中设置父容器背景色为浅红色，子容器背景色为浅蓝色，通过 hover-class="view-hover"为标签增加属性，点击态均设置为点击后背景色更新为红色，第一组不阻止点击态传递给父容器，在第二组子类容器中通过 hover-stop-propagation 来阻止点击态传递给父容器，并设置属性 hover-start-time="3000"，hover-stay-time="4000"，当点击子容器时，3s 后出现点击状态，当手指松开 4s 后，子容器背景色变为初始颜色。

图 3.1(a)为页面初始效果；图 3.1(b)为点击第 1 组的子容器后，父子容器背景色均变为红色；图 3.1(c)为点击第 2 组的子容器后，仅有子容器背景色变为红色。

3.2.2　scroll-view

scroll-view 容器为可滚动的视图容器，允许用户通过手指在容器上滑动来改变显示区域，常见的滑动方向有水平滑动和垂直滑动。其属性如表 3.3 所示。

表 3.3　<scroll-view>组件属性

属 性 名	类 型	默认值	说 明
scroll-x	boolean	false	允许横向滑动
scroll-y	boolean	none	允许纵向滑动
upper-threshold	number	50	距顶部/左边多远时(单位：px)，触发 scrolltoupper 事件
lower-threshold	number	50	距底部/右边多远时(单位：px)，触发 scrolltolower 事件
scroll-top	number		设置纵向滚动条位置
scroll-left	number		设置横向滚动条位置
scroll-into-view	string		值应为某子元素 id。设置哪个方向可滚动，则在哪个方向滚动到该元素
scroll-with-animation	boolean	false	在设置滚动条位置时使用动画过渡
enable-back-to-top	boolean	false	iOS 下单击顶部状态、Android 双击标题栏滚动条返回顶部，仅支持纵向
bindscrolltoupper	eventhandle		滚动到顶部/左边，会触发 scrolltoupper 事件
bindscrolltolower	eventhandle		滚动到底部/右边，会触发 scrolltolower 事件
bindscroll	eventhandle		滚动时触发，event. detail = {scrollLeft, scrollTop, scrollHeight, scrollWidth, deltaX, deltaY}

注意：在使用纵向滚动时，需要为< scroll-view >设置一个固定宽度。

例 3-2　本例设计一个纵向 scroll-view 组件，运行效果如图 3.2 所示。

视频讲解

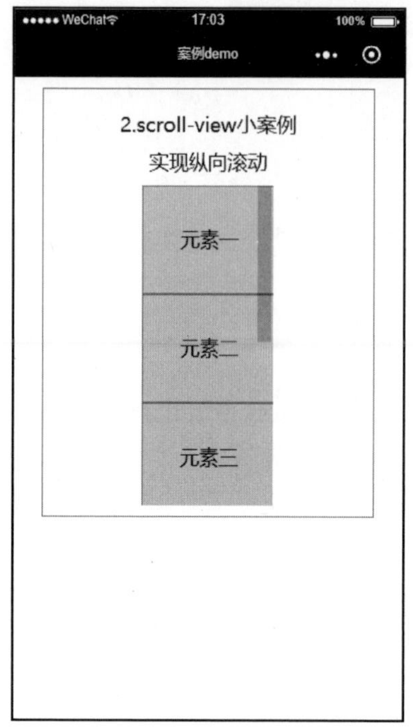

(a) 页面初始效果

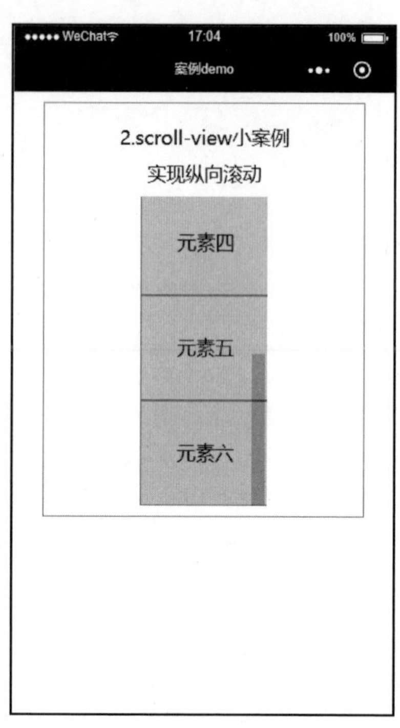

(b) scroll-view滚动后效果

图 3.2　组件 scroll-view 小案例

pages/scroll-view/scroll-view. wxml 文件代码如下：

```
< view class = "demo - box">
< view class = "title"> 2. scroll - view 小案例</view >
< view class = "title">实现纵向滚动</view >
< scroll - view scroll - y >
< view class = "scroll - item - y">元素一</view >
< view class = "scroll - item - y">元素二</view >
< view class = "scroll - item - y">元素三</view >
< view class = "scroll - item - y">元素四</view >
< view class = "scroll - item - y">元素五</view >
< view class = "scroll - item - y">元素六</view >
</scroll - view >
</view >
```

pages/scroll-view/scroll-view. wxss 文件代码如下：

```
scroll - view {
  height: 600rpx; width: 250rpx; margin: 0 auto;
}
.scroll - item - y {
  height: 200rpx; line - height: 200rpx;
  text - align: center; background - color: skyblue; border: 1px solid gray;
```

```
}
```

【代码讲解】 本例在 scroll-view. wxml 文件中放置< scroll-view >组件，通过设置属性 scroll-y，允许组件上下滑动，在 scroll-view. wxss 文件中设置其高度为 600rpx，使得 scroll-view 组件能够纵向滚动，在< scroll-view >中嵌套 6 组< view >用于显示滚动效果，内部元素宽度均为 250rpx。

图 3. 2(a)为页面初始效果；图 3. 2(b)为< scroll-view >组件滑动到底部后的效果。

在图 3. 3 中，虚线框是< scroll-view >组件在京东小程序分类页中的应用。

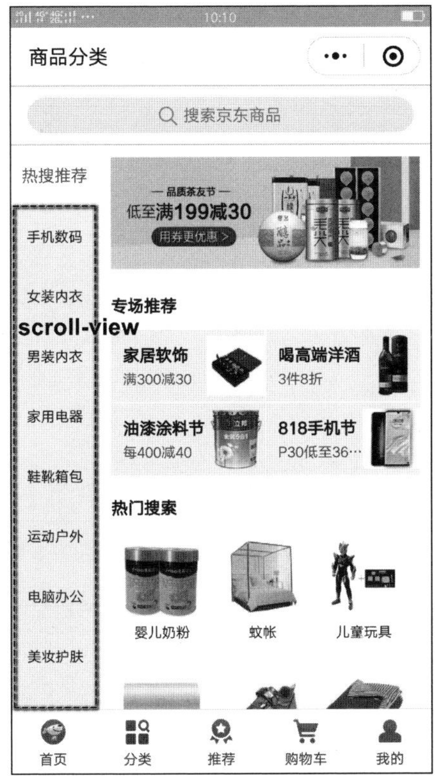

图 3. 3　scroll-view 组件应用实例

3.2.3　swiper

< swiper >组件为滑块视图容器，通常用于图片之间的切换播放，被形象地称为轮播图。其属性如表 3. 4 所示。

表 3. 4　< swiper >组件属性

属 性 名	类 型	默 认 值	说 明
indicator-dots	boolean	false	是否显示面板指示点
indicator-color	color	rgba(0,0,0,0. 3)	指示点颜色
indicator-active-color	color	＃000000	当前选择的指示点颜色

续表

属性名	类型	默认值	说明
autoplay	boolean	false	是否自动切换
current	number	0	当前所在滑块的 index
current-item-id	string	" "	当前所在滑块的 item-id，不能与 current 被同时指定
interval	number	5000	自动切换时间间隔，单位：ms
duration	number	500	滑动动画时长，单位：ms
circular	boolean	false	是否采用衔接滑动
vertical	boolean	false	滑动方向是否为纵向
previous-margin	string	"0px"	前边距，可用于露出前一项的一小部分，接受 px 和 rpx 值
next-margin	string	"0px"	后边距，可用于露出后一项的一小部分，接受 px 和 rpx 值
display-multiple-items	number	1	同时显示的滑块数量
skip-hidden-item-layout	boolean	false	是否跳过未显示的滑块布局，设为 true 可优化复杂情况下的滑动性能，但会丢失隐藏状态滑块的布局信息
bindchange	eventhandle		current 改变时会触发 change 事件，event.detail = {current：current，source：source}
bindtransition	eventhandle		swiper-item 的位置发生改变时会触发 transition 事件，event.detail = {dx：dx，dy：dy}
bindanimationfinish	eventhandle		动画结束时会触发 animationfinish 事件，event.detail 同上

例 3-3 swiper 组件小案例，运行效果如图 3.4 所示。pages/swiper/swiper.wxml 文件代码如下：

视频讲解

```
< view class = "demo - box">
  < view class = "title"> 3. swiper 小案例</view >
  < view class = "title">图片进行翻页切换</view >
  < swiper indicator - dots autoplay interval = "3000">
    < swiper - item >
      < image src = "/images/cat1.jpg"></image >
    </swiper - item >
    < swiper - item >
      < image src = "/images/cat2.jpg"></image >
    </swiper - item >
    < swiper - item >
      < image src = "/images/cat3.jpg"></image >
    </swiper - item >
  </swiper >
</view >
```

(a) 页面初始效果

(b) 切换到第2张照片

(c) 切换到第3张照片

图 3.4　swiper 组件小案例

pages/swiper/swiper.wxss 文件代码如下：

```
swiper {
  height: 350rpx;
}
```

【代码讲解】　本例在 swiper.wxml 文件中放置<swiper>组件，设置属性 autoplay 允许自动切换图片，设置属性 interval="3000"，图片每隔 3s 发生一次切换，属性 indicator-dots 用于显示面板指示点，<swiper>组件中嵌套 3 组<swiper-item>，swiper 容器的高度设置为 300rpx。

图 3.4(a)为页面初始效果，此时默认显示第一张图片；图 3.4(b)和图 3.4(c)分别显示第二张照片和第三张照片，照片数据来自本地，保存在 images 文件夹下。

在图 3.5 中，虚线框是<swiper>组件在携程小程序中的应用。

图 3.5　swiper 组件应用实例

3.3　基础内容组件

3.3.1　icon

<icon>为图标组件，常用于页面装饰，开发者可以自定义其类型、大小和颜色。其属性

如表 3.5 所示。

表 3.5 <icon>组件属性

属性名	类　型	默认值	说　明
type	string	false	icon 的类型,有效值：success,success_no_circle, info, warn,waiting,cancel,download, search, clear
size	number/string	23px	icon 的大小,单位：px(基础库 2.4.0 起支持 rpx)
color	color	#000000	icon 的颜色,同 CSS 的 color

例如,自定义一个绿色、40px 大小的 success 图标。

WXML 中的代码如下：

```
< icon type = "success" size = "40" color = "green" />
```

如果有多个图标需要批量生成,利用 wx:for 循环精简代码,在 JS 文件的 data 中存放数据,然后在 WXML 文件中使用< block >标签进行列表渲染。

批量生成不同大小的 success 图标的示例代码如下。

WXML 中的代码如下：

```
< view >
  < block wx:for = "{{iconSize}}">
    < icon type = "success" size = "{{item}}" />
  </block >
</view >
```

JS 文件代码如下：

```
Page({
  data: {
    iconSize: ["20", "25", "30"]
  }
})
```

上述代码生成的图标大小分别为 20rpx、25rpx 和 30rpx。

例 3-4　icon 组件小案例,运行效果如图 3.6 所示。

pages/icon/icon.wxml 文件代码如下：

视频讲解

```
< view class = "demo - box">
  < view class = "title"> 4. icon 小案例</view >
  < view class = "title">(1)实现大小变化</view >
  < block wx:for = "{{iconSize}}">
    < icon type = "success" size = "{{item}}" />
  </block >
  < view class = "title">(2)实现内容变化</view >
  < block wx:for = "{{iconType}}">
    < icon type = "{{item}}" size = "40" />
  </block >
  < view class = "title">(3)实现颜色变化</view >
  < block wx:for = "{{iconColor}}">
```

```
    <icon type="success" size="40" color="{{item}}" />
  </block>
</view>
```

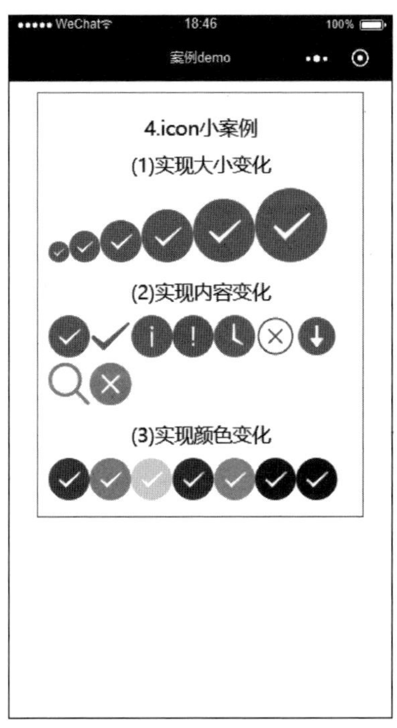

图 3.6　icon 组件小案例

pages/icon/icon.js 文件代码如下：

```
Page({
  data: {
    iconSize: [20, 30, 40, 50, 60, 70],
    iconType: [
      "success", "success_no_circle", "info", "warn", "waiting", "cancel", "download",
"search", "clear"
    ],
    iconColor: [
      "red", "orange", "yellow", "green", "red", "blue", "purple"
    ],
  }
})
```

【代码讲解】　本例在 icon.js 文件中的 data 中设置 3 个数组，分别为 iconSize、iconType 和 iconColor，用于设置图标的大小、图标的类型和图标的颜色。在 icon.wxml 文件中使用<block>标签配合 wx:for 批量生成多个标签组件。

3.3.2　text

<text>为文本组件，用于文字的显示，小程序的文本组件支持转义字符。其属性如

表 3.6 所示。

表 3.6 <text>组件属性

属性名	类型	默认值	说明	最低版本
selectable	boolean	false	文本是否可选	1.1.0
space	string		显示连续空格	1.4.0
decode	boolean	false	是否解码	1.4.0

例 3-5 组件 text 小案例，运行效果如图 3.7 所示。

pages/text/text.wxml 文件代码如下：

视频讲解

```
< view class = "demo - box">
  < view class = "title"> 5.text 小案例</view >
  < view class = "title">用于文本的显示</view >
  < text >{{text}}</text >
  < button bindtap = "add">增加一行</button >
  < button bindtap = "reduce">删除一行</button >
</view >
```

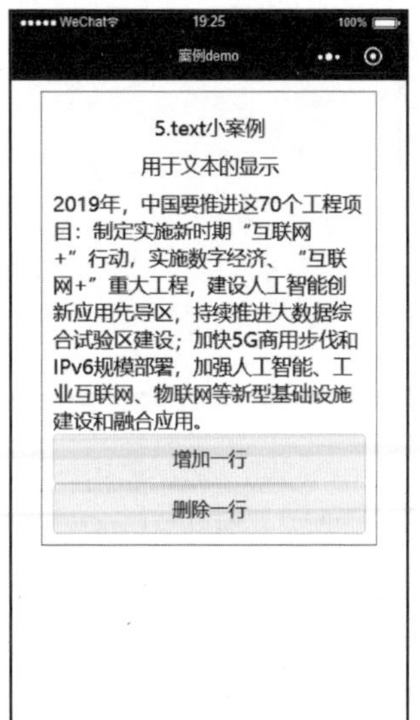

(a) 页面初始效果

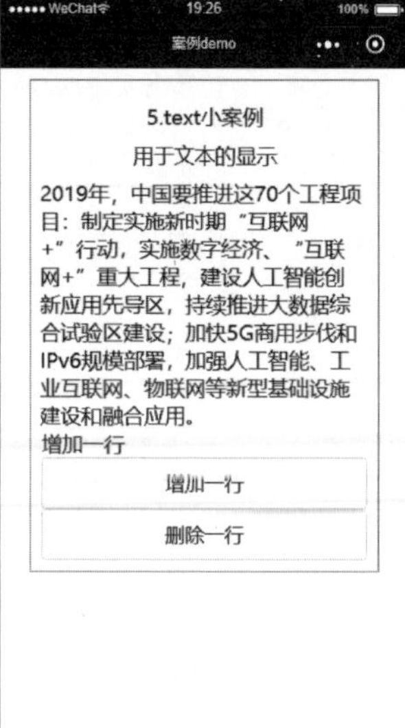

(b) 增加一行文字

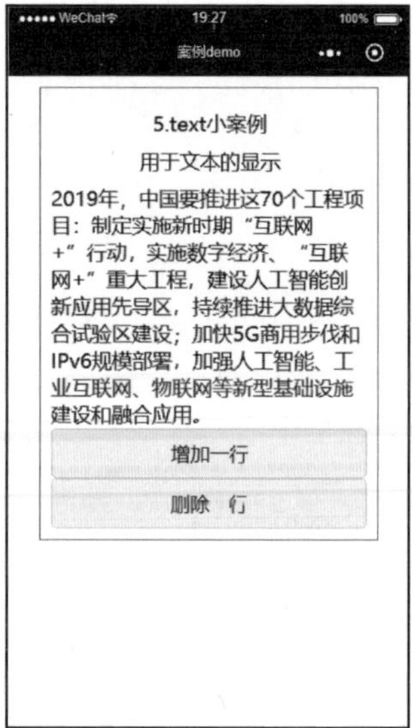

(c) 删除一行文字

图 3.7 text 组件小案例

pages/text/text.js 文件代码如下：

```
var initData = "2019 年,中国要推进这 70 个工程项目: 制定实施新时期"互联网 + "行动,实施数字经济、"互联网 + "重大工程,建设人工智能创新应用先导区,持续推进大数据综合试验区建设; 加快5G 商用步伐和 IPv6 规模部署,加强人工智能、工业互联网、物联网等新型基础设施建设和融合应用。"
var extraLine = [];                                    //创建一个空数组
Page({
  data: {
```

```
      text: initData
    },
    add: function(e) {
      extraLine.push("增加一行")                    //增加一行
      this.setData({
        text: initData + '\n' + extraLine.join('\n')      //更新数组值
      })
    },
    reduce: function(e) {
      if(extraLine.length > 0) {
        extraLine.pop()                              //删除一行
        this.setData({
          text: initData + '\n' + extraLine.join('\n')    //更新数组值
        })
      }
    }
})
```

【代码讲解】 本例在 text.wxml 文件中通过< text >组件存放文字，以及增加两个< button >按钮，分别绑定了点击事件，用于实现增加一行和删除一行的操作，对应在 text.js 文件中自定义 add()和 reduce()两个函数。

图 3.7(a)为初始页面；图 3.7(b)为点击"增加一行"按钮后，在文字内容下方增加一行；图 3.7(c)为点击"删除一行"按钮后，新增加的一行消失，回到页面初始状态。

3.3.3　progress

< progress >为进度条组件，用于进度的显示，长度单位默认为 px。其属性如表 3.7 所示。

表 3.7　< progress >组件属性

属性名	类型	默认值	说明
percent	number		百分比 0%～100%
show-info	boolean	false	在进度条右侧显示百分比
border-radius	number/string	0	圆角大小
font-size	number/string	16	右侧百分比字体大小
stroke-width	number/string	6	进度条线的宽度
color	string	#09BB07	进度条颜色(请使用 activeColor)
activeColor	string	#09BB07	已选择的进度条的颜色
backgroundColor	string	#EBEBEB	未选择的进度条的颜色
active	boolean	false	进度条从左往右的动画
active-mode	string	backwards	backwards：动画从头播；forwards：动画从上次结束点接着播

例如，自定义一个当前进度为 30，宽度为 10rpx 的进度条，示例代码如下：

```
<progress percent = "30" stroke - width = "10rpx"></progress>
```

其运行效果如图 3.8 所示。

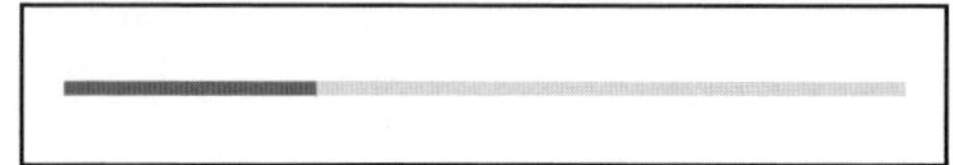

图 3.8　进度条图示

例 **3-6**　progress 组件小案例，运行效果如图 3.9 所示。

视频讲解

pages/progress/progress.wxml 文件代码如下：

```
<view class = "demo - box">
  <view class = "title"> 6.progress 小案例</view>
  <view class = "title">增加和减少进度</view>
  <progress percent = "{{progress}}" stroke - width = "10rpx" show - info ></progress>
  <button bindtap = "add">增加 10 % </button>
  <button bindtap = "reduce">减少 10 % </button>
</view>
```

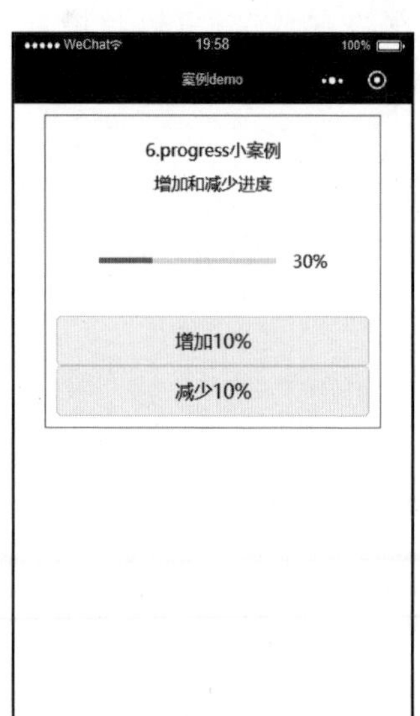

(a) 页面初始效果

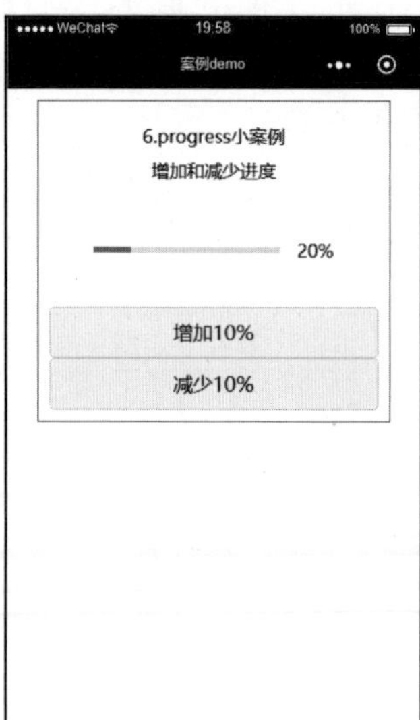

(b) 减少10%

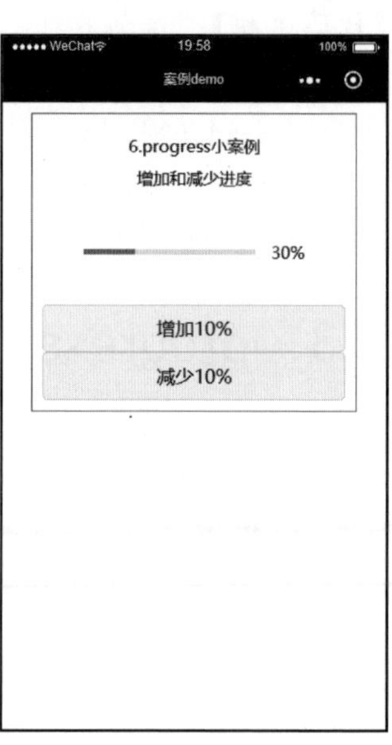

(c) 增加10%

图 3.9　progress 组件小案例

pages/progress/progress.js 文件代码如下：

```
var per = 30;
Page({
  data: {
    progress: per,                          //数据初始化
  },
  add: function(e) {
    per += 10;                              //增加 10 %
    if (per > 100) {
      per = 100;                            //进度超过 100 % 不增加
    }
```

```
      this.setData({
        progress: per                                //更新进度值
      })
    },
    reduce: function(e) {
      per -= 10;                                     //减少 10%
      if (per < 0) {
        per = 0;                                     //进度小于 0%不减少
      }
      this.setData({
        progress: per                                //更新进度值
      })
    }
  })
```

pages/progress/progress.wxss 文件代码如下：

```
progress {
  padding: 80rpx;
}
```

【代码讲解】 本例在 progress.wxml 文件中放置< progress >组件，通过设置属性 show-info 让进度条右边显示当前进度值，以及增加两个< button >按钮，分别绑定了点击事件，用于实现增加 10%的进度值和减少 10%的进度值，在 progress.js 文件中自定义 add()和 reduce()两个函数。

图 3.9(a)为初始页面；图 3.9(b)为点击“减少 10%”按钮后，进度值从 30%变为 20%；图 3.9(c)为点击“增加 10%”按钮后，进度值从 20%变为 30%。

3.4 表单组件

表单组件在网页开发中十分常见，对于小程序而言，当用户需要设计注册、登录等页面时可以使用表单。表单组件是多类组件的统称。

3.4.1 button

< button >为按钮组件，是常用的表单组件之一，用于事件的触发以及表单的提交。其属性如表 3.8 所示。

表 3.8 < button >组件属性

属 性 名	类型	默认值	说 明	最低版本
size	string	default	按钮的大小	
type	string	default	按钮的样式类型	
plain	boolean	false	按钮是否镂空，背景色透明	
disabled	boolean	false	是否禁用	
loading	boolean	false	名称前是否带 loading 图标	

续表

属 性 名	类型	默认值	说 明	最低版本
form-type	string		用于< form >组件，点击分别会触发< form >组件的 submit/reset 事件	
open-type	string	button-hover	微信开放能力	1.1.0
hover-class	string	button-hover	指定按钮按下去的样式类。当 hover-class="none"时，没有点击态效果	
hover-stop-propagation	boolean	false	指定是否阻止本节点的祖先节点出现点击态	1.5.0
hover-start-time	number	20	按住后多久出现点击态，单位：ms	
hover-stay-time	number	70	手指松开后点击态保留时间，单位：ms	

注意：button-hover 默认为{background-color：rgba(0，0，0，0.1)；opacity：0.7；}。

size 属性值如下。

(1) default：默认按钮，宽度与手机屏幕宽度保持一致。

(2) mini：迷你按钮，尺寸小于默认按钮。

示例代码如下：

```
<button>默认按钮</button>
<button size="mini">迷你按钮 </button>
```

其运行效果如图 3.10 所示。

type 属性值如下。

(1) primary：主要按钮，颜色为绿色。

(2) default：默认按钮，颜色为灰白色。

(3) warn：警告按钮，颜色为红色。

示例代码如下：

```
<button type="primary">primary 按钮</button>
<button type="default">default 按钮</button>
<button type="warn">warn 按钮</button>
```

其运行效果如图 3.11 所示。

图 3.10　默认按钮和迷你按钮

图 3.11　type 属性值对应的按钮

form-type 属性值如下。

(1) submit：提交表单。

(2) reset：重置表单。

示例代码如下：

```
< button form - type = "submit">提交按钮</button >
< button form - type = "reset">重置按钮</button >
```

其运行效果如图 3.12 所示。

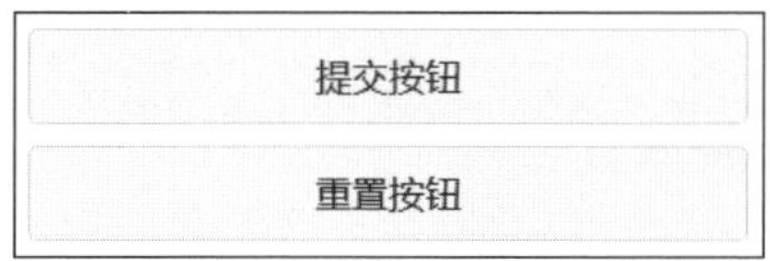

图 3.12　form-type 属性值对应的按钮

视频讲解

例 3-7　button 组件小案例，运行效果如图 3.13 所示。

图 3.13　button 组件小案例

pages/button/button.wxml 文件代码如下：

```
< view class = "demo - box">
  < view class = "title"> 7.button 小案例</view >
  < view class = "title">(1)迷你按钮</view >
  < button size = "mini" type = "primary">主要按钮</button >
  < button size = "mini" type = "default">次要按钮</button >
  < button size = "mini" type = "warn">警告按钮</button >
  < view class = "title">(2)按钮状态</view >
  < button >普通按钮</button >
  < button disabled >禁用按钮</button >
  < button loading >加载按钮</button >
  < view class = "title">(3)增加按钮事件</view >
  < button bindgetuserinfo = "getUserDetail" open - type = "getUserInfo">点我获取用户信息
```

```
</button>
</view>
```

pages/button/button.js 文件代码如下：

```
Page({
  getUserDetail: function(e) {
    console.log(e.detail.userInfo)
  }
})
```

pages/button/button.wxss 文件代码如下：

```
button {
  margin: 10rpx;
}
```

【代码讲解】 在 button.wxml 文件中设置 3 组效果，分别为迷你按钮、普通按钮和带点击事件按钮。第 1 组设置相同的 size 属性和不同的 type 属性实现 3 种不同类型的迷你按钮；第 2 组设置属性 disabled 和 loading 实现按钮禁用和加载动画效果；第 3 组通过 bindgetuserinfo="getUserDetail"为按钮增加事件，并追加 open-type="getUserInfo"状态，然后在 JS 文件中定义点击事件。

3.4.2 checkbox

<checkbox>为复选框组件，常用于在表单中进行多项数据的选择。复选框的<checkbox-group>为父控件，其内部嵌套若干个<checkbox>子控件。

<checkbox-group>组件只有一个属性，如表 3.9 所示。

表 3.9 <checkbox-group>组件属性

属性名	类型	默认值	说明
bindchange	eventhandle		<checkbox-group>选中项发生改变时触发 change 事件，detail = {value:[选中的 checkbox 的 value 数组]}

<checkbox>组件的属性如表 3.10 所示。

表 3.10 <checkbox>组件属性

属性名	类型	默认值	说明
value	string		<checkbox>标识，选中时触发<checkbox-group>的 change 事件，并携带<checkbox>的 value
disabled	boolean	false	是否禁用
checked	boolean	false	当前是否选中，可用来设置默认选中
color	color		checkbox 的颜色，同 CSS 的 color

示例代码如下：

```
<checkbox-group>
```

```
    <checkbox value="tiger" checked="true" />老虎
    <checkbox value="elephant" disabled="true" />大象
    <checkbox value="lion" />狮子
    <checkbox value="penguin" />企鹅
</checkbox-group>
```

其运行效果如图 3.14 所示。

图 3.14　复选框图示

视频讲解

如图 3.14 所示,“老虎”选项被选中,“大象”选项禁止选择,“狮子”和“企鹅”选项均未选中。

例 3-8　checkbox 组件小案例,运行效果如图 3.15 所示。

(a) 多个选项被选中

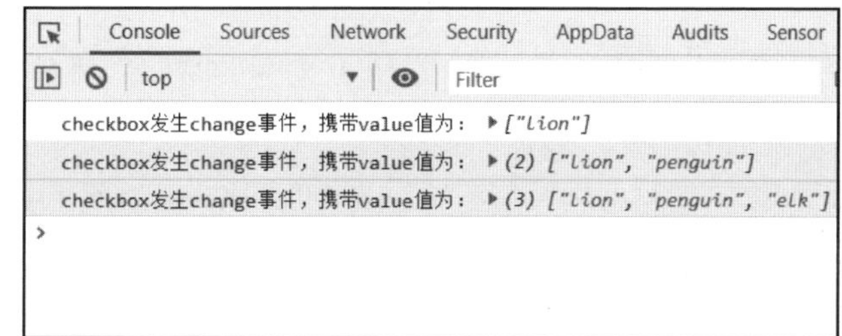

(b) 多个选项被选中时Console输出的内容

图 3.15　checkbox 组件小案例

pages/checkbox/checkbox.wxml 文件代码如下:

```
<view class="demo-box">
  <view class="title">8.checkbox 小案例</view>
  <view class="title">利用 for 循环批量生成</view>
  <checkbox-group bindchange="checkboxChange">
    <label wx:for="{{items}}">
      <checkbox value="{{item.name}}" checked="{{item.checked}}" />{{item.value}}
    </label>
  </checkbox-group>
</view>
```

pages/checkbox/checkbox.js 文件代码如下:

```
Page({
  data: {
    items: [
      { name: "tiger", value: "老虎" },
      { name: "elephant", value: "大象" },
      { name: "lion", value: "狮子", checked: "true" },
      { name: "penguin", value: "企鹅" },
      { name: "elk", value: "麋鹿" },
      { name: "swan", value: "天鹅" },
    ]
```

```
    },
    checkboxChange:function(e) {
      console.log("checkbox 发生 change 事件,携带 value 值为：", e.detail.value)
    }
})
```

【代码讲解】 本例首先在 checkbox.js 文件中的 data 中定义一个数组 items,用于记录复选框的名称(name)、值(value)以及选中情况,并在 check.wxml 文件中使用<checkbox-group>标签包裹<checkbox>,使用<label>标签配合 wx:for 实现批量生成多个 checkbox 组件；其次在<checkbox-group>标签上绑定监听事件,在 checkbox.js 文件中自定义 checkboxChange()函数,以达到每次被触发后都在 Console 控制台输出最新选中的值。

图 3.15(a)为多个选项被选中后的效果；图 3.15(b)为 Console 控制台输出的内容,显示被选中的选项所携带的值。

3.4.3 input

<input>为输入框组件,常用于文本(如姓名、年龄等信息)的输入。其属性如表 3.11 所示。

表 3.11 <input>组件属性

属性名	类型	默认值	说明
value	string		输入框的初始内容
type	string	"text"	input 的类型
password	boolean	false	是否是密码类型
placeholder	string		输入框为空时的占位符
placeholder-style	string		指定 placeholder 的样式
placeholder-class	string		"input-placeholder"指定 placeholder 的样式类
disabled	boolean	false	是否禁用
maxlength	number	140	最大输入长度,设置为-1 的时候不限制最大长度
cursor-spacing	number	0	指定光标与键盘的距离,单位：px(基础库 2.4.0 起支持 rpx)。取 input 距离底部的距离和 cursor-spacing 指定距离的最小值作为光标与键盘的距离
auto-focus	boolean	false	(即将废弃,请直接使用 focus)自动聚焦,拉起键盘
focus	boolean	false	获取焦点
confirm-type	string	"done"	设置键盘右下角按钮的文字,仅在 type='text'时生效
confirm-hold	boolean	false	点击键盘右下角按钮时是否保持键盘不收起
cursor	number		指定 focus 时的光标位置
selection-start	number	-1	光标起始位置,自动聚集时有效,需与 selection-end 搭配使用
selection-end	number	-1	光标结束位置,自动聚集时有效,需与 selection-start 搭配使用
adjust-position	boolean	true	键盘弹起时,是否自动上推页面
bindinput	eventhandle		键盘输入时触发,event.detail = {value, cursor, keyCode},keyCode 为键值,基础库 2.1.0 起支持,处理函数可以直接返回一个字符串,将替换输入框的内容

续表

属　性　名	类　　型	默认值	说　　明
bindfocus	eventhandle		输入框聚焦时触发，event. detail ＝ {value, height}，height 为键盘高度，在基础库 1.9.90 起支持
bindblur	eventhandle		输入框失去焦点时触发，event. detail ＝ {value：value}
bindconfirm	eventhandle		点击“完成”按钮时触发，event. detail ＝ {value：value}

type 属性值如下。

(1) text：文本输入键盘。

(2) Number：数字输入键盘。

(3) idcard：身份证输入键盘。

(4) digit：带小数点的数字键盘。

confirm-type 属性值如下。

(1) send：右下角按钮为“发送”。

(2) search：右下角按钮为“搜索”。

(3) next：右下角按钮为“下一个”。

(4) go：右下角按钮为“前往”。

(5) done：右下角按钮为“完成”。

视频讲解

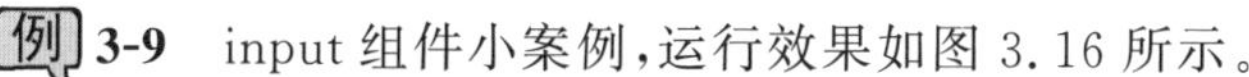

例 3-9　input 组件小案例，运行效果如图 3.16 所示。

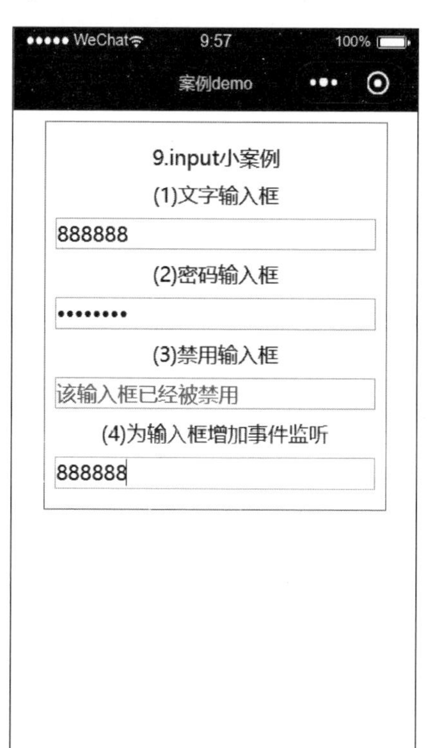

(a) 模拟器效果

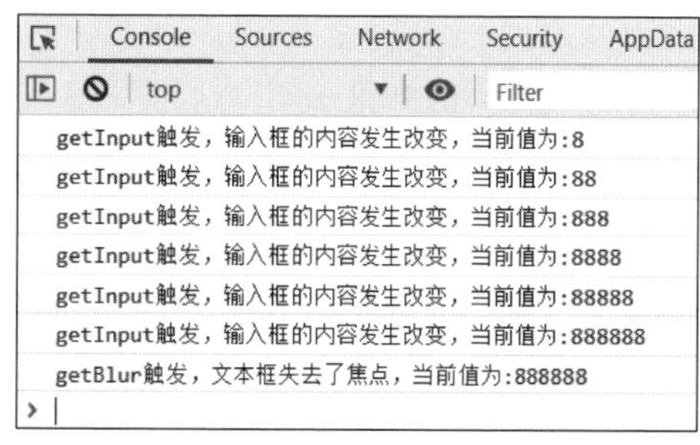

(b) 控制台效果

图 3.16　input 组件小案例

pages/input/input.wxml 文件代码如下：

```
<view class = "demo - box">
  <view class = "title">9.input 小案例</view>
  <view class = "title">(1)文字输入框</view>
  <input type = "text" maxlength = "10" placeholder = "这里最多只能输入 10 个字" />
  <view class = "title">(2)密码输入框</view>
  <input type = "password" placeholder = "请输入密码" />
  <view class = "title">(3)禁用输入框</view>
  <input disabled placeholder = "该输入框已经被禁用" />
  <view class = "title">(4)为输入框增加事件监听</view>
  <input bindinput = "getInput" bindblur = "getBlur" placeholder = "这里输入的内容将会被监
听"/>
</view>
```

pages/input/input.js 文件代码如下：

```
Page({
  getInput: function(e) {
    console.log("getInput 触发,输入框的内容发生改变,当前值为:" + e.detail.value);
  },
  getBlur: function(e) {
    console.log("getBlur 触发,文本框失去了焦点,当前值为:" + e.detail.value);
  },
})
```

pages/input/input.wxss 文件代码如下：

```
input {
  margin: 20rpx auto; border: 1px solid silver;
}
```

【代码讲解】 本例包含 4 个输入框 input 组件，分别为最大字符长度限时为 10、密码输入框、禁用输入框以及带有监听事件的输入框。在 input.wxml 文件中前三组分别通过设置属性 maxlength、password、disabled 来实现效果；第四组通过属性 bindinput 和 bindfocus 为输入框增加当键盘输入时和失去焦点时所触发的事件，并在 input.js 文件中自定义两个事件函数 getInput()和 getBlur()。

图 3.16(a)是模拟器上的程序效果；图 3.16(b)是输入数据和失去焦点时程序在 Console 控制台打印的结果。

3.4.4 label

<label>是标签组件，label 组件不会呈现任何效果，但是可以用来改进表单组件的可用性。当用户在 label 元素内点击文本时，就会触发此控件，即当用户选择该标签时，事件会传递到和标签相关的表单控件上，可以使用 for 属性绑定 id，也可以将控件放在该标签内部。该组件对应的属性如表 3.12 所示。

表 3.12 <label>组件属性

属 性 名	类 型	说 明
for	String	绑定控件的 id

注意：目前可以绑定的控件有<button>、<checkbox>、<radio>、<switch>。

这里以复选框<checkbox>为例，使用<label>标签的 for 属性绑定对应复选框，示例代码如下：

```
<checkbox-group>
  <checkbox id="tiger" value="tiger" checked="true" />
  <label for="tiger">老虎</label>
</checkbox-group>
```

也可以使用<label>包裹<checkbox>标签：

```
<checkbox-group>
  <label>
    <checkbox value="tiger" checked="true" />老虎
  </label>
</checkbox-group>
```

视频讲解

上述两种做法效果一样，当用户点击"老虎"内容区域时，<checkbox>组件均被选中。

例 3-10 label 组件小案例，运行效果如图 3.17 所示。

pages/label/label.wxml 文件代码如下：

```
<view class="demo-box">
  <view class="title">10.label 小案例</view>
  <view class="title">(1)利用 for 属性</view>
  <checkbox-group>
    <checkbox id="tiger" checked />
    <label for="tiger">老虎</label>
    <checkbox id="elephant" />
    <label for="elephant">大象</label>
    <checkbox id="lion" />
    <label for="lion">狮子</label>
  </checkbox-group>
  <view class="title">(2)label 包裹组件</view>
  <checkbox-group>
    <label>
      <checkbox checked />老虎
    </label>
    <label>
      <checkbox/>大象
    </label>
    <label>
      <checkbox/>狮子
    </label>
```

```
  </checkbox - group>
</view>
```

pages/label/label.wxss 文件代码如下：

```
checkbox - group {
  margin: 0 80rpx ;
}
```

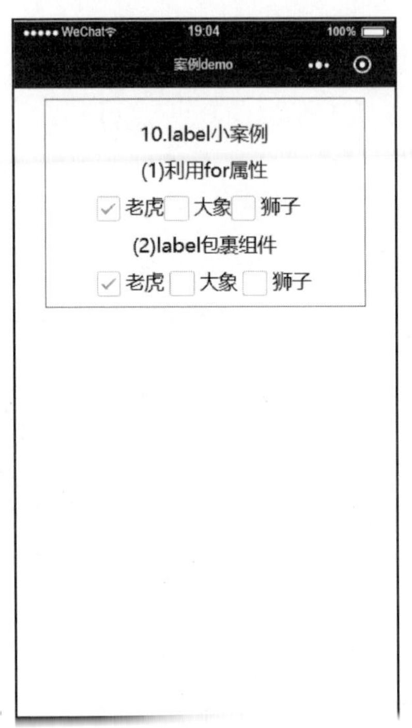

(a) “老虎”选项被默认选中

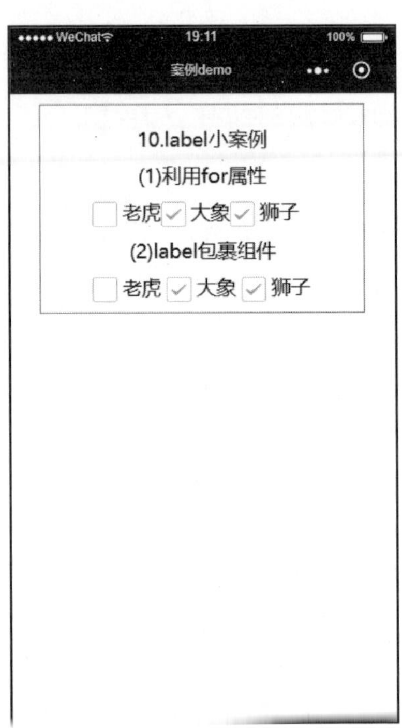

(b) 点击标签后的效果

图 3.17　label 组件小案例

【代码讲解】 本例在 label.wxml 文件中放置两组标签组件，第一组使用 for 属性绑定 id，第二组让 label 包裹<checkbox>组件。

如图 3.17 所示，两种方法效果相同，当分别单击每个文字内容后，如果左边复选框是选中状态，就会变成未选中状态；如果左边复选框是未选中状态，就会变成选中状态。

3.4.5　form

<form>为表单控件组件，用于提交表单组件中的内容。<form>控件组件内部可以嵌套多种组件，具体组件类型如下所示。

(1) <input>：输入框组件；

(2) <button>：按钮组件；

(3) <checkbox>：复选框组件；

(4) <switch>：开关选择器；

(5) < radio >：单项框组件；
(6) < picker >：滚动选择器；
(7) < slider >：滑动选择器；
(8) < textarea >：多行输入框；
(9) < label >：标签组件。

< form >控件组件的属性如表 3.13 所示。

表 3.13　< form >控件组件属性

属　性　名	类　　型	说　　明
report-submit	boolean	是否返回 formId 用于发送模板消息
bindsubmit	eventhandle	携带 form 中的数据触发 submit 事件，event.detail = {value：{'name'：'value'}，formId：''}
bindreset	eventhandle	表单重置时会触发 reset 事件

注意：当表单组件< form >需要提交二级内容表单组件(如< input >)的内容时，< form >组件中需添加 bindsubmit 事件，< button >组件需添加 form-type 属性并赋值为 submit，此时的二级内容表单组件(如< input >)还需要设置 name 属性。当需要在用户提交表单之后发送模板消息时，表单组件< form >需设置 report-submit 属性并赋值为 true。

例 3-11　form 控件组件小案例，运行效果如图 3.18 所示。

视频讲解

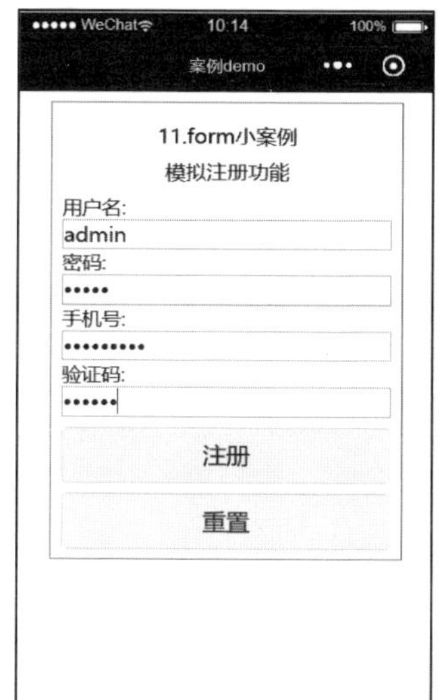

(a) 页面初始效果

Console　Sources　Network　Security　AppData　Audits　Sensor　Storage　Trace
top　Filter　Default levels
form发生了submit事件，携带数据为：
{username: "admin", password: "admin", phonenumber: "18888888888", code: "123456"}
code: "123456"
password: "admin"
phonenumber: "18888888888"
username: "admin"
__proto__: Object
form发生了reset事件，表单已被重置

(b) Console控制台打印的信息

图 3.18　form 组件小案例

pages/form/form.wxml 文件代码如下：

```
<view class = "demo - box">
  <view class = "title"> 11.form 小案例</view>
  <view class = "title">模拟注册功能</view>
  <form bindsubmit = "onSubmit" bindreset = "onReset">
```

```
    <text>用户名:</text>
    <input name = "username" type = "text" placeholder = "请输入你的用户名"></input>
    <text>密码:</text>
    <input name = "password" type = "password" placeholder = "请输入你的密码"></input>
    <text>手机号:</text>
    <input name = "phonenumber" type = "password" placeholder = "请输入你的手机号"></input>
    <text>验证码:</text>
    <input name = "code" type = "password" placeholder = "请输入验证码"></input>
    <button form - type = "submit">注册</button>
    <button form - type = "reset">重置</button>
  </form>
</view>
```

pages/form/form.js 文件代码如下：

```
Page({
  onSubmit(e) {
    console.log("表单被注册")
    console.log("form 发生了 submit 事件,携带数据为: ", e.detail.value)
  },
  onReset() {
    console.log("form 发生了 reset 事件,表单已被重置")
  }
})
```

pages/form/form.wxss 文件代码如下：

```
input {
  border: 1px solid silver;
}
button {
  margin - top: 20rpx;
}
text {
  font - size: 36rpx;
}
```

【代码讲解】 本示例在 form.wxml 文件中放置一个<form>控件组件，为其绑定监听事件 bindsubmit="onSubmit"和 bindreset="onReset"，用于监听表单的提交和重置。在<form>控件组件内部嵌套 4 个<input>组件，设置属性 type="text"以及 type="password"用于用户名和密码的输入。页面底部放置两个<button>组件，设置属性 form-type="submit"和 form-type="reset"用于提交和重置表单，并在 form.js 文件中自定义 onSubmit()函数和 onReset()函数。

图 3.18(a)为输入数据时模拟器的效果；图 3.18(b)是 Console 控制台打印的注册信息效果。

3.4.6 picker

<picker>为滚动选择器，从页面底部弹出供用户选择。根据 mode 属性值的不同共有 5 种选择器，分别为普通选择器、多列选择器、时间选择器、日期选择器和省市区选择器。

1. 普通选择器

当 mode ="selector"时为普通选择器效果，相关属性如表 3.14 所示。

表 3.14 <picker mode="selector">组件属性

属 性 名	类 型	默认值	说 明
range	Array/ObjectArray	[ ]	mode 为 selector 或 multiSelector 时，range 有效
range-key	string		当 range 是一个 Object Array 时，通过 range-key 来指定 Object 中 key 的值作为选择器显示内容
value	number	0	表示选择了 range 中的第几个(下标从 0 开始)
bindchange	eventhandle		value 改变时触发 change 事件，event. detail = {value: value}
disabled	boolean	false	是否禁用
bindcancel	eventhandle		取消选择或点遮罩层收起 picker 时触发

2. 多列选择器

当 mode="multiSelector"时为多列选择器效果，其属性如表 3.15 所示。

表 3.15 <picker mode="multiSelector">组件属性

属 性 名	类 型	默认值	说 明
range	Array/ObjectArray	[ ]	mode 为 selector 或 multiSelector 时，range 有效
range-key	string		当 range 是一个 Object Array 时，通过 range-key 来指定 Object 中 key 的值作为选择器显示内容
value	number	0	表示选择了 range 中的第几个(下标从 0 开始)
bindchange	eventhandle		value 改变时触发 change 事件，event. detail = {value: value}
bindcolumnchange	eventhandle		列改变时触发

3. 时间选择器

当 mode="time"时为时间选择器效果，其属性如表 3.16 所示。

表 3.16 <picker mode="time">组件属性

属 性 名	类 型	默认值	说 明
value	string		表示选中的时间，格式为"hh:mm"
start	string		表示有效时间范围的开始，字符串格式为"hh:mm"
end	string		表示有效时间范围的结束，字符串格式为"hh:mm"
bindchange	eventhandle		value 改变时触发 change 事件，event. detail = {value}
bindcolumnchange	eventhandle		列改变时触发

4. 日期选择器

当 mode="date"时为日期选择器效果，其属性如表 3.17 所示。

表 3.17 <picker mode="date">组件属性

属性名	类型	默认值	说明
value	string	0	表示选中的日期，格式为"YYYY-MM-DD"
start	string		表示有效日期范围的开始，字符串格式为"YYYY-MM-DD"
end	string		表示有效日期范围的结束，字符串格式为"YYYY-MM-DD"
fields	string	day	有效值 year，month，day，表示选择器的粒度
bindchange	eventhandle		value 改变时触发 change 事件，event. detail = {value}

5. 省市区选择器

当 mode="region"时为省市区选择器效果，其属性如表 3.18 所示。

表 3.18 <picker mode="region">组件属性

属性名	类型	默认值	说明
value	array	[]	表示选中的省市区，默认选中每一列的第一个值
custom-item	string		可为每一列的顶部添加一个自定义的项
bindchange	eventhandle		value 改变时触发 change 事件，event. detail = {value, code, postcode}，其中字段 code 是统计用区划代码，postcode 是邮政编码

例 3-12 picker 组件小案例，运行效果如图 3.19 所示。

视频讲解

pages/picker/picker. wxml 文件代码如下：

```
<view class = "demo - box">
  <view class = "title"> 12.表单组件 picker 的简单应用</view>
  <view class = "title">(1)普通选择器</view>
  <picker mode = "selector" range = "{{oneItems}}" bindchange = "selectorChange">
    <view>当前选择: {{selector}}</view>
  </picker>
  <view class = "title">(2)多列选择器</view>
  <picker mode = "multiSelector" range = "{{doubleItems}}"
    bindchange = "multiSelectorChange">
    <view>当前选择: {{multiSelector}}</view>
  </picker>
  <view class = "title">(3)时间选择器</view>
  <picker mode = "time" bindchange = "timeChange">
    <view>当前选择: {{time}}</view>
  </picker>
  <view class = "title">(4)日期选择器</view>
  <picker mode = "date" bindchange = "dateChange">
    <view>当前选择: {{date}}</view>
  </picker>
  <view class = "title">(5)省市区选择器</view>
  <picker mode = "region" bindchange = "regionChange">
    <view>当前选择: {{region}}</view>
```

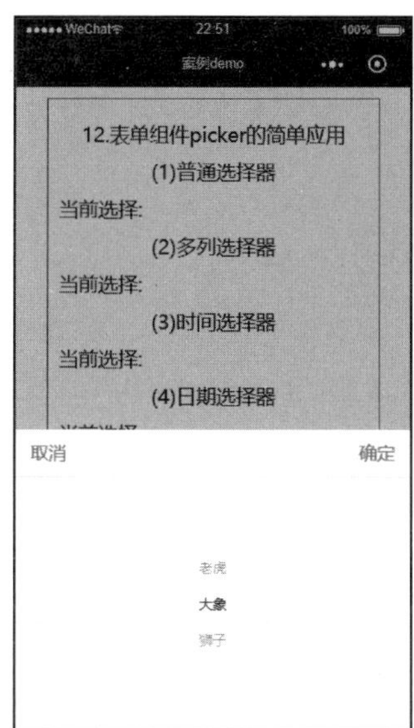

(a) 普通选择器

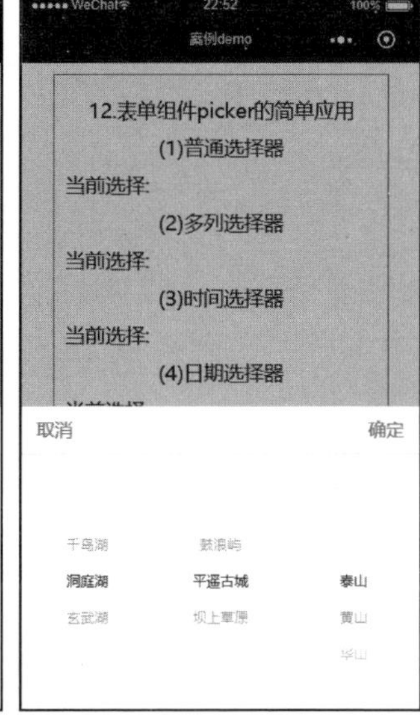

(b) 单击多列选择器

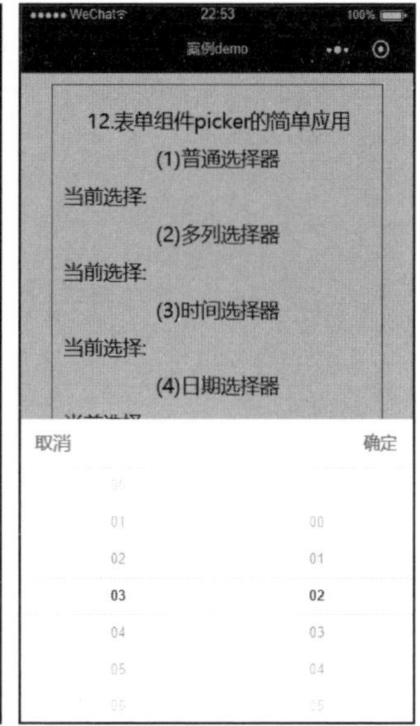

(c) 时间选择器

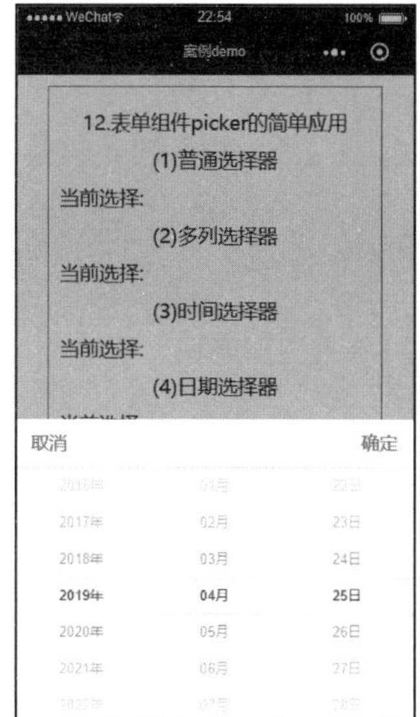

(d) 日期选择器

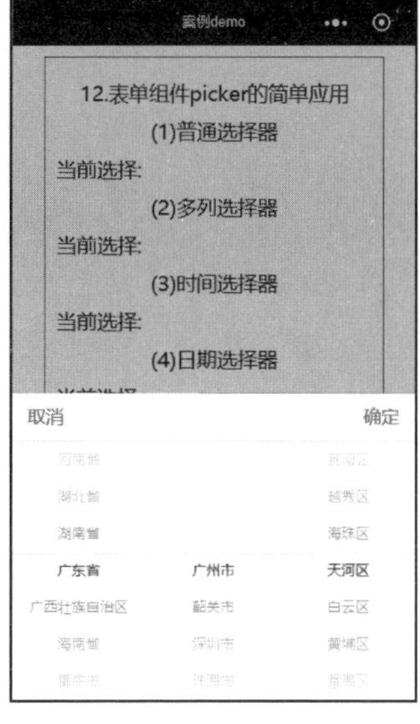

(e) 省市区选择器

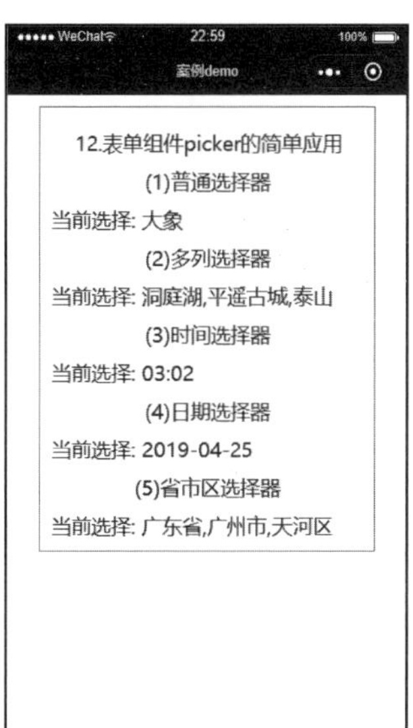

(f) 选择器确定区域后的效果

图 3.19　picker 组件小案例

```
  </picker>
</view>
```

pages/picker/picker.js 文件代码如下：

```
Page({
  //页面的初始数据
  data: {
    oneItems: ["老虎", "大象", "狮子"],
    doubleItems: [
```

```
        ["千岛湖", "洞庭湖", "玄武湖"],
        ["鼓浪屿", "平遥古城", "坝上草原"],
        ["泰山", "黄山", "华山"]
      ]
    },
    selectorChange: function(e) {
      var i = e.detail.value;                     //获得数组的下标
      var value = this.data.oneItems[i];          //获得选项的值
      this.setData({
        selector: value                           //将用户选择的值更新赋给 selector
      });
    },
    multiSelectorChange: function(e) {
      var arrayIndex = e.detail.value;            //获得数组的下标
      var value = new Array();                    //声明一个空数组,用于存放用户选择的值
      for (var i = 0; i < arrayIndex.length; i++) {
        var m = arrayIndex[i];                    //通过数组的遍历,获得第 i 个数组元素的下标
        var n = this.data.doubleItems[i][m];      //获得第 i 个数组的元素值
        value.push(n);                            //往数组中追加新的值
      }
      this.setData({
        multiSelector: value                      //将用户选择的值更新赋给 multiSelector
      });
    },
    timeChange: function(e) {
      var value = e.detail.value;                 //获得选择的时间
      this.setData({
        time: value                               //将用户选择的值更新赋给 time
      });
    },
    dateChange: function(e) {
      var value = e.detail.value;                 //获得选择的日期
      this.setData({
        date: value                               //将用户选择的值更新赋给 date
      });
    },
    regionChange: function(e) {
      var value = e.detail.value;                 //获得选择的省市区
      this.setData({
        region: value                             //将用户选择的值更新赋给 region
      });
    },
  })
```

【代码讲解】 本示例在 picker.wxml 文件中设置了 5 组不同效果的选择器,分别为普通选择器、多列选择器、时间选择器、日期选择器和省市区选择器。在普通选择器中设置属性 mode="selector"并为绑定监听事件,在 JS 文件中自定义选项数组和函数 selectorChange(),在 WXML 文件中通过{{selector}}动态获取用户选择的内容;在多列选择器中设置属性 mode= "selector" 并为组件绑定监听事件,在 JS 文件中自定义选项数组和函数

multiSelectorChange()，在 WXML 文件中通过{{multiSelector}}动态获取用户选择的内容；在时间选择器中设置属性 mode="time"并为组件绑定监听事件，在 JS 文件中自定义函数 timeChange()，在 WXML 文件中通过{{time}}动态获取用户选择的时间；在日期选择器中设置属性 mode="date"并为组件绑定监听事件，在 JS 文件中自定义函数 dateChange()，在 WXML 文件中通过{{date}}动态获取用户选择的日期；在省市区选择器中设置属性 mode="region"并为组件绑定监听事件，在 JS 文件中自定义函数 regionChange()，在 WXML 文件中通过{{region}}动态获取用户选择的地区。

3.4.7　picker-view

<picker-view>是嵌入页面的滚动选择器，其中只可放置 picker-view-column 组件，其属性如表 3.19 所示。

表 3.19　<picker-view>组件属性

属性名	类型	默认值	说明
value	Array.<number>	[]	数组中的数字依次表示 picker-view 内的 picker-view-column 选择的第几项(下标从 0 开始)，数字大于 picker-view-column 可选项长度时，选择最后一项
indicator-style	string		设置选择器中间选中框的样式
indicator-class	string		设置选择器中间选中框的类名
mask-style	string		设置蒙层的样式
mask-class	string		设置蒙层的类名
bindchange	eventhandle		滚动选择时触发 change 事件，event.detail = {value}；value 为数组，表示 picker-view 内的 picker-view-column 当前选择的是第几项(下标从 0 开始)

例 3-13　picker-view 组件小案例，运行效果如图 3.20 所示。

视频讲解

pages/picker-view/picker-view.wxml 文件代码如下：

```
<view class = "demo - box">
  <view class = "title"> 13.picker - view 小案例</view>
  <view class = "title">旅游计划</view>
  <view class = "title">{{travel}}</view>
  <picker - view value = "{{value}}" bindchange = "pickerviewChange">
    <picker - view - column>
      <view wx:for = "{{touristone}}">{{item}}</view>
    </picker - view - column>
    <picker - view - column>
      <view wx:for = "{{touristtwo}}">{{item}}</view>
    </picker - view - column>
    <picker - view - column>
      <view wx:for = "{{touristthree}}">{{item}}</view>
    </picker - view - column>
  </picker - view>
</view>
```

pages/picker-view/picker-view.js 文件代码如下：

```
Page({
  //页面的初始数据
  data: {
    touristone: ["五台山", "普陀山", "峨眉山"],
    touristtwo: ["莫高窟", "云冈石窟", "龙门石窟"],
    touristthree: ["法门寺", "佛光寺", "大相国寺"],
    value: [0, 0, 0],                        //设置默认的每个选项的数组下标
  },
  pickerviewChange: function(e) {
    var v = e.detail.value;                  //获得数组的下标
    var travel = [];                         //声明一个空数组,用于存放用户选择的值
    travel.push(this.data.touristone[v[0]]); //追加用户选择第一个数组的元素
    travel.push(this.data.touristtwo[v[1]]); //追加用户选择第二个数组的元素
    travel.push(this.data.touristthree[v[2]]);//追加用户选择第三个数组的元素
    this.setData({
      travel: travel                         //将用户选择的值更新赋给 travel
    });
  },
})
```

pages/picker-view/picker-view.wxss 文件代码如下：

```
picker - view {
  width: 100 %;
  height: 400rpx;
}
```

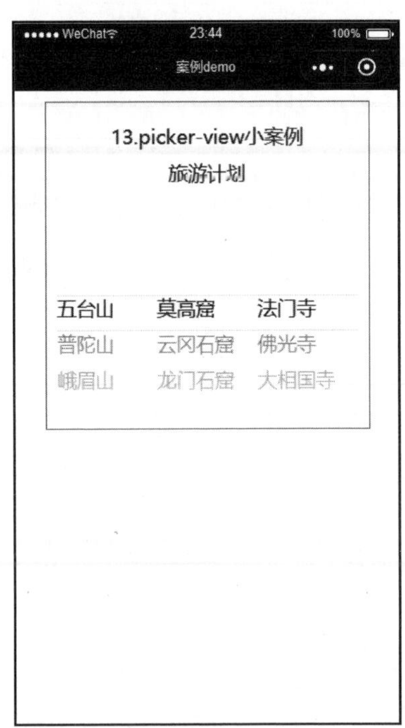

(a) 页面初始效果

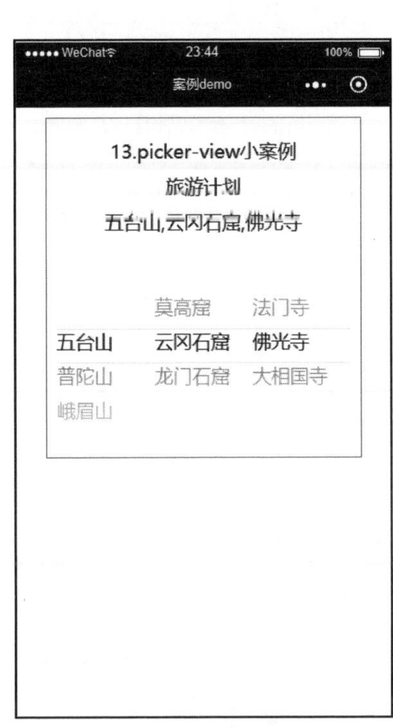

(b) 更改后的效果

图 3.20 picker-view 组件小案例

【代码讲解】< picker-view-column >组件是< picker-view >组件的二级组件，本例中 picker-view. wxml 文件< picker-view >组件包含了 3 个< picker-view-column >组件，分别用于渲染 3 个景点。< picker-view >组件和< picker >组件可以达到同样的表现效果，读者可以根据自己的喜好选择使用。

图 3. 20(a)为页面初始效果；图 3. 20(b)为用户选择完每个选项后，内容会显示在旅游计划下方。

3.4.8　radio

< radio >为单选框组件，往往需要配合< radio-group >组件来使用，< radio >标签嵌套在< radio-group >当中。< radio-group >组件只有一个属性，如表 3. 20 所示。

表 3. 20　< radio-group >组件属性

属　性　名	类　　型	说　　明	备　　注
bindchange	eventhandle	< radio-group >选中项发生变化时触发 change 事件，event. detail = {value:选中项 radio 的 value}	携带值为 vent. detail = {value:选中项 radio 的 value}

< radio >组件的属性如表 3. 21 所示。

表 3. 21　< radio >组件属性

属　性　名	类　　型	默认值	说　　明
value	string		< radio >标识。当该< radio >选中时，< radio-group >的 change 事件会携带< radio >的 value
checked	boolean	false	当前是否选中
disabled	boolean	false	是否禁用
color	color		radio 的颜色，同 css 的 color

示例代码如下：

```
<radio - group>
  <radio value = "tiger" checked = "true">老虎</radio>
  <radio value = "elephant" disabled = "true">大象</radio>
  <radio value = "lion">狮子</radio>
  <radio value = "penguin">企鹅</radio>
</radio - group>
```

其效果如图 3. 21 所示。

图 3. 21　单选框图示

如图 3. 21 所示，“老虎”选项被选中，“大象”选项禁止选择，“狮子”和“企鹅”选项均未选中。< radio-group >组件不允许多选，当前状态有且仅有一个选项被选中，< checkbox-group >组件允许用户多选。

视频讲解

例 3-14　radio 组件小案例，运行效果如图 3. 22 所示。

pages/radio/radio. wxml 文件代码如下：

```
<view class = "demo - box">
  <view class = "title"> 14. radio 小案例</view>
  <view class = "title">利用 for 循环批量生成</view>
  <radio - group bindchange = "radioChange">
    <block wx:for = "{{radioItems}}">
      <radio value = "{{item.name}}" checked = "{{item.checked}}" />{{item.value}}
    </block>
  </radio - group>
</view>
```

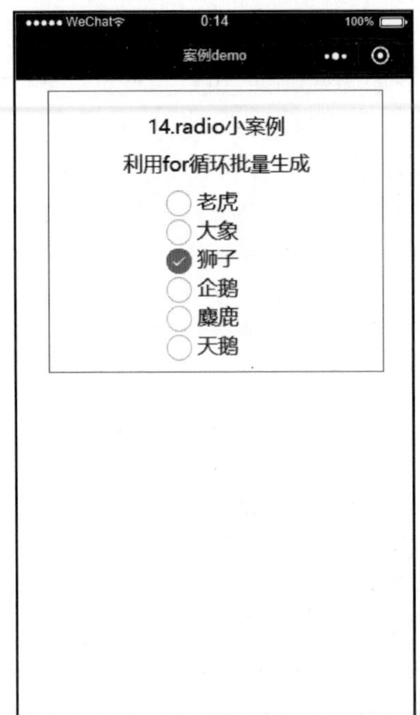

(a) “狮子” 选项被选中

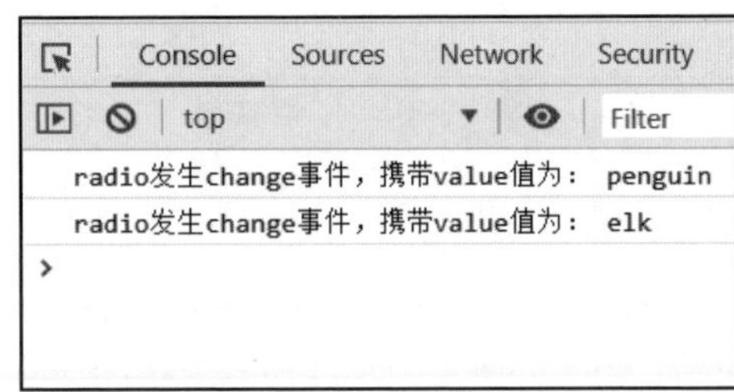

(b) 新选项被选中时Console控制台输出的内容

图 3.22　radio 组件小案例

pages/radio/radio.js 文件代码如下：

```
Page({
  data: {
    radioItems: [
      { name: 'tiger', value: '老虎' },
      { name: 'elephant', value: '大象' },
      { name: 'lion', value: '狮子', checked: 'true' },
      { name: 'penguin', value: '企鹅' },
      { name: 'elk', value: '麋鹿' },
      { name: 'swan', value: '天鹅' },
    ]
  },
  radioChange:function(e) {
    console.log("radio 发生 change 事件,携带 value 值为: ", e.detail.value)
  }
```

```
})
```

pages/radio/radio.wxss 文件代码如下：

```
radio - group {
  margin: 0 200rpx;
}
```

【代码讲解】 本例首先在 radio.js 文件中的 data 中定义一个数组 radioitems，用于记录单选框的名称 name、值 value 及选中情况，并在 radio.wxml 文件中使用<radio-group>标签包裹<radio>，使用<label>标签配合 wx：for 实现批量生成多个 radio 组件；其次在<radio-group>标签上绑定监听事件，在 radio.js 文件中自定义 checkboxChange()函数，以达到每次被触发后都在 Console 控制台输出最新选中的所有值。图 3.22(a)为模拟器效果；图 3.22(b)为控制台效果。

3.4.9 slider

<slider>为滑动选择器，用于可视化地动态改变某变量的取值。其属性如表 3.22 所示。

表 3.22 <slider>组件属性

属性名	类型	默认值	说明
min	number	0	最小值
max	number	100	最大值
step	number	1	步长，取值必须大于 0，并且可被(max－min)整除
disabled	boolean	false	是否禁用
value	number	0	当前取值
color	color	#e9e9e9	背景条的颜色(使用 backgroundColor)
selected-color	color	#1aad19	已选择的颜色(使用 activeColor)
backgroundColor	color	#e9e9e9	背景条的颜色
block-size	number	28	滑块的大小，取值范围为 12～28
block-color	color	#ffffff	滑块的颜色
show-value	boolean	false	是否显示当前 value
bindchange	eventhandle		完成一次拖动后触发的事件，event.detail = {value:valuc}
bindchanging	eventhandle		拖动过程中触发的事件，event.detail = {value: value}

例如，设置一个自定义滑动条，最小值为 10、最大值为 100，在右侧显示当前数值，示例代码如下：

```
< slider min = "10" max = "100" show - value />
```

其运行效果如图 3.23 所示。

图 3.23 滑动条图示

滑动条上的滑块向右边滑动时，右侧的数值会逐渐增大。

视频讲解

【例】3-15 slider 组件小案例，运行效果如图 3.24 所示。

pages/slider/slider.wxml 文件代码如下：

```
<view class = "demo - box">
  <view class = "title"> 14.slider 小案例</view>
  <view class = "title">(1)滑动条右侧显示当前进度值</view>
  <slider min = "0" max = "100" value = "30" step = "10" show - value = "true" />
  <view class = "title">(2)自定义滑动条颜色与滑块样式</view>
  <slider min = "0" max = "100" value = "30" block - size = "20" block - color = "gray"
activeColor = "skyblue" />
  <view class = "title">(3)禁用滑动条</view>
  <slider min = "0" max = "100" value = "30" disabled = "true" />
  <view class = "title">(4)增加滑动条监听事件</view>
  <slider min = "0" max = "100" value = "30" bindchange = "sliderChange" />
</view>
```

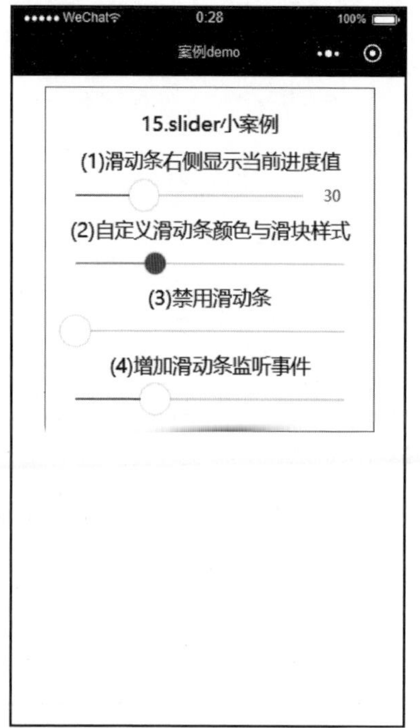

(a) 页面初始效果

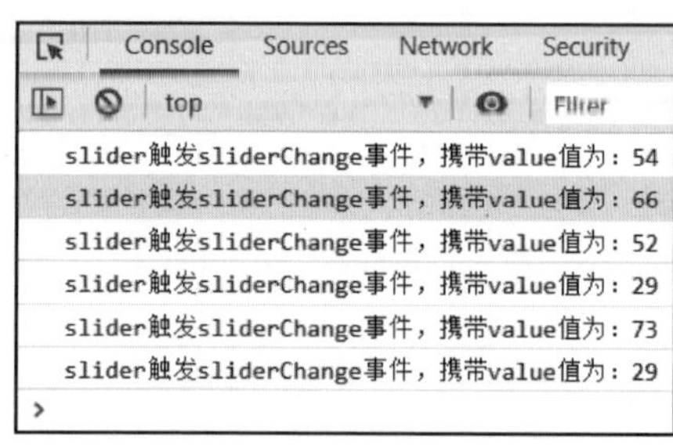

(b) Console控制台打印的slider变化

图 3.24 slider 组件小案例

pages/slider/slider.js 文件代码如下：

```
Page({
  sliderChange(e) {
    console.log("slider 触发 sliderChange 事件，携带 value 值为：" + e.detail.value)
  }
})
```

【代码讲解】 本例在 slider.wxml 文件中设置 4 种情况，分别为滑动条右侧显示当前进度值、自定义滑动条颜色和滑块样式、禁用滑动条(无法改变当前数值)、增加滑动条监听事件。

图 3.24(a)是页面初始状态，第 4 个滑动条为其滑块增加监听事件并在 slider.js 文件中自定义函数；图 3.24(b)所示，当第 4 个滑动条被拖动时，会在 Console 控制台上输出 slider 的最新值。

3.4.10　switch

<switch>为开关选择器，常用于表单上的开关功能，其属性如表 3.23 所示。

表 3.23　<switch>组件属性

属性名	类型	默认值	说明
checked	boolean	false	是否选中
disabled	boolean	false	是否禁用
type	string	switch	样式，有效值：switch，checkbox
bindchange	eventhandle		value 改变时触发 change 事件，event.detail = {value: value}
color	color		switch 的颜色，同 CSS 的 color

示例代码如下：

```
< switch checked = "true" />选中
< switch/>未选中
```

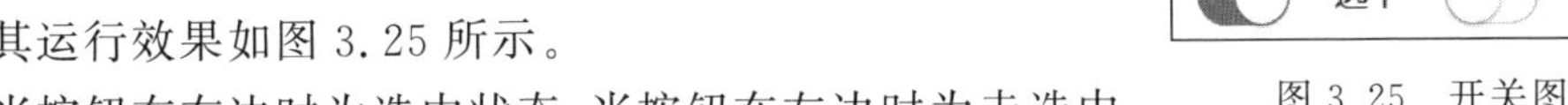

其运行效果如图 3.25 所示。

图 3.25　开关图示

当按钮在右边时为选中状态，当按钮在左边时为未选中状态。

视频讲解

例 3-16　switch 组件小案例，运行效果如图 3.26 所示。

pages/switch/switch.wxml 文件代码如下：

```
< view class = "demo - box">
  < view class = "title"> 16.swtich 小案例</view >
  < view class = "title">增加 switch 事件监听</view >
  < switch checked bindchange = "switch1Change"></switch >
  < switch bindchange = "switch2Change"></switch >
</view >
```

pages/switch/switch.js 文件代码如下：

```
Page({
  switch1Change1: function(e) {
    console.log("switch1 触发 switch1Change 事件,携带 value 值为: " + e.detail.value)
  },
  switch2Change2: function(e) {
    console.log("switch2 触发 switch2Change 事件,携带 value 值为: " + e.detail.value)
  }
})
```

(a) 页面初始效果

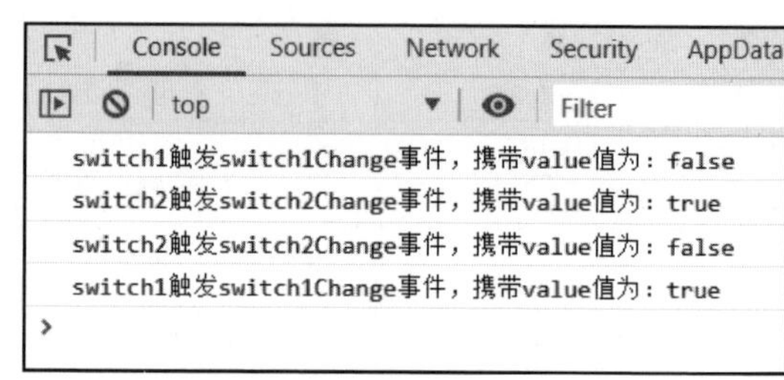

(b) Console控制台输出的switch状态变化

图 3.26　switch 组件小案例

【代码讲解】 本例在 switch.wxml 文件中放置两组< switch >标签，第 1 组设置属性 checked 实现选中状态，第 2 组默认为未选中状态，并为两个组件分别绑定监听事件；在 switch.js 文件中自定义 switch1Change()函数和 switch2Change()函数，当 switch 组件被触发后会在 Console 控制台输出当前状态。

图 3.26(a)为页面初始效果图；图 3.26(b)为 Console 控制台输出当前两组 switch 的选中状态。

3.4.11　textarea

< textarea >为多行输入框，常用于多行文字的输入。其属性如表 3.24 所示。

表 3.24　< textarea >组件属性

属性名	类　型	默认值	说　明
value	string		输入框的内容
placeholder	string		输入框为空时占位符
placeholder-style	string	switch	指定 placeholder 的样式，目前仅支持 color、font-size 和 font-weight
placeholder-class	string	textarea-placeholder	指定 placeholder 的样式类
disabled	boolean	false	是否禁用
maxlength	number	140	最大输入长度，设置为－1 的时候不限制最大长度

续表

属性名	类型	默认值	说明
auto-focus	boolean	false	自动聚焦，拉起键盘
focus	boolean	false	获取焦点
auto-height	boolean	false	是否自动增高，设置 auto-height 时，style.height 不生效
fixed	boolean	false	如果 textarea 是在一个 position：fixed 的区域，需要显示指定属性 fixed 为 true
cursor-spacing	number/string	0	指定光标与键盘的距离，单位：px（基础库 2.4.0 起支持）。取 textarea 底部的距离和 cursor-spacing 指定的距离的最小值作为光标与键盘的距离
cursor	number		指定 focus 时的光标位置（基础库 1.5.0 起支持）
show-confirm-bar	boolean	true	是否显示键盘上方带有"完成"按钮那一栏（基础库 1.6.0 起支持）
selection-start	number	−1	光标起始位置，自动聚集时有效，需与 selection-end 搭配使用（基础库 1.9.0 起支持）
selection-end	number	−1	光标结束位置，自动聚集时有效，需与 selection-start 搭配使用（基础库 1.9.0 起支持）
adjust-position	boolean	true	键盘弹起时，是否自动上推页面（基础库 1.9.0 起支持）
bindfocus	eventhandle		输入框聚焦时触发，event.detail＝{value，height}，height 为键盘高度（基础库 1.9.0 起支持）
bindblur	eventhandle		输入框失去焦点时触发，event.detail ＝ {value，cursor}
bindlinechange	eventhandle		输入框行数变化时调用，event.detail ＝ {height：0 heightRpx：0，lineCount：0}
bindinput	eventhandle		当键盘输入时，触发 input 事件，event.detail ＝ {value，cursor，keyCode}，keyCode 为键值，目前工具还不支持返回 keyCode 参数。bindinput 处理函数的返回值并不会反映到 textarea 上
bindconfirm	eventhandle		点击完成时，触发 confirm 事件，event.detail ＝ {value：value}

例 3-17 textarea 组件小案例，运行效果如图 3.27 所示。

视频讲解

pages/textarea/textarea.wxml 文件代码如下：

```
<view class = "demo - box">
  <view class = "title"> 17.textarea 小案例</view>
  <view class = "title">(1)文本框自动变高</view>
```

```
    <textarea auto-height placeholder="允许自动变高" />
    <view class="title">(2)禁用文本框</view>
    <textarea placeholder="禁用该文本框无法输入内容" disabled/>
    <view class="title">(3)自定义占位符样式</view>
    <textarea placeholder="占位符颜色是天蓝色并加粗" placeholder-style="color:skyblue;
font-weight:bold" />
</view>
```

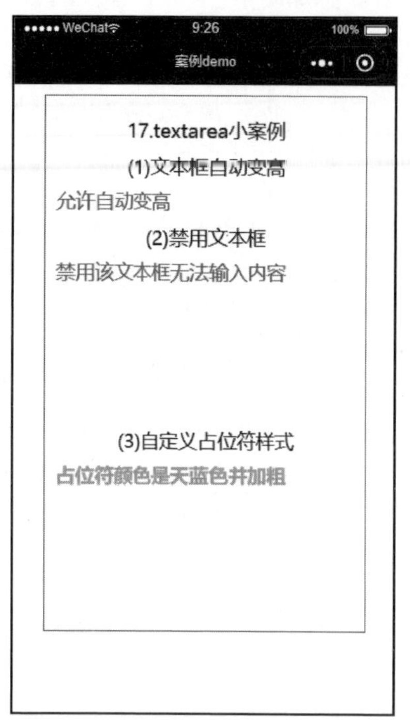

(a) 页面初始效果

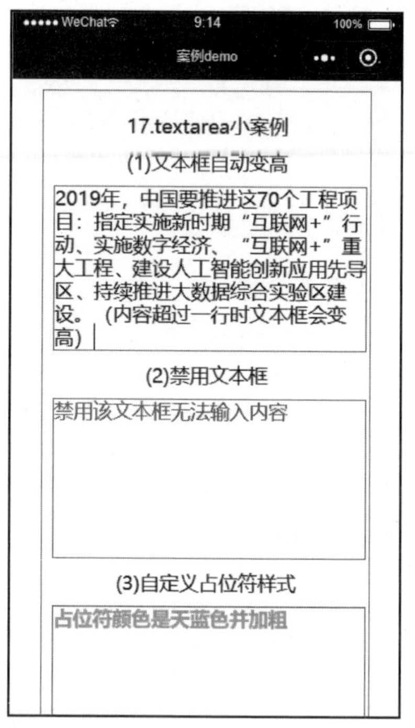

(b) 第一组文本框在输入的内容超过一行时自动变高

图 3.27　textarea 组件小案例

【代码讲解】 本例在 switch.wxml 文件中放置 3 组<textarea>标签：第 1 组通过设置属性 auto-height，允许文本框在输入内容超过一行时自动变高；第 2 组通过设置属性 disabled，文本框被禁用无法进行输入；第 3 组通过设置属性"color:skyblue;font-weight:bold"，使文本框中的占位符呈现天蓝色并加粗。

图 3.27(a)为页面初始效果；图 3.27(b)中第 1 组文本框在输入的内容超过一行时，文本框高度会自动发生变化，第 2 组文本框被禁，第 3 组文本框的占位符样式发生改变。

3.5 导航组件

<navigator>导航组件用于页面之间的跳转，是使用频繁的组件之一。其属性如表 3.25 所示。

表 3.25 <navigator>组件属性

属性名	类型	默认值	说明
target	string	self	是在哪个目标上发生跳转,默认为当前小程序
url	string	false	当前小程序内的跳转链接
open-type	navigate	switch	跳转方式

其中,open-type 属性对应 6 种取值。其属性如表 3.26 所示。

表 3.26 open-type 属性

属性名	说明
navigate	对应 wx. navigateTo 或 wx. navigateToMiniProgram 的功能
redirect	对应 wx. redirectTo 的功能
switchTab	navigate 跳转方式
reLaunch	对应 wx. reLaunch 的功能
navigateBack	对应 wx. navigateBack 的功能
exit	退出小程序,target="miniProgram"时生效

例 3-18 navigator 组件小案例,运行效果如图 3.28 所示。

视频讲解

pages/navigator/navigator. wxml 文件代码如下:

```
<view class = "demo - box">
  <view class = "title"> 18. 导航组建 navigator 小案例</view>
  <view class = "title">(1)点击打开新页面</view>
  <navigator url = "../new/new" open - type = "navigate">
    <button type = "primary">点击按钮跳转到新页面</button>
  </navigator>
  <view class = "title">(2)点击重定向到新页面</view>
  <navigator url = "../redirect/redirect" open - type = "redirect">
    <button type = "primary">点击按钮跳转到当前页面</button>
  </navigator>
</view>
```

在 app. js 文件中通过 pages/new/new、pages/redirect/redirect 注册两个新页面。

pages/new/new. wxml 文件代码如下:

```
<text>这里是新窗口打开的新页面,点击左上角图标可以返回上一页</text>
```

pages/redirect/redirect. wxml 文件中代码如下:

```
<text>重定向的新页面,无法返回到 navigator.wxml 页面</text>
```

【代码讲解】 本例创建 3 个页面:navigator 页面是项目首页;new 和 redirect 页面是 navigator 页面中两种不同的跳转方式的着陆页。

图 3.28(a)为页面初始效果;图 3.28(b)页面利用 open-type="navigate"方式跳转而来,着陆页左上角的返回图标可以返回上一页;图 3.28(c)页面由 open-type="redirect"方式跳转而来,着陆页左上角没有返回图标。

(a) 页面初始效果　(b) 点击第1个按钮跳转到新页面　(c) 点击第2个按钮在当前页打开

图 3.28　navigator 组件小案例

3.6 媒体组件

媒体组件用于播放媒体内容，常用的有音频组件、图片组件和视频组件。

3.6.1 audio

<audio>为音频组件，用于网络音频的播放。其属性如表 3.27 所示。

表 3.27　<audio>组件属性

属性名	类型	默认值	说明
id	string		audio 组件的唯一标识符
src	string	false	要播放音频的资源地址
loop	boolean	false	是否循环播放
controls	boolean	false	是否显示默认控件
poster	string		默认控件上的音频封面的图片资源地址，如果 controls 属性值为 false，则设置 poster 无效
name	string	未知音频	默认控件上的音频名字，如果 controls 属性值为 false，则设置 name 无效
author	string	未知作者	默认控件上的作者名字，如果 controls 属性值为 false，则设置 author 无效
binderror	eventhandle		当发生错误时触发 error 事件，detail = {errMsg: MediaError.code}

续表

属　性　名	类　　型	默认值	说　　明
bindplay	eventhandle		当开始/继续播放时触发 play 事件
bindpause	eventhandle		当暂停播放时触发 pause 事件
bindtimeupdate	eventhandle		当播放进度改变时触发 imeupdate 事件，detail = {currentTime，duration}
bindended	eventhandle		当播放到末尾时触发 ended 事件

其中，binderror 属性触发后的返回值 MediaError. code 共有 4 种取值：

(1) 获取资源被用户禁止；

(2) 网络错误；

(3) 解码错误；

(4) 不合适资源。

视频讲解

例 3-19　audio 组件小案例，运行效果如图 3.29 所示。

(a) 页面初始效果　　(b) 音频正在播放　　(c) 音频回到10秒处

图 3.29　audio 组件小案例

pages/audio/audio. wxml 文件代码如下：

```
<view class = "demo - box">
  <view class = "title"> 19.audio 小案例</view>
  <view class = "title">网络音频播放</view>
  <view>
    <audio poster = "{{poster}}" name = "{{name}}" author = "{{author}}"
  src = "{{src}}" id = "myAudio" controls loop>
    </audio>
    <button size = "mini" bindtap = "audioPlay">播放</button>
```

```
    <button size = "mini" bindtap = "audioPause">暂停</button>
    <button size = "mini" bindtap = "audio10">设置当前播放时间为 10 秒</button>
  </view>
</view>
```

pages/audio/audio.js 文件代码如下：

```
Page({
  onReady: function(e) {
    //使用 wx.createAudioContext 获取 audio 上下文 context
    this.audioCtx = wx.createAudioContext('myAudio')
  },
  data: {
    poster: 'http://y.gtimg.cn/music/photo_new/T002R300x300M00000
            3rsKF44GyaSk.jpg?max_age = 2592000',
    name: '此时此刻',
    author: '许巍',
    src: 'http://ws.stream.qqmusic.qq.com/M500001VfvsJ21xFqb.mp3?guid =
        ffffffff82def4af4b12b3cd9337d5e7&uin = 346897220&vkey = 6292F51E1E384
        E06DCBDC9AB7C49FD713D632D313AC4858BACB8DDD29067D3C601481D36E62053
        BF8DFEAF74C0A5CCFADD6471160CAF3E6A&fromtag = 46',
  },
  audioPlay: function() {
    this.audioCtx.play()                        //播放音乐
  },
  audioPause: function() {
    this.audioCtx.pause()                       //暂停播放
  },
  audio10: function() {
    this.audioCtx.seek(10)                      //回到 10s
  },
})
```

【代码讲解】 本例先在 audio.wxml 文件中放置<audio>组件，然后设置 3 组<button>按钮，分别用于控制音频的播放、暂停以及定位到 10s 处的播放位置。<audio>组件的属性值在 audio.js 文件中定义，包括音频封面图片、歌曲名、演唱者和音频来源，并为 3 组按钮绑定点击事件，在 audio.js 文件中自定义 audioPlay()函数、audioPause()函数和 audio10()函数。

图 3.29(a)为页面初始状态；图 3.29(b)为点击“播放”按钮后，网络音频呈现播放状态，音频处于 0 分 23 秒；图 3.29(c)为网络音频正在播放时，点击“设置当前播放时间为 10 秒”按钮后，音频回到 0 分 10 秒处播放。

若程序中音频的 src 地址失效，请参考 8.6 节获取新的 src 地址。

3.6.2 image

<image>为图片组件，常用于浏览图片，通过设置属性 mode 能够对图片进行裁剪和缩放，共有 13 种模式，其中 4 种是缩放模式，9 种是裁剪模式，其属性如表 3.28 所示。

表 3.28 <image>组件属性

属性名	类型	默认值	说明
src	string		图片资源地址，支持云文件 ID(基础库 2.2.3 起支持)
mode	string	'scaleToFill'	图片裁剪、缩放的模式
lazy-load	boolean	false	图片懒加载，在即将进入当前屏幕可视区域时才开始加载(基础库 1.5.0 起支持)
binderror	eventhandle		当错误发生时，发布到 AppService 的事件名，事件对象 event.detail = {errMsg: 'something wrong'}
bindload	eventhandle		当图片载入完毕时，发布到 AppService 的事件名，事件对象 event.detail = {height:'图片高度 px', width:'图片宽度 px'}

image 组件默认宽度 300px、高度 225px。image 组件中二维码/小程序码图片不支持长按识别，仅在 wx.previewImage()中支持长按识别。

mode 有效值的属性名及说明如表 3.29 所示。

表 3.29 mode 有效值的属性名及说明

属性名		说明
缩放模式	scaleToFill	不保持纵横比例缩放图片，使图片的宽、高完全拉伸至填满 image 元素
	aspectFit	保持纵横比例缩放图片，使图片的长边能完全显示出来。也就是说，可以完整地将图片显示出来
	aspectFill	保持纵横比例缩放图片，只保证图片的短边能完全显示出来。也就是说，图片通常只在水平或垂直方向是完整的，另一个方向将会发生截取
	widthFix	宽度不变，高度自动变化，保持原图宽、高比不变
裁剪模式	top	不缩放图片，只显示图片的顶部区域
	bottom	不缩放图片，只显示图片的底部区域
	center	不缩放图片，只显示图片的中间区域
	left	不缩放图片，只显示图片的左边区域
	right	不缩放图片，只显示图片的右边区域
	top left	不缩放图片，只显示图片的左上边区域
	top right	不缩放图片，只显示图片的右上边区域
	bottom left	不缩放图片，只显示图片的左下边区域
	bottom right	不缩放图片，只显示图片的右下边区域

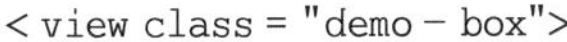

例 3-20 image 组件小案例。

视频讲解

pages/image/image.wxml 文件代码如下：

```
<view class = "demo - box">
  <view class = "title"> 20.audio 小案例</view>
  <view class = "title">(1)不保持纵横比缩放图片,使图片完全适应</view>
  <image src = "{{src}}" mode = "scaleToFill"></image>
  <view class = "title">(2)保持纵横比缩放图片,使图片的长边能完全显示出来</view>
  <image src = "{{src}}" mode = "aspectFit"></image>
  <view class = "title">(3)保持纵横比缩放图片,只保证图片的短边能完全显示出来</view>
  <image src = "{{src}}" mode = "aspectFill"></image>
  <view class = "title">(4)宽度不变,高度自动变化,保持原图宽高比不变</view>
```

```
    <image src="{{src}}" mode="widthFix"></image>
    <view class="title">(5)不缩放图片,只显示图片的顶部区域</view>
    <image src="{{src}}" mode="top"></image>
    <view class="title">(6)不缩放图片,只显示图片的底部区域</view>
    <image src="{{src}}" mode="bottom"></image>
    <view class="title">(7)不缩放图片,只显示图片的中间区域</view>
    <image src="{{src}}" mode="center"></image>
    <view class="title">(8)不缩放图片,只显示图片的左边区域</view>
    <image src="{{src}}" mode="left"></image>
    <view class="title">(9)不缩放图片,只显示图片的右边区域</view>
    <image src="{{src}}" mode="right"></image>
    <view class="title">(10)不缩放图片,只显示图片的左上边区域</view>
    <image src="{{src}}" mode="top left"></image>
    <view class="title">(11)不缩放图片,只显示图片的右上边区域</view>
    <image src="{{src}}" mode="top right"></image>
    <view class="title">(12)不缩放图片,只显示图片的左下边区域</view>
    <image src="{{src}}" mode="bottom left"></image>
    <view class="title">(13)不缩放图片,只显示图片的右下边区域</view>
    <image src="{{src}}" mode="bottom right"></image>
</view>
```

pages/image/image.wxss 文件代码如下：

```
image {
  width: 300rpx;
  height: 300rpx;
  margin-left: 150rpx;
}
```

pages/image/image.js 文件代码如下：

```
Page({
  data: {
    src: "/images/xingkong.jpg"
  }
})
```

本例在 images 文件夹下存放素材图片 xingkong.jpg,图片选择梵高的名画《星空》,原图如图 3.30 所示。

图 3.30　图片素材

运行效果如图 3.31 所示。

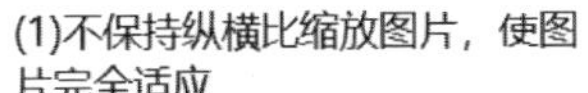

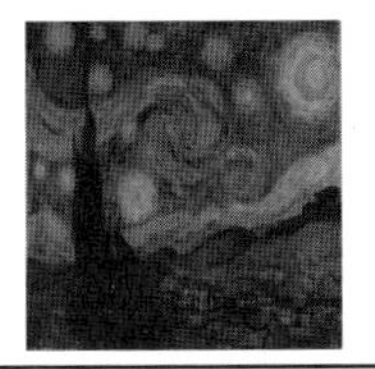

(a) 缩放模式：scaleToFill

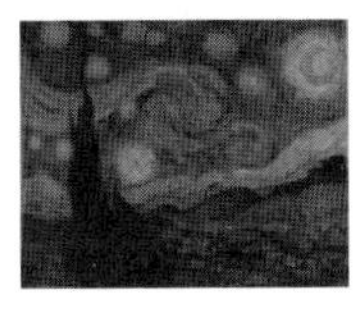

(b) 缩放模式：aspectFit

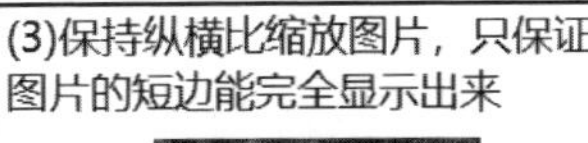

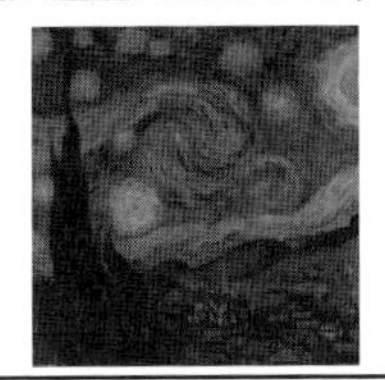

(c) 缩放模式：aspectFill

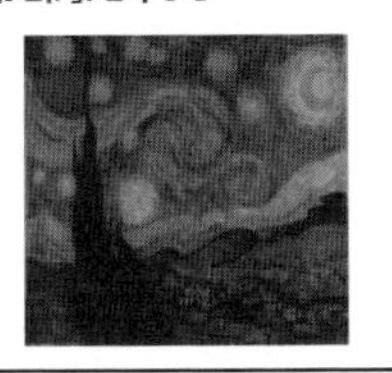

(d) 缩放模式：widthFix

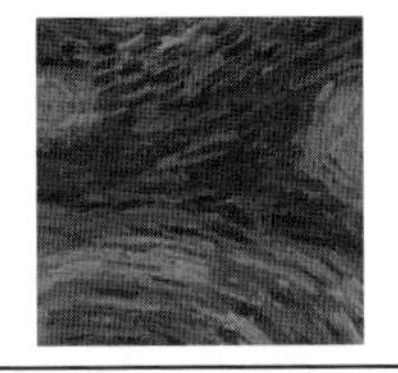

(e) 裁剪模式：top

(f) 裁剪模式：bottom

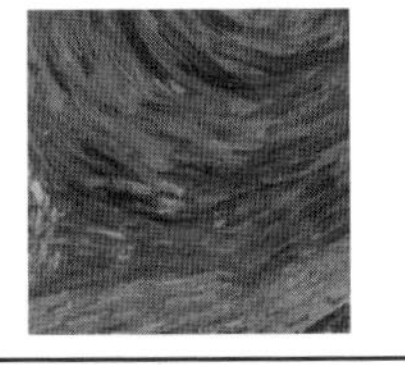

(g) 裁剪模式：center

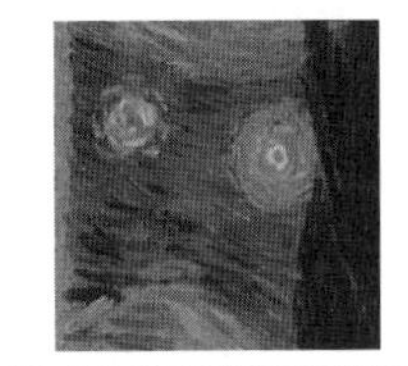

(h) 裁剪模式：left

(i) 裁剪模式：right

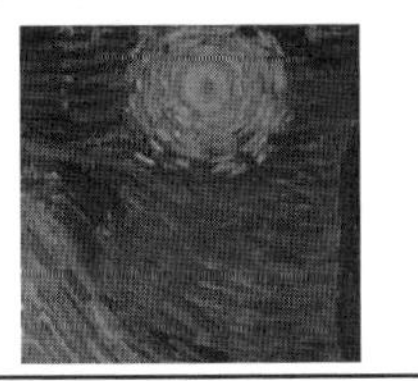

(j) 裁剪模式：top left

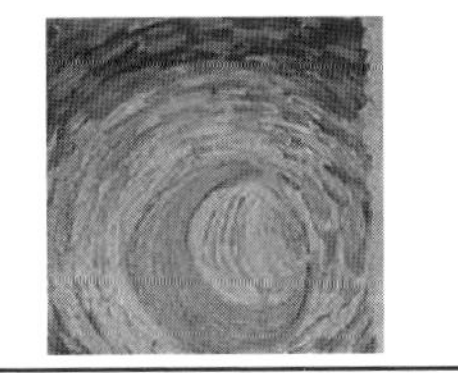

(k) 裁剪模式：top right

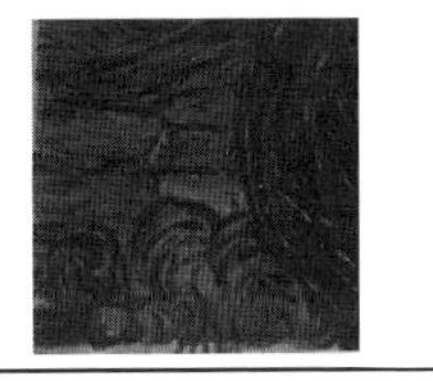

(l) 裁剪模式：bottom left

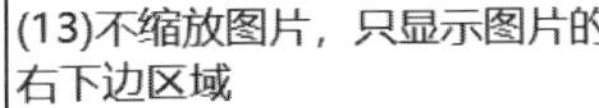

(m) 裁剪模式：bottom right

图 3.31　image 组件小案例

【代码讲解】 本例在 image. wxml 文件中放置了 13 组<image>组件，通过设置不同的 mode 属性值，实现对图片的缩放和剪裁。

3.6.3 video

<video>为视频组件，用于播放本地或网络的视频资源，视频的默认宽度为 300px、高度为 225px，视频组件提供弹幕功能，其属性如表 3.30 所示。

表 3.30 <video>组件属性

属性名	类型	默认值	说明
src	string		要播放视频的资源地址，支持云文件 ID (2.3.0)
duration	number		指定视频时长
controls	boolean	true	是否显示默认播放控件（播放/暂停按钮、播放进度、时间）
danmu-list	Array.<object>		弹幕列表
enable-danmu	boolean	false	是否展示弹幕，只在初始化时有效，不能动态变更
autoplay	boolean	false	是否自动播放
loop	boolean	false	是否循环播放
muted	boolean	false	是否静音播放
initial-time	number		指定视频初始播放位置
page-gesture	boolean	false	在非全屏模式下，是否开启亮度与音量调节手势
direction	number		设置全屏时视频的方向，不指定则根据宽、高比自动判断
show-progress	boolean	true	若不设置，宽度大于 240 时才会显示
show-fullscreen-btn	boolean	true	是否显示全屏按钮
show-play-btn	boolean	true	是否显示视频底部控制栏的播放按钮
show-center-play-btn	boolean	true	是否显示视频中间的播放按钮
enable-progress-gesture	boolean	true	是否开启控制进度的手势
object-fit	string	contain	当视频大小与 video 容器大小不一致时，视频的表现形式
poster	string		视频封面的图片网络资源地址或云文件 ID。若 controls 属性值为 false 则设置 poster 无效
show-mute-btn	boolean	false	是否显示静音按钮
title	string		视频的标题，全屏时在顶部展示
play-btn-position	string	bottom	播放按钮的位置
enable-play-gesture	boolean	false	是否开启播放手势，即双击切换播放/暂停
auto-pause-if-navigate	boolean	true	当跳转到其他小程序页面时，是否自动暂停本页面的视频

续表

属 性 名	类 型	默认值	说 明
auto-pause-if-open-native	boolean	true	当跳转到其他微信原生页面时，是否自动暂停本页面的视频
vslide-gesture	boolean	false	在非全屏模式下，是否开启亮度与音量调节手势(同 page-gesture)
vslide-gesture-in-fullscreen	boolean	true	在全屏模式下，是否开启亮度与音量调节手势
bindplay	eventhandle		当开始/继续播放时触发 play 事件
bindpause	eventhandle		当暂停播放时触发 pause 事件
bindended	eventhandle		当播放到末尾时触发 ended 事件
bindtimeupdate	eventhandle		播放进度变化时触发，event. detail = {currentTime，duration}。触发频率 250ms 一次
bindfullscreenchange	eventhandle		视频进入和退出全屏时触发，event. detail = {fullScreen，direction}，direction 有效值为 vertical 或 horizontal
bindwaiting	eventhandle		视频出现缓冲时触发
binderror	eventhandle		视频播放出错时触发
bindprogress	eventhandle		加载进度变化时触发，只支持一段加载。event. detail = {buffered}，百分比

例 3-21 video 组件小案例，运行效果如图 3.32 所示。

视频讲解

pages/video/video. wxml 文件代码如下：

```
<view class = "demo - box">
  <view class = "title"> 21. video 小案例</view>
  <video id = "myVideo" src = "{{src}}" danmu - list = "{{danmuList}}"
    enable - danmu controls ></video>
  <view class = "weui - label">弹幕内容:</view>
  <input bindblur = "bindInputBlur" type = "text" placeholder = "在此处输入弹幕内容" />
  <button size = "mini" bindtap = "bindPlay">播放</button>
  <button size = "mini" bindtap = "bindPause">暂停</button>
  <button bindtap = "bindSendDanmu" size = "mini" formType = "submit">发送弹幕</button>
</view>
```

pages/video/video. js 文件代码如下：

```
Page({
  onReady: function(res) {
    this.videoContext = wx.createVideoContext('myVideo')
  },
  inputValue: '',                          //创建一个空字符串用于保存弹幕内容
  data: {
    src: "http://wxsnsdy.tc.qq.com/105/20210/snsdyvideodownload?filekey =
30280201010421301f0201690402534804102ca905ce620b1241b726bc41dcff44e002040128
82540400&bizid = 1023&hy = SH&fileparam = 302c020101042530230204136ffd93020457e3c4
```

```
ff02024ef202031e8d7f02030f42400204045a320a0201000400",
    danmuList: [{
        text: "第 1s 出现的弹幕",
        color: "#ff0000",
        time: 1
      },
      {
        text: "第 2s 出现的弹幕",
        color: "#ff00ff",
        time: 2
      }
    ]
  },
  bindInputBlur: function(e) {
    this.inputValue = e.detail.value              //获取输入内容
  },
  bindSendDanmu: function() {
    this.videoContext.sendDanmu({
      text: this.inputValue,                      //发送弹幕
    })
  },
  bindPlay: function() {
    this.videoContext.play()                      //播放视频
  },
  bindPause: function() {
    this.videoContext.pause()                     //暂停视频
  },
})
```

(a) 页面初始状态

(b) 视频播放状态

(c) 发送弹幕

图 3.32　video 组件小案例

pages/video/video. wxss 文件代码如下：

```
input {
  border: 1px solid gray;
}
```

【代码讲解】 本例在 video. wxml 文件中放置< video >组件并设置其相关属性，< input >组件用来输入弹幕内容，通过 3 组< button >组件分别绑定点击事件 bindPlay()、bindPause()、bindSendDanmu()来实现视频播放、暂停和发送弹幕，并在 video. js 文件中定义视频播放的资源地址及自定义事件函数。

图 3. 32(a)为初始化界面，呈现暂停状态；图 3. 32(b)为点击"播放"按钮后，视频呈现播放状态；图 3. 32(c)为在输入框输入内容"12345678"并点击"发送弹幕"按钮后，输入框中的内容会呈现在屏幕上。

3.7 地图组件

< map >为地图组件，用于地图的展示。小程序使用的地图来自腾讯地图，地图组件为用户提供视角中心点地位、缩放层级的设置、标记物的增加以及内部组件的事件绑定，其属性如表 3. 31 所示。

表 3. 31 < map >组件属性

属 性 名	类 型	默认值	说 明
longitude	number		中心经度
latitude	number		中心纬度
scale	number	16	缩放级别，取值范围为 3～20
marker	Array. < marker >		弹幕列表(基础库 1. 0. 0 起支持)
cover	Array. < cover >		即将移除，请使用 markers
polyline	Array. < polyline >		路线
circle	Array. < circles >		圆
control	Array. < control >	false	控件(即将废弃，建议使用 cover-view 代替)
include-point	Array. < point >		缩放视野以包含所有给定的坐标点
show-location	boolean	false	显示带有方向的当前定位点
bindtap	eventhandle		点击地图时触发
bindmarkertap	eventhandle		点击标记点时触发，会返回 marker 的 id
bindcontroltap	eventhandle		点击控件时触发，会返回 control 的 id
bindcallouttap	eventhandle		点击标记点对应的气泡时触发，会返回 marker 的 id
bindupdated	eventhandle		视野发生变化时触发
bindregionchange	eventhandle		是否开启控制进度的手势(基础库 1. 9. 0 起支持)
bindpoitap	eventhandle		点击地图 poi 点时触发

其中，marker 为标记点，用于在地图上显示标记的位置，其属性如表 3.32 所示。

表 3.32 marker 属性

属性名	说　明	类　型	备　注
id	标记点 id	number	marker 点击事件回调会返回此 id。建议为每个 marker 设置 number 类型 id，保证更新 marker 时有更好的性能
latitude	纬度	number	浮点数，范围为－90°～90°
longitude	经度	number	浮点数，范围为－180°～180°
title	标注点名	string	value 改变时触发 change 事件，event.detail＝{value：value}
iconPath	显示的图标	string	项目目录下的图片路径，支持相对路径写法，以'/'开头则表示相对小程序根目录；也支持临时路径和网络图片
rotate	旋转角度	number	顺时针旋转的角度，范围为 0°～360°，默认为 0
alpha	标注的透明度	number	默认为 1，无透明，范围为 0～1
width	标注图标宽度	number/string	默认为图片实际宽度
height	标注图标高度	number/string	默认为图片实际高度
callout	自定义标记点上方的气泡窗口	Object	支持的属性见表 3.33，可识别换行符
label	为标记点旁边增加标签	Object	支持的属性见表 3.34，可识别换行符
anchor	经度和纬度在标注图标的锚点，默认底边中点		{x，y}，x 表示横向(0～1)，y 表示纵向(0～1)。{x：.5，y：1}表示底边中点

marker 上的气泡 callout，其属性如表 3.33 所示。

表 3.33 callout 属性

属　性　名	说　明	类　型	最低版本
content	文本	string	1.2.0
color	文本颜色	string	1.2.0
fontSize	文字大小	number	1.2.0
borderRadius	边框圆角	number	1.2.0
borderWidth	边框宽度	number	1.2.0
borderColor	边框颜色	string	1.2.0
bgColor	背景色	string	1.2.0
padding	文本边缘留白	number	1.2.0
display	'BYCLICK'：点击显示；'ALWAYS'：常显	string	1.2.0
textAlign	文本对齐方式，有效值：left、right、center、string	string	1.6.0

marker 上的气泡 label，其属性如表 3.34 所示。

表 3.34 label 属性

属性名	说明	类型	最低版本
content	文本	string	1.2.0
color	文本颜色	string	1.2.0
fontSize	文字大小	number	1.2.0
borderRadius	边框圆角	number	1.2.0
borderWidth	边框宽度	number	2.3.0
borderColor	边框颜色	string	2.3.0
bgColor	背景色	string	1.2.0
padding	文本边缘留白	number	1.2.0
display	'BYCLICK'：点击显示；'ALWAYS'：常显	string	1.2.0
textAlign	文本对齐方式，有效值：left、right、center	string	1.6.0

polyline 用于指定一系列坐标点，从数组第一项连线至最后一项，其属性如表 3.35 所示。

表 3.35 polyline 属性

属性名	说明	类型	备注
points	经度和纬度数组	array	[{latitude：0，longitude：0}]
color	线的颜色	string	十六进制
width	线的宽度	number	
dottedLine	带箭头的线	boolean	默认值为 false
arrowLine	边框宽度	number	默认值为 false，开发者工具暂不支持该属性
arrowIconPath	更换箭头图标	string	在 arrowLine 为 true 时生效
borderColor	线的边框颜色	string	
borderWidth	线的厚度	number	

circle 属性用于在地图上显示圆，通过设定中心点的经度、纬度和半径来绘制一个圆形图案作为地图上的标记物，其属性如表 3.36 所示。

表 3.36 circle 属性

属性名	说明	类型	备注
latitude	纬度	number	浮点数，范围为－90°～90°
longitude	经度	number	浮点数，范围为－180°～180°
color	描边的颜色	string	十六进制
fillColor	填充颜色	string	十六进制
radius	半径	number	
strokeWidth	描边的宽度	number	

例 3-22 map 组件小案例，程序运行效果如图 3.33 所示。

视频讲解

pages/map/map.wxml 文件代码如下：

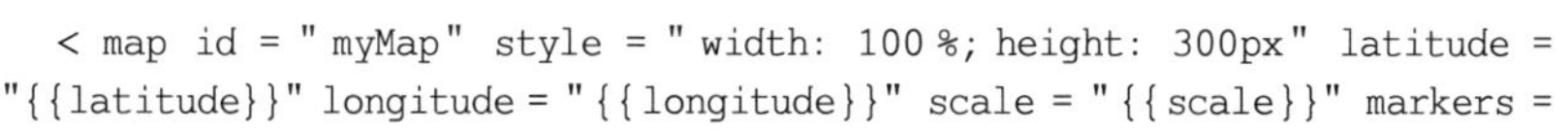
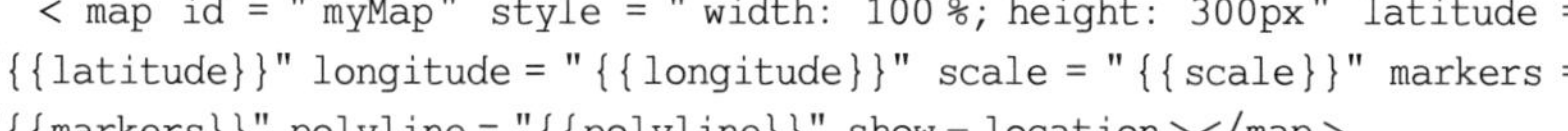

```
<map id="myMap" style="width: 100%; height: 300px" latitude="{{latitude}}" longitude="{{longitude}}" scale="{{scale}}" markers="{{markers}}" polyline="{{polyline}}" show-location></map>
```

```
<view class = "content">
  <button size = "mini" bindtap = "reduce"> - </button>
  <button size = "mini" bindtap = "default">默认缩放比例</button>
  <button size = "mini" bindtap = "add"> + </button>
  </view>
  <view class = "content">
   <button size = "mini" bindtap = "includePoints">按 includePoints 缩放视野</button> </view>
```

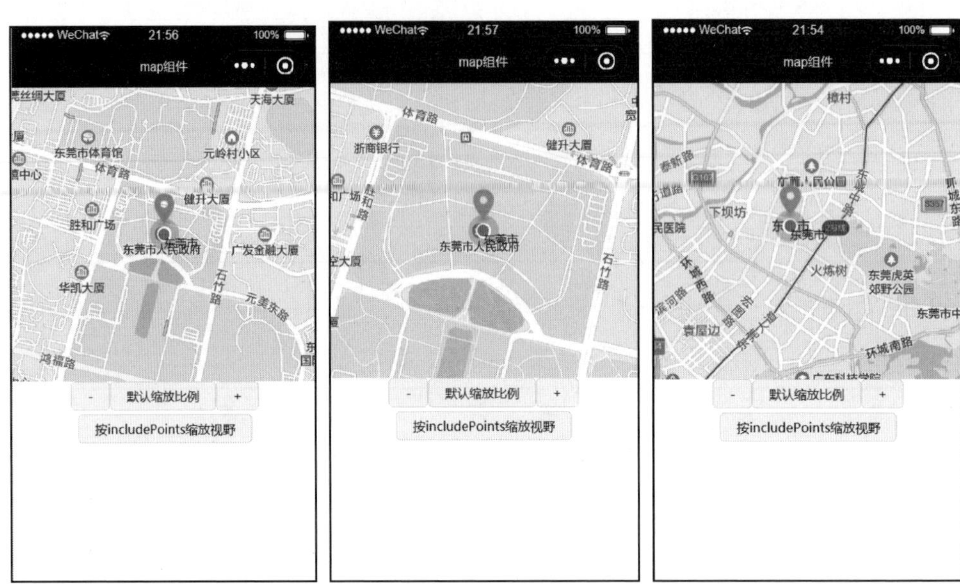

(a) 16倍缩放比例效果　　(b) 17倍缩放比例效果　　(c) 包含两个点时的缩放比例效果

图 3.33　map 组件小案例

pages/map/map.js 文件代码如下：

```
Page({
  data: {
    latitude: 23.020670, longitude: 113.751790,
    scale:16,
    markers: [{
      latitude: 23.020670, longitude: 113.751790,
      iconPath: "/images/location.png",
      label: {
        content: "东莞市" }
    }],
    polyline: [{
      points: [{
        longitude: 113.3245211,
        latitude: 23.10229
      }, {
        longitude: 113.324520, latitude: 23.21229
      }],
      color: "#FF0000DD",
      width: 2,
      dottedLine: true
    }]
  },
```

```
    onReady: function(e) {
      this.mapCtx = wx.createMapContext("myMap")
    },
    default:function()
    {
      this.setData({ scale: 16, })
    },
    reduce: function() {
      this.setData({
        scale: this.data.scale - 1,
      })
    },
    add: function() {
      this.setData({
        scale: this.data.scale + 1,
      })
    },
    includePoints: function() {                          //缩放视野
      this.mapCtx.includePoints({
        padding: [10],
        points: [{
          latitude: 23.0403, longitude: 113.7446,
        }, {
          latitude: 22.9983, longitude: 113.7724,
        }]
      })
    }
  })
```

【代码讲解】 本案例共定义 4 个函数：default、reduce、add 和 includePoints，其中，default 函数把地图设置为默认的 16 倍缩放比例，reduce 和 add 两个函数实现缩放比例的加一和减一，includePoints 函数指定(23.0403，113.7446)和(22.9983，113.7724)两个点，此时 map 地图缩放比例的依据就是需要在地图中包含这两个点，即这两个点需作为 map 地图 4 个顶点中的两个。

图 3.33(a)为默认的 16 倍缩放比例效果；图 3.33(b)为 17 倍缩放比例效果；图 3.33(c)为包含(23.0403，113.7446)和(22.9983，113.7724)两个点时的缩放比例效果。

3.8 实训项目——问卷调查

本实训项目是对微信小程序学习的问卷调查，知识点主要涉及表单组件的使用和表单数据在逻辑端 JS 中的获取。项目中涉及的表单组件有<form><radio-group><radio><checkbox-group><checkbox><label><slider><textarea>和<button>等。

视频讲解

项目在模拟器中的效果如图 3.34 所示；图 3.35 是 Console 控制台上显示的 JS 接收到的用户输入数据。

pages/survey/survey.wxml 文件代码如下：

```
<view class = "content">
```

```
  <form bindsubmit="onSubmit" bindreset="onReset">
    <view class="title">1.你现在大几?</view>
    <radio-group bindchange="universityChange">
      <radio value="大一"/>大一
      <radio value="大二"/>大二
      <radio value="大三"/>大三
      <radio value="大四"/>大四
    </radio-group>
    <view class="title">2.你使用手机最大的用途是什么?</view>
    <checkbox-group bindchange="mobilChange">
      <label><checkbox value="社交"/>社交</label>
      <label>
        <checkbox value="网购"/>网购</label>
      <label>
        <checkbox value="学习"/>学习</label><label>
        <checkbox value="其他"/>其他</label>
    </checkbox-group>
    <view class="title">3.平时每天使用手机多少小时?</view>
    <slider min="0" max="24" show-value bindchange="timechange" />
     <view class="title">4.你之前使用过微信小程序吗?</view>
    <radio-group bindchange="programChange">
      <radio value="无"/>无
      <radio value="有"/>有
    </radio-group>
    <view class="title">5.谈谈你对微信小程序未来发展的看法</view>
    <textarea auto-height placeholder="请输入你的看法" name="textarea" />
    <button size="mini" form-type="submit">提交</button>
    <button size="mini" form-type="reset">重置</button>
  </form>
</view>
```

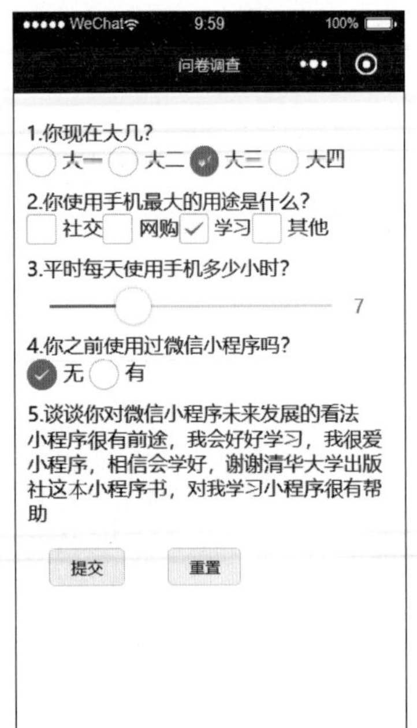

图 3.34　问卷调查示例图

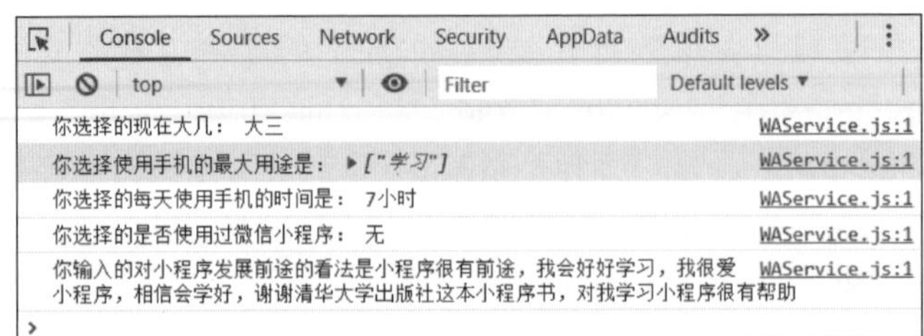

图 3.35　Console 控制台显示数据

pages/survey/survey.js 文件代码如下：

```
Page({
  universityChange: function(e) {
    console.log("你选择的现在大几：", e.detail.value)
  },
  mobilChange: function(e) {
    console.log("你选择使用手机的最大用途是：", e.detail.value)
  },
  timechange: function(e) {
    console.log("你选择的每天使用手机的时间是：", e.detail.value + "小时")
  },
  programChange: function(e) {
    console.log("你选择的是否使用过微信小程序：", e.detail.value)
  },
  onSubmit(e) {
    console.log("你输入的对小程序发展前途的看法是" + e.detail.value.textarea)
  },
  onReset() {
    console.log("表单已被重置")
  }
})
```

pages/survey/survey.wxss 文件代码如下：

```
.content {
  padding: 30rpx;
}
button {
  margin: 40rpx;
}
.title {
  margin - top: 20rpx;
}
```

【代码讲解】 本项目是常用的问卷调查页面，“你现在大几?”和“你之前使用过微信小程序吗?”两项使用了< radio-group >和< radio >组件；“你使用手机最大的用途是什么?”使用了< checkbox-group >和< checkbox >组件；“平时每天使用手机多少个小时?”使用了< slider >组件，“谈谈你对微信小程序未来发展的看法”使用了< textarea >组件；“提交”“重置”按钮使用了< button >组件；而< form >和< label >组件属于控制类组件，没有在页面中显示效果。读者需认真学习 JS 页面获取表单数据的方法，为后续编程做准备。

第4章 样式与布局

对于不熟悉CSS的程序员来说，开发微信小程序面临比较困难的问题就是界面的排版。微信小程序的排版跟WXML和WXSS有关，其中，WXML指定了界面的框架结构，而WXSS指定了界面的框架及元素的显示样式。本章结合小程序编程的特点介绍其样式与布局。

本章主要目标

- 了解样式与布局的基本概念；
- 掌握样式中盒子模型、选择器和常见的样式属性；
- 熟练掌握flex布局、layer布局和float布局；
- 具备设计美观小程序界面的能力。

4.1 小程序样式

要想设计出美观的小程序，需要努力学习小程序样式和布局。

WXML(WeiXin Markup Language)是小程序框架提供的标签语言，结合基础组件以构建出页面结构。WXSS(WeiXin Style Sheets)是小程序框架提供的样式语言，用于设置组件的样式。

4.1.1 定义样式

WXSS具有CSS的大部分特性，其书写规则由两部分组成：选择器和声明(属性:属性值)，如图4.1所示。

选择器(selector)：表明该样式将作用于哪些对象，这些对象可以是某个标签、指定的class或id值等。在解析这个样式时，根据选择器来渲染对象的显示效果。

属性(property)：样式的选择项。

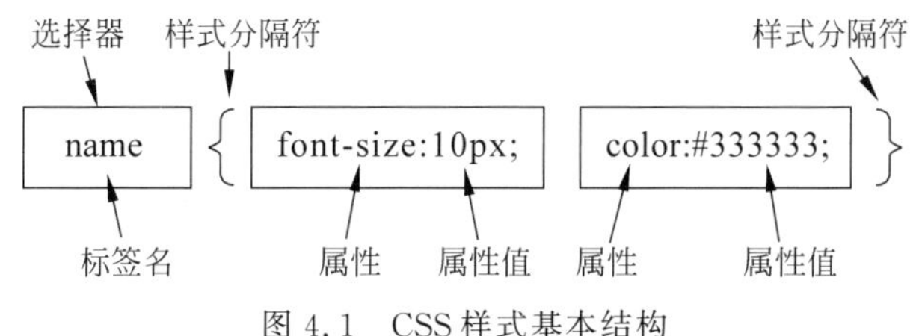

图 4.1　CSS 样式基本结构

属性值(value)：决定样式的参数。

自定义文字的字体为 12px,颜色为红色,示例代码如下：

```
text {
  font - size: 12px;            /* 定义字体大小为 12px */
  color: #f00;                  /* 定义字体颜色为红色 */
}
```

上述代码中 text 为选择器,font-size 和 color 为两个属性,12px 和 #f00 为对应的属性值。WXSS 注释以"/ * "开始,以" * /"结束。关于选择器的介绍详见 4.2 节。

4.1.2　使用样式

1. 内联样式

在 WXML 文件中使用 style 可以直接设置组件的样式,示例代码如下：

```
<view style = "width:150rpx; height:150rpx">我是 view 容器 <view>
```

上述代码中通过 style 对< view >组件设置属性,宽度和高度均设置为 150rpx。

2. 外部样式

外部样式分为自动引用和样式导入。

1) 自动引用

在小程序目录结构中,page 文件夹中的样式为局部样式,只作用于对应的页面。app.wxss 文件中的样式为全局样式,会作用于所有页面中具有相同选择器的组件,当局部样式出现与全局样式同名的选择器时,局部样式会覆盖全局样式。

2) 样式导入

使用@import 语句可以将其他路径中的样式引入当前文件中,@import 后面需要注明外联样式表的相对路径,用"；"表示语句结束。

假设有一个公共样式 common.wxss,示例代码如下：

```
.bg - gray {
  background - color: gray;
}
```

在其他的页面中都可以通过@import 语句引入外部文件样式表,示例代码如下：

```
@import "common.wxss";
.bg - blue {
```

```
  background - color: blue;
}
```

通过导入外部样式,.bg-gray 和.bg-blue 样式均可以被当前页面使用。

4.2 选择器

选择器(selector)是标签与样式的纽带,在 WXSS 文件中常用的选择器有基础选择器和复合选择器。

4.2.1 基础选择器

1. 标签选择器

给标签设置样式,会自动指向该标签。语法:

```
标签选择器名{属性:属性值;}
```

示例代码如下。

WXML 文件中:

```
< text >这是一段文字< text >
```

WXSS 文件中:

```
text {
 color: red;
}
```

其中,< text >是 WXML 中的文本组件,text 是 WXSS 文件中的样式选择器,color 是 text 选择器的属性。

2. class 选择器

语法:

```
.类选择器名{属性: 属性值;}
```

选择器的名称自定义,名称应当醒目且具有意义,页面中一旦设置了类别选择器,页面中该标签都具有相同的样式。示例代码如下。

WXML 文件中:

```
< text class = "title">这是一行文字</text >
< text class = "title">这是一行文字</text >
< text class = "title">这是一行文字</text >
```

WXSS 文件中:

```
.title {
  color: red;
}
```

其中,.title 为类选择器,color 为属性,其值为红色,页面中这类标签的元素属性相同。

3. id 选择器

语法:

```
#选择器名{属性: 属性值;}
```

选择器的名称自定义,名称应当醒目且有意义。示例代码如下。

WXML 文件中:

```
<text id="p-style">这是一行文字</text>
<text>这是一行文字</text>
```

WXSS 文件中:

```
#p-style {
  color: red;
}
```

其中,#p-style 为 id 选择器,对特定的<text>标签设置属性; color 为属性,其值为红色,只对第一行中的<text>标签生效,其他<text>标签不受影响。id 选择器比 class 选择器更具有针对性。

4.2.2 复合选择器

1. 多元素选择器

多元素选择器用于同时设置多个标签具有共同的样式。语法:

```
选择器,选择器,选择器,…{共有属性: 属性值;}
```

选择器之间用逗号隔开。示例代码如下。

WXML 文件中:

```
<view>我是 view 组件</view>
<text>我是 text 组件</text>
```

WXSS 文件中:

```
view, text {
  background-color: red;
}
```

其中,"view, text"为多元素选择器,同时设置<view>和<text>组件背景颜色为红色。

2. 后代元素选择器

后代元素选择器为标签的后代设置样式。语法:

```
选择器 1 选择器 2 选择器 3{属性: 属性值;}
```

选择器之间用空格隔开。示例代码如下。

WXML 文件中:

```
<view>
  <image>存放图片</image>
</view>
```

WXSS 文件中：

```
view image{
  width: 50rpx; height: 100rpx;
}
```

其中，view image 为后代元素选择器，设置< view >标签嵌套中的< image >图片大小，宽度为 50rpx，高度为 100rpx，后代元素选择器用于对特定元素进行设置。

4.3 基础样式

在 Web 页面设计中使用的样式繁多，而微信小程序的界面中使用的样式有所减少，读者可以通过查询 CSS 手册获取详细信息。现列举在小程序开发中经常使用的样式。

4.3.1 文本样式

文本样式决定页面中文字的排版，可以设置字符缩进、文字颜色、字符间距、文本对齐方式和装饰文字等，常用属性如表 4.1 所示。

表 4.1 文本样式属性

属　　性	说　　明	CSS 版本
color	设置文本的颜色	1
letter-spacing	设置字符间距	1
line-height	设置行高	1
text-align	规定文本的水平对齐方式	1
text-decoration	规定添加到文本的装饰效果	1
text-indent	规定文本块首行的缩进	1
text-transform	控制文本的大小写	1
vertical-align	设置元素的垂直对齐方式	1
white-space	设置如何给元素控件留白	1
word-spacing	设置单词间距	1
text-align-last	当 text-align 设置为 justify 时，最后一行的对齐方式	3
text-justify	当 text-align 设置为 justify 时，指定分散对齐的方式	3
text-outline	设置文字的轮廓	3
text-overflow	指定文本溢出包含的元素，应该发生什么	3
text-shadow	为文本添加阴影	3
text-wrap	指定文本换行规则	3
word-break	指定文字的断行规则	3
word-wrap	是否对过长的单词进行换行	3

示例代码如下。

WXML 文件中：

```
<view class = "text">对段落首行缩进、文字颜色、字符间距、行高、水平对齐方式和装饰效果进行设置.</view>
```

WXSS 文件中：

```
.text {
  text - indent: 2em; color: #00f; letter - spacing: 20rpx;
  line - height: 60rpx; text - align: left; text - decoration: underline;
}
```

运行效果如图 4.2 所示。

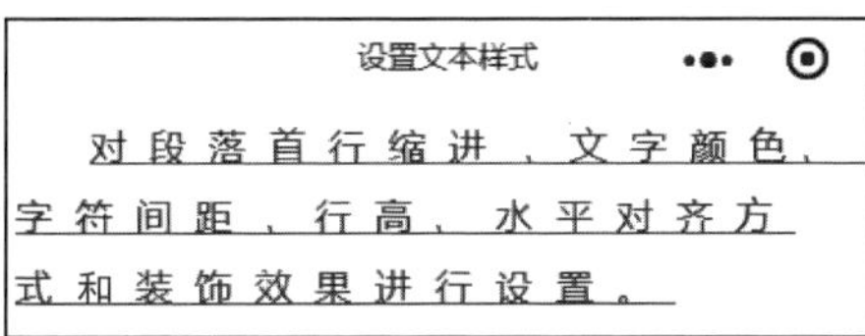

图 4.2　设置文本样式

在 WXSS 文件中对文本样式进行设置，文本首行缩进 2em、字体颜色蓝色、字符间距 20rpx、行高 60rpx、文本左对齐、文本增加下画线修饰。

4.3.2　字体样式

字体样式用于设置字体的属性，常用属性如表 4.2 所示。

表 4.2　字体样式属性

属　　性	说　　明	CSS 版本
font	复合属性，设置字体相关属性	1
font-family	规定文本的字体系列	1
font-size	规定文本的字体尺寸	1
font-style	规定文本的字体样式	1
font-variant	规定文本的字体大小写	1
font-weight	规定字体的粗细	1

示例代码如下。

WXML 文件中：

```
<text>设置字体样式</text>
```

WXSS 文件中：

```
text {
font - family: "Microsoft YaHei"; font - size: 50rpx;
```

```
font - style:oblique; font - weight:bold;
}
```

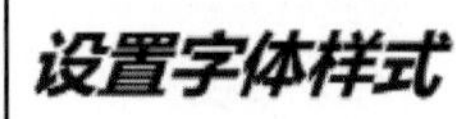

图 4.3　设置字体样式

在 WXSS 文件中对文字的样式进行设置，字体雅黑、字体大小 50rpx、样式斜体、字体加粗，运行效果如图 4.3 所示。

4.4　盒子模型

4.4.1　盒子模型概述

当对一个文档进行布局(layout)时，浏览器的渲染引擎会根据标准之一的 CSS 基础盒子模型(CSS basic box model)，将所有元素表示为一个个矩形的盒子。CSS 决定这些盒子的大小、位置及属性(例如颜色、背景、边框尺寸等)。盒子模型规定了内部处理的元素内容(content)、内边距(padding)、边框(border)和外边距(margin)的样式，其模型结构如图 4.4 所示。

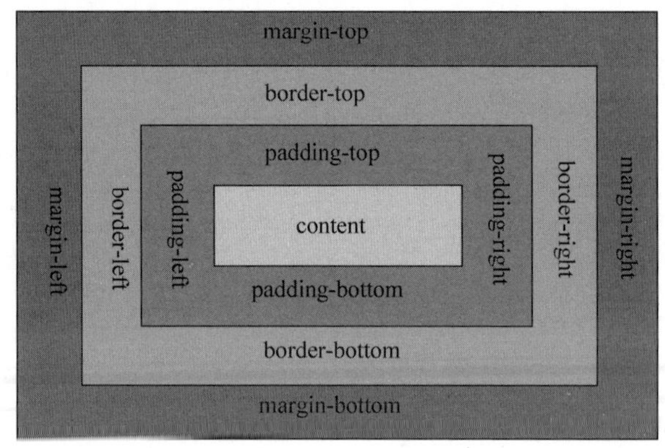

图 4.4　盒子模型

在盒子模型结构中，最内层为盒子的内容(content)，向外依次是内边距(padding-top、padding-right、padding-bottom、padding-left)、边框(border-top、border-right、border-bottom、border-left)以及外边距(margin-top、margin-right、margin-bottom、margin-left)。内边距、边框和外边距分别都有上、右、下、左 4 个属性，这 4 个属性可以同时应用于一个元素，也可以单独或部分应用于同一个元素。

在盒子模型中，很多样式需要指定方向，通常会有 4 个属性值，在 WXSS 文件中常遵循 TRBL 原则和相同合并原则来简化代码。

1. TRBL 原则

TRBL 原则按照 top、right、bottom、left 的方向来设置属性。

示例代码如下：

```
padding - top:1rpx;
padding - right:2rpx;
```

```
padding - bottom:3rpx;
padding - left:4rpx;
```

上述代码遵循了 TRBL 原则，按照上、右、下、左的顺序顺时针的方向将代码合并成一行，其效果等同于下方代码：

```
padding:1rpx 2rpx 3rpx 4rpx;
```

2. 相同合并原则

当属性值相同的时候，可以按照相同合并原则简化代码。

(1) 上、下、左、右的属性值相同时，简化为 1 个值。

示例代码如下：

```
padding:1rpx 1rpx 1rpx 1rpx;
padding:1rpx;
```

(2) 上、下和左、右的属性值分别相同时，简化为 2 个值。

示例代码如下：

```
padding:1rpx 2rpx 1rpx 2rpx;
padding:1rpx 2rpx;
```

(3) 左、右的属性值相同时，简化为 3 个值。

示例代码如下：

```
padding:1rpx 2rpx 3rpx 2rpx;
padding:1rpx 2rpx 3rpx;
```

4.4.2 盒子模型属性

1. width 与 height

盒子模型使用 width 和 height 定义内容区域的大小。除此之外还可以通过 max-height、min-height 设置最大高度和最小高度，属性如表 4.3 所示。

表 4.3 width 和 height 属性

属　　性	说　　明	CSS 版本
height	设置元素的高度	1
max-height	设置元素的最大高度	2
max-width	设置元素的最大宽度	2
min-height	设置元素的最小高度	2
min-width	设置元素的最小宽度	2
width	设置元素的宽度	1

2. padding 与 margin

padding 指内边距，即内容与边框之间的部分。内边距的属性有 4 种，分别为 padding-top、padding-right、padding-bottom 和 padding-left，其属性值可以是像素，也可以是百分比，通过设置内边距可以控制内容与边框的间隔。

margin 指外边距，主要用于设置元素之间的距离。外边距的属性有 4 种，分别为 margin-top、margin-right、margin-bottom、margin-left，其使用方法与内边距类似。padding 与 margin 的属性如表 4.4 所示。

表 4.4　padding 和 margin 属性

属　　性	说　　明	CSS 版本
padding	在一个声明中设置所有内边距属性	1
padding-top	设置元素的上内边距	1
padding-right	设置元素的右内边距	1
padding-bottom	设置元素的下内边距	1
padding-left	设置元素的左内边距	1
margin	在一个声明中设置所有外边距属性	1
margin-top	设置元素的上外边距	1
margin-right	设置元素的右外边距	1
margin-bottom	设置元素的下外边距	1
margin-left	设置元素的左外边距	1

3. border

边框是内容与外部填充的边界，使用边框属性可以设置边框的样式，常见属性如表 4.5 所示。

表 4.5　border 属性

属　　性	说　　明	CSS 版本
border	在一个声明中设置所有的边框属性	1
border-color	设置四条边框的颜色	1
border-style	设置元素四条边框的样式	1
border-width	设置四条边框的宽度	1
border-radius	设置四条边框的边角形状	3
outline	在一个声明中设置所有的轮廓属性	2

4.5　元素类别

小程序页面布局是指页面元素在页面中的显示位置，而页面元素与其他页面元素之间的位置关系会决定该页面元素在页面中的显示位置。所以在学习小程序的页面布局之前，开发者需要掌握页面元素的分类。根据展示特点的不同，页面元素分为块级元素、行内元素、内联块级元素 3 种。

4.5.1 块级元素

块级元素(block element)是一种以块状方式展示的容器,高度和宽度均可设置,默认宽度为 100 %,默认占一行的高度。在一行中两个块级元素无法并列排放(float 浮动后除外,详见 4.8 节),两个块级元素连续出现时,会在页面中自动换行显示。块级元素常作为页面布局的组织结构,内部可以嵌套块级元素和行内元素,小程序中< view >容器默认为块级元素,内部可嵌入< text >< image >等。通过设置属性 display: block 能够将元素设置为块级元素。

例 **4-1** 块级元素展示特点小案例,运行效果如图 4.5 所示。

视频讲解

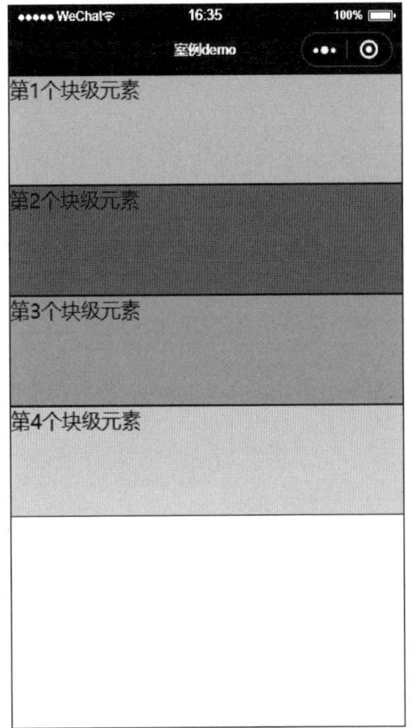

图 4.5 块级元素展示特点

pages/block/block.wxml 文件代码如下:

```
<view class = "red">第 1 个块级元素</view>
<view class = "blue">第 2 个块级元素</view>
<view class = "green">第 3 个块级元素</view>
<view class = "gray">第 4 个块级元素</view>
```

pages/block/block.wxss 文件代码如下:

```
.red {
  background-color: red;
}
.blue {
```

```
  background-color: blue;
}
.green {
  background-color: green;
}
.gray {
  background-color: gray;
}
view {
  display: block; height: 200rpx; border: 1px solid black;
}
```

【代码讲解】 通过设置属性 display: block,使得< view >容器为块级元素(默认为块级元素可省略),宽度设置为 200rpx,每个< view >容器独占一行,默认宽度 100%和父元素宽度一致,后面的< view >容器位于下一行。

4.5.2 行内元素

行内元素(inline element)一般嵌套在块级元素中使用,它只能容纳文本或者其他行内元素,高度和宽度均不可设置,WXML 的< text >就是行内元素。

例 4-2 行内元素展示特点小案例,运行效果如图 4.6 所示。

视频讲解

图 4.6 行内元素展示特点

pages/inline/inline. wxml 文件代码如下：

```
<view class = "red">第一个行内元素</view>
<view class = "blue">第二个行内元素</view>
<view class = "green">第三个行内元素</view>
<view class = "gray">第四个行内元素</view>
```

pages/inline/inline. wxss 文件代码如下：

```
.red {
  background - color: red;
}
.blue {
  background - color: blue;
}
.green {
  background - color: green;
}
.gray {
  background - color: gray;
}
view {
  display: inline; border: 1px solid black;
}
```

【代码讲解】 通过设置属性 display: inline，使< view >容器变为行内元素，宽度为内部文字的宽度，每个< view >容器并排在同一行，当超过父容器宽度时会换行显示。

4.5.3 内联块级元素

内联块级元素(inline-block element)同时具备块级元素和行内元素的特点，多个内联块级元素可以并列在同一行，并且高度和宽度都可以设置。例如，WXML 中的< image >是内联块级元素。通过设置属性 display:inline-block 能够将元素设置为内联块级元素。

例 4-3 内联块级元素展示特点小案例，运行效果如图 4.7 所示。

pages/inline-block/inline-block. wxml 文件代码如下：

视频讲解

```
<view class = "red">第一个内联块级元素</view>
<view class = "blue">第二个内联块级元素</view>
```

pages/inline-block/inline-block/block. wxss 文件代码如下：

```
.red {
 background - color: red;
}
.blue {
      background - color: blue;
}
view{
```

```
    display: inline-block; height: 300rpx; width: 300rpx;
    border: 1px solid black;
}
```

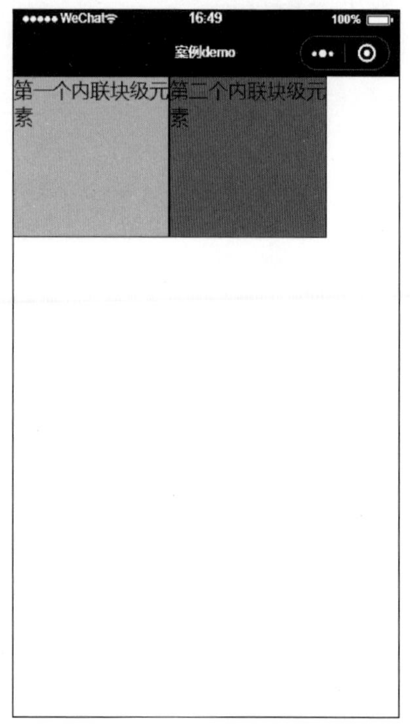

图 4.7　内联块级元素展示特点

【代码讲解】 通过设置属性 display：inline-block，使< view >容器变为内联块级元素；宽度和高度均设置为 300rpx，当不超过父容器宽度时，每个< view >容器并列排在同一行；同时具备行内元素及块级元素的特点。

4.6　flex 布局

4.6.1　flex 基本概念

flex 是 Flexible Box 的缩写，意为"弹性布局"，为盒子模型提供最大的灵活性。任何一个容器都可以指定为 flex 布局。示例代码如下：

```
.view {
    display: flex;
}
```

上述代码中将< view >组件设置为 flex 布局后，子元素的 float、clear 和 vertical-align 属性将失效。采用 flex 布局的容器，称为 flex 容器（flex container）。容器内部包含的子容器，称为 flex 项目(flex item),简称"项目"。flex 布局发生在父容器和子容器之间,父容器需要有 flex 的环境,只有设置其属性 display: flex 后,子容器才能根据自身的属性来布局。简单地说,就是如果父容器有 flex 的环境,子容器就可以瓜分父容器的空间。如果父容器没有 flex 的环境,那么子容器就无法使用 flex 的规则来使用父容器的空间。flex 模型如图 4.8 所示。

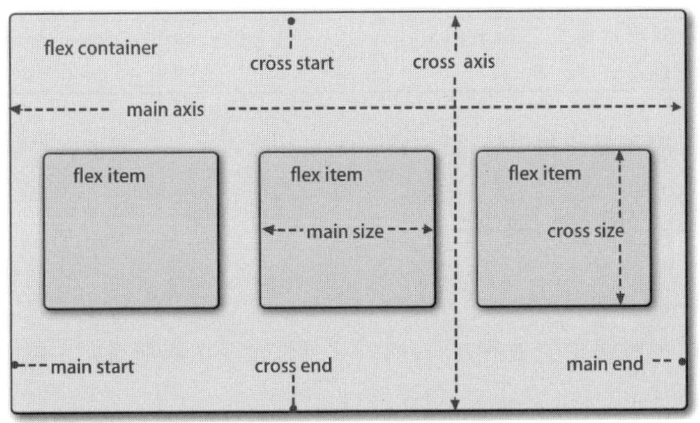

图 4.8　flex 模型

在图 4.8 中,父容器默认存在两根轴: 水平的主轴(main axis)和垂直的交叉轴(cross axis)。主轴的开始位置(与边框的交叉点)叫作 main start,结束位置叫作 main end; 交叉轴的开始位置叫作 cross start,结束位置叫作 cross end; 单个项目占据的主轴空间叫作 main size,占据的交叉轴空间叫作 cross size。项目默认沿主轴排列,默认属性为 flex-direction: row,水平轴从左向右,交叉轴为垂直方向自上而下; 通过设置属性 flex-direction: column 能够将主轴与交叉轴的位置进行互换,即主轴变为垂直方向的自上而下,交叉轴变为水平方向的从左向右。在 flex 布局中,通过设置容器属性来控制内部项目的排列与对齐方式。

flex 属性如表 4.6 所示。

表 4.6　flex 属性

属　　性	说　　明	默认值	其他属性值
flex-direction	项目的排列方向	row	row-reverse | column | column-reverse
flex wrap	项目是否换行	nowrap	wrap | wrap-reverse
justify-content	设置项目在主轴方向上的对齐方式	flex-start	flex-end | center | space-between | space-around | space-evenly
align-items	设置项目在交叉轴方向上的对齐方式	stretch	center | flex-end | baseline | flex-start
align-content	当多行排列时,设置行项目在交叉轴方向上的对齐方式	stretch	flex-start | center | flex-end | space-between | space-around | space-evenly
order	设置项目在主轴上的排列顺序	0	<integer>
flex-shrink	设置项目在主轴上溢出的收缩比率	1	<number>

续表

属　　性	说　　明	默认值	其他属性值
flex-grow	扩张在主轴方向上还有空间的项目	0	<number>
flex-basis	代替项目宽/高的属性	auto	<length>
align-self	设置项目在交叉轴上的对齐方式	auto	flex-start | center | flex-end | baseline |stretch

flex 属性分为容器属性和项目属性，将在 4.6.2 节和 4.6.3 节依次讲解。

4.6.2 flex 容器属性

1. flex-direction 属性

flex-direction 属性决定主轴的方向，即 flex 弹性盒子内部项目在主轴的排列方向。其语法格式如下：

```
.container {
  flex - direction: row(默认值)| row - reverse | column | column - reverse;
}
```

对应的属性值如下。

- row(默认值)：主轴为水平方向，起点在左端。
- row-reverse：主轴为水平方向，起点在右端。
- column：主轴为垂直方向，起点在顶端。
- column-reverse：主轴为垂直方向，起点在底端。

视频讲解

例 4-4　flex-direction 属性小案例。

pages/flex-direction/flex-direction. wxml 文件代码如下：

```
<view class = "flex - container">
  <view class = "item">元素 1 </view>
  <view class = "item">元素 2 </view>
  <view class = "item">元素 3 </view>
</view>
```

pages/flex-direction/flex-direction. wxss 文件代码如下：

```
.flex - container {
  margin: 50rpx auto; width: 740rpx;
  height: 600rpx; border: 1px solid #000;
  display: flex;
  flex - direction: row; /* 更改 flex - direction 属性，从 row 依次替换为 row - reverse | column
| column - reverse */
}
.item {
  width: 200rpx; height: 150rpx;
  line - height: 150rpx; border: 1px solid #000;
  text - align: center; background - color: skyblue;
}
```

将 WXSS 文件中的 flex-direction 属性依次设置为 row、row-reverse、column、column-reverse，4 种属性值对应的效果如图 4.9 所示。

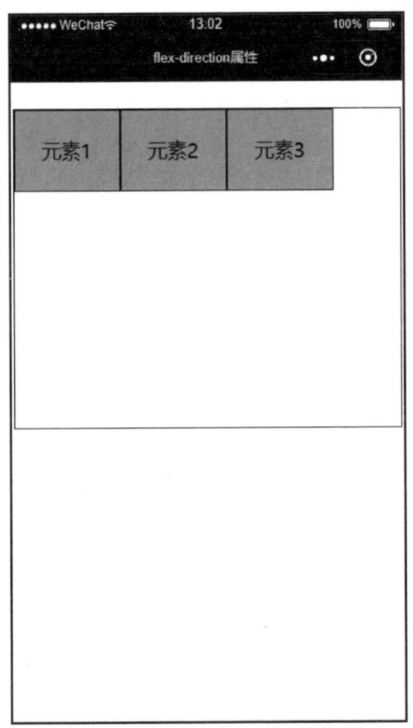

(a) row(默认)

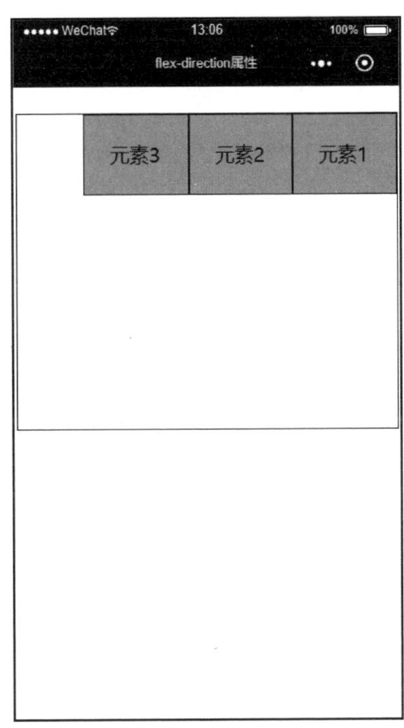

(b) row-reverse

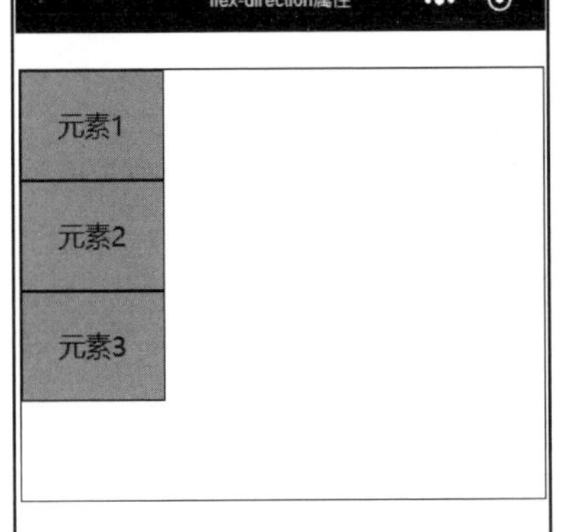

(c) column

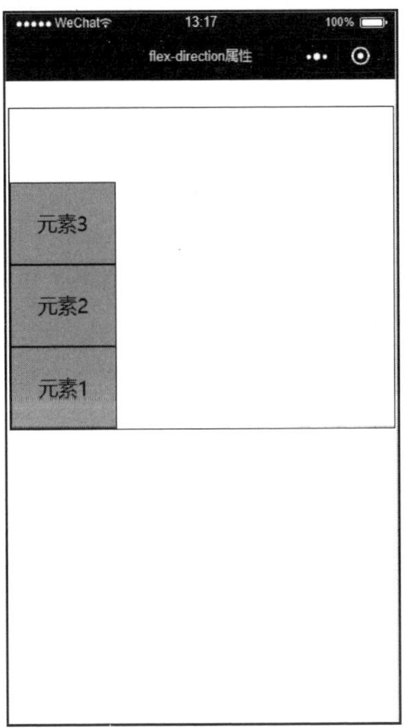

(d) column-reverse

图 4.9　flex-direction 属性示例图

2. flex-wrap 属性

flex-wrap 属性用于设置 flex 弹性盒子内部的项目是否允许项目换行以及换行时的方向，默认情况项目不换行。

其语法格式如下：

```
.container {
  flex - wrap: nowrap(默认值)| wrap | wrap - reverse;
}
```

对应的属性值如下。

- nowrap：不允许换行，当容器中所有项目的宽度超过父容器时，可能会被压缩。
- wrap：当容器中所有项目的宽度超过父容器时，允许换行排列。
- wrap-reverse：当容器中所有项目的宽度超过父容器时，换行的方向与 wrap 反向。

视频讲解

例 4-5 flex-wrap 属性小案例。

pages/flex-wrap/flex-wrap. wxml 文件代码如下：

```
<view class = "flex - container">
  <view class = "item">元素 1 </view>
  <view class = "item">元素 2 </view>
  <view class = "item">元素 3 </view>
  <view class = "item">元素 4 </view>
</view>
```

pages/flex-wrap/flex-wrap. wxss 文件代码如下：

```
.flex - container {
  margin: 50rpx auto;  width: 740rpx;
  height: 600rpx;  border: 1px solid #000;
  display: flex;  flex - direction: row;
  flex - wrap: nowrap; /* 更改 flex - wrap 属性,把 nowrap 依次替换为 wrap|wrap - reverse */
}
.item {
  width: 200rpx;  height: 150rpx;
  line - height: 150rpx;  border: 1px solid #000;
  text - align: center;  background - color: skyblue;
}
```

将 WXSS 文件中的 flex-wrap 属性依次替换为 nowrap、wrap、wrap-reverse，3 种属性值对应的效果如图 4.10 所示。

3. justify-content 属性

justify-content 属性用于设置 flex 弹性盒子内部的项目在主轴方向上的对齐方式。

其语法格式如下：

```
.container {
  justify - content: flex - start(默认值)| center | flex - end | space - between | space - around
| space - evenly;
}
```

对应的属性值如下。

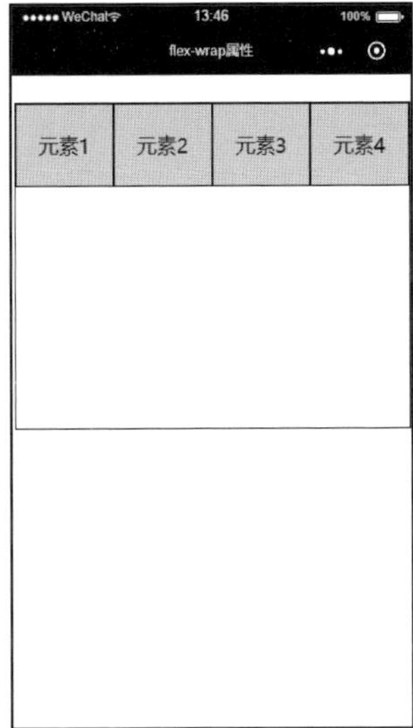

(a) nowrap (默认)

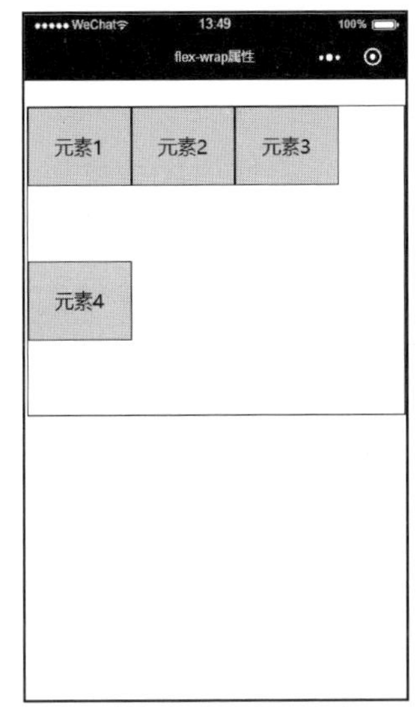

(b) wrap

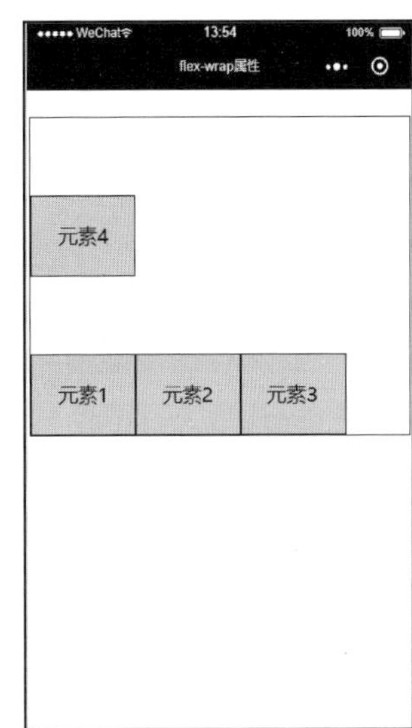

(c) wrap-reverse

图 4.10　flex-wrap 属性示例图

- flex-start：项目对齐于主轴起点，项目之间不留空隙。
- center：项目在主轴上居中排列，位于容器的中心，项目之间不留空隙。
- flex-end：项目对齐于主轴终点，项目之间不留空隙。
- space-between：项目间距相等，第一个和最后一个项目分别靠在主轴的起点和终点。
- space-around：第一个项目离主轴的起点和最后一个项目离主轴的终点的距离是中间相邻项目间距的一半。
- space-evenly：第一个项目离主轴起点以及最后一个项目离主轴的终点的距离以及中间相邻项目间距均相等。

例 4-6　justify-content 属性小案例。

视频讲解

pages/justify-content/justify-content.wxml 文件代码如下：

```
<view class = "flex - container">
  <view class = "item">元素 1 </view>
  <view class = "item">元素 2 </view>
  <view class = "item">元素 3 </view>
</view>
```

pages/justify-content/justify-content.wxss 文件代码如下：

```
.flex - container {
  margin: 50rpx auto; width: 740rpx;
  height: 600rpx; border: 1px solid #000;
  display: flex; flex - direction: row;
  flex - wrap: wrap;
  justify - content: flex - start; /* 更改 justify - content 属性,从 flex - start 依次替换为
center | flex - end | space - between | space - around | space - evenly */
}
.item {
  width: 200rpx; height: 150rpx;
```

```
  line-height: 150rpx; border: 1px solid #000;
  text-align: center; background-color: skyblue;
}
```

将 WXSS 文件中的 justify-content 属性依次替换为 flex-start、center、flex-end、space-between 和 space-around、space-evenly，6 种属性值对应的效果如图 4.11 所示。

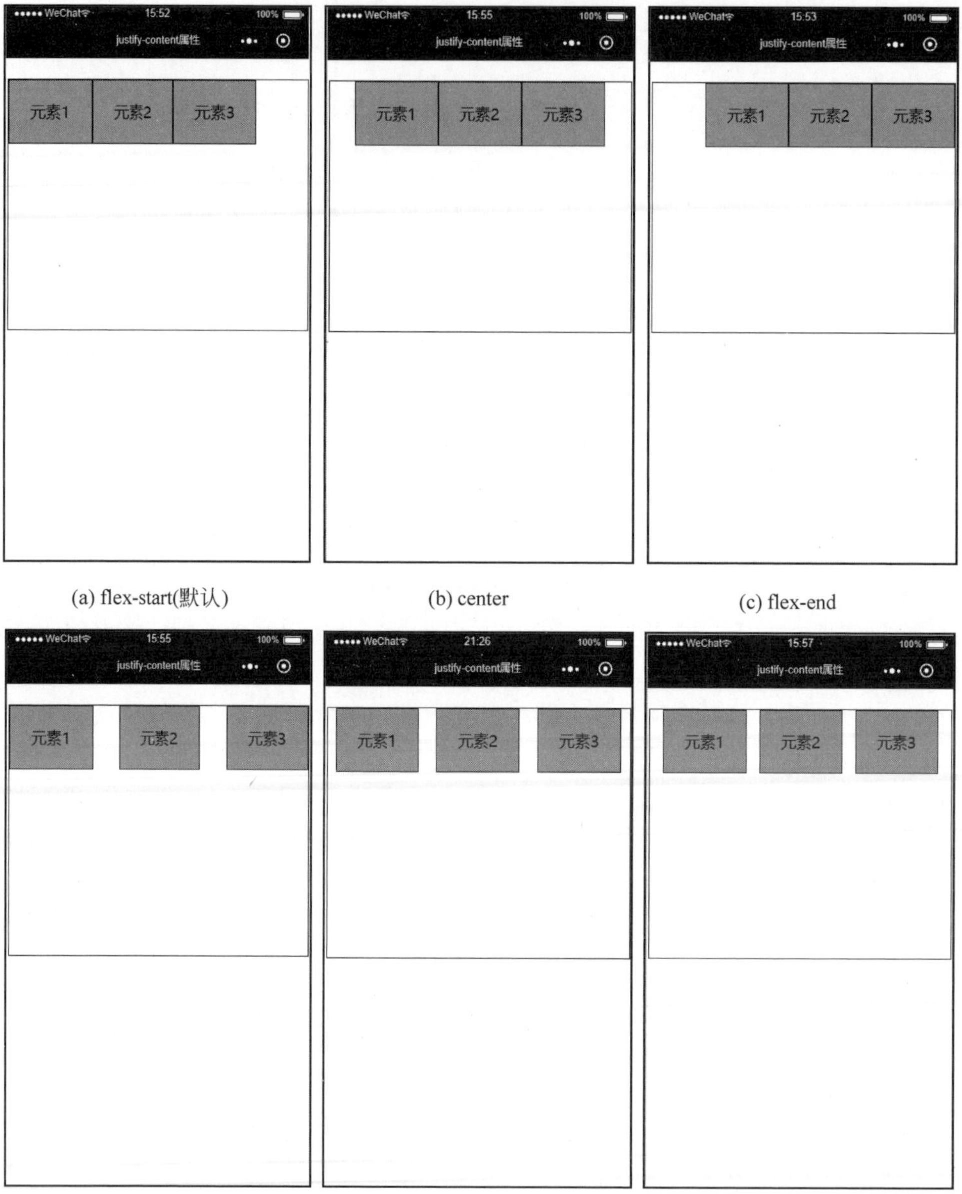

(a) flex-start(默认)　(b) center　(c) flex-end

(d) space-between　(e) space-around　(f) space-evenly

图 4.11　justify-content 属性示例图

4. align-items 属性

align-items 属性用于设置 flex 弹性盒子内部的项目在交叉轴方向上的对齐方式。其语法格式如下：

```
.container {
  align-items: stretch(默认值)| flex-start | center |flex-end | baseline;
}
```

对应的属性值如下。

- stretch：当项目大小未设置时项目会被拉伸至填满交叉轴。
- flex-start：项目对齐于交叉轴起点。
- center：项目在交叉轴居中对齐。
- flex-end：项目对齐于交叉轴终点。
- baseline：项目对齐于基线上，未设置基线时等同于 flex-start。

例 4-7 align-items 属性小案例。

pages/align-items/align-items. wxml 文件代码如下：

视频讲解

```
<view class="flex-container">
  <view class="item">元素 1 </view>
  <view class="item">元素 2 </view>
  <view class="item">元素 3 </view>
</view>
```

pages/align-items/align-items. wxss 文件代码如下：

```
.flex-container {
  margin: 50rpx auto;
  width: 740rpx;
  height: 600rpx;
  border: 1px solid #000;
  display: flex; flex-direction: row;
  flex-wrap: no-wrap;
  align-items: stretch       /* 更改 align-items 属性,从 stretch 依次替换为 flex-start |
    center | flex-end */
}
.item {
  width: 200rpx;
  height: 100rpx;            /* 当 align-items 的属性值为 stretch,删掉 height 属性 */
  line-height: 100rpx;
  border: 1px solid #000;
  text-align: center; background-color: skyblue;
}
```

将 WXSS 文件中的 align-items 属性依次替换为 stretch、flex-start、center、flex-end，当替换为 stretch 时需要将设置项目尺寸的代码删除，不设置项目的高度。4 种属性值对应的效果如图 4.12 所示。

5. align-content 属性

align-content 属性用于设置 flex 弹性盒子内部的项目进行多行排列时，在交叉轴方向上的对齐方式。

其语法格式如下：

```
.container {
```

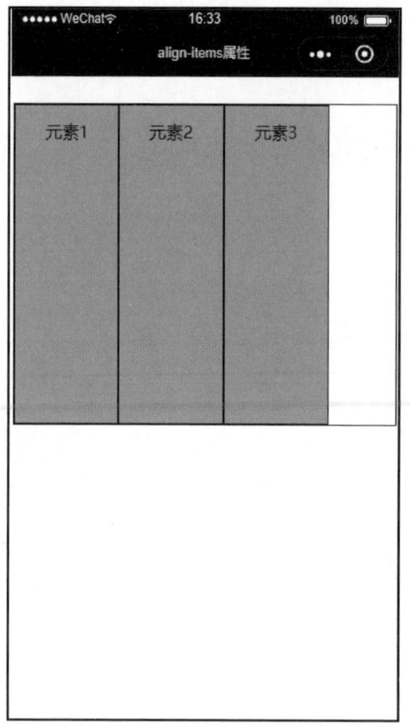

(a) stretch(默认)

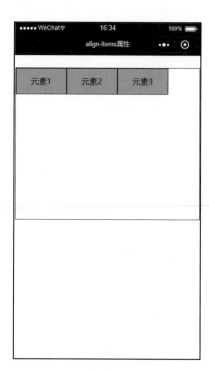

(b) flex-start

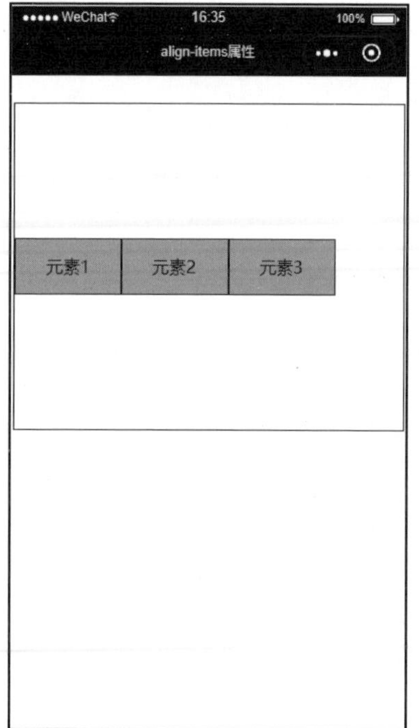

(c) center

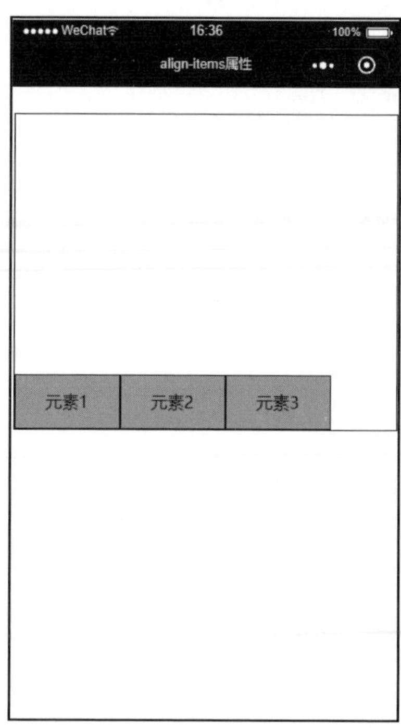

(d) flex-end

图 4.12　align-items 属性示例图

```
  align-content: | stretch(默认值)| flex-start | center | flex-end | space-between | space-around | space-evenly;
}
```

对应的属性值如下。

- stretch：未设置项目大小时将项目拉伸至填满交叉轴。
- flex-start：项目在交叉轴起点对齐。
- center：项目在交叉轴居中对齐。
- flex-end：项目在交叉轴终点对齐。
- space-between：行间距相等，首行与尾行靠在交叉轴起点和交叉轴终点。
- space-around：行间距相等，首行离交叉轴起点和尾行离交叉轴终点的距离为行间距的一半。
- space-evenly：首行离交叉轴起点和尾行离交叉轴终点的距离与行间距相等。

注意：多行排列时需要设置 flex-wrap 属性值为 wrap 允许换行。

例 4-8　align-content 属性小案例。

pages/align-content/align-content.wxml 文件代码如下：

视频讲解

```
<view class = "flex-container">
  <view class = "item a">元素 1 </view>
  <view class = "item b">元素 2 </view>
  <view class = "item c">元素 3 </view>
  <view class = "item b">元素 4 </view>
  <view class = "item a">元素 5 </view>
</view>
```

pages/align-content/align-content.wxss 文件代码如下：

```
.flex-container {
  margin: 50rpx auto;
  width: 740rpx;
  height: 600rpx;
  border: 1px solid #000;
  display: flex; flex-direction: row;
  flex-wrap: wrap;
  align-content: stretch;   /* 更改 align-content 属性，从 Stretch 依次替换为 flex-start|center | flex-end | space-between | space-around | space-evenly */
}
.item {
  height: 100rpx; /* 当 align-items 的属性值替换为 stretch，删掉 height 属性 */
  line-height: 100rpx;
  border: 1px solid #000;
  text-align: center;
  background-color: skyblue;
}
.a {
  width: 300rpx;
}
.b {
```

```
  width: 350rpx;
}
.c {
  width: 400rpx;
}
```

将 WXSS 文件中的 align-content 属性依次替换为 stretch、flex-start、center、flex-end、space-between、space-around 和 space-evenly，当替换为 stretch 时需要将 height 属性删掉，不设置项目的高度。7 种属性值对应的效果如图 4.13 所示。

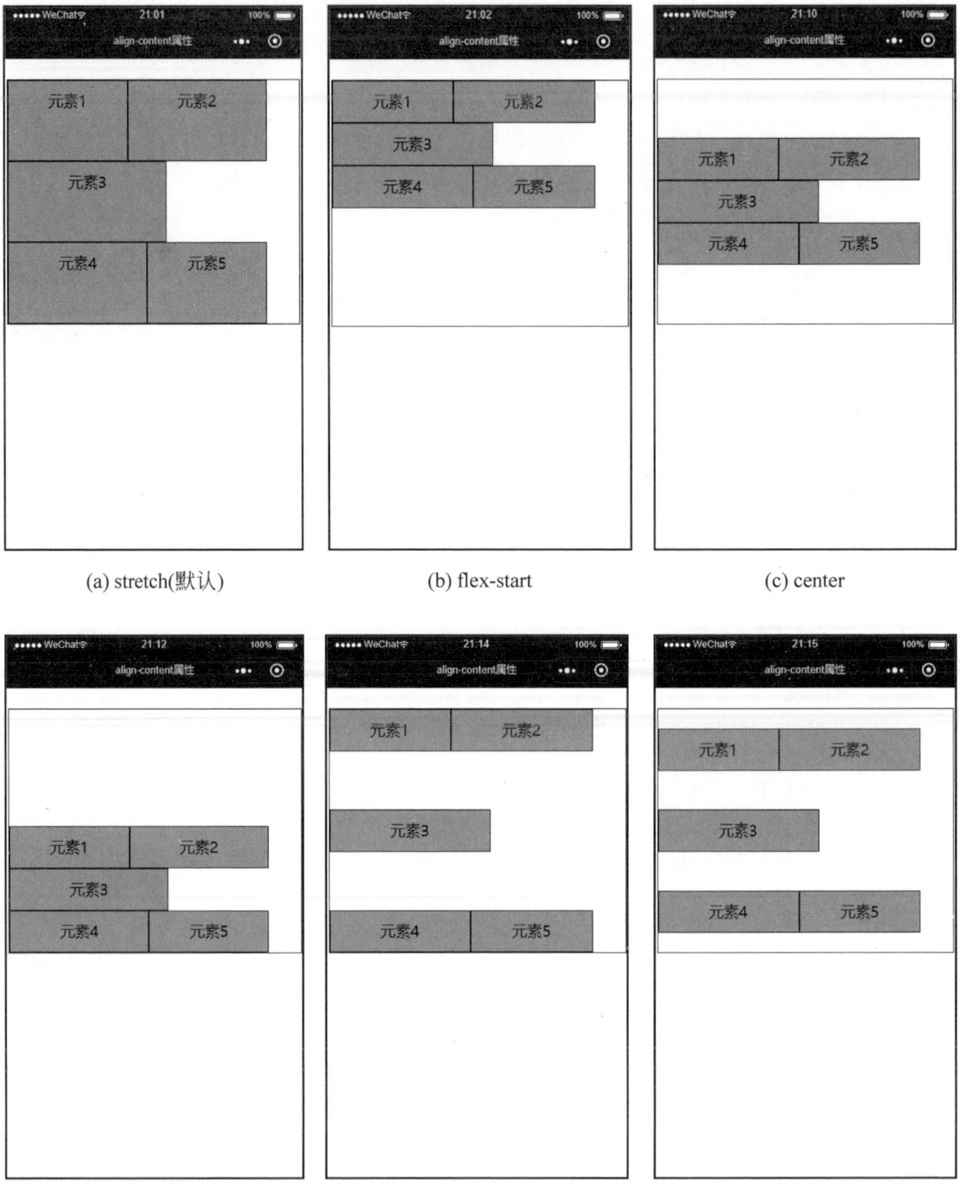

(a) stretch(默认)　(b) flex-start　(c) center

(d) flex-end　(e) space-between　(f) space-around

图 4.13　align-content 属性示例图

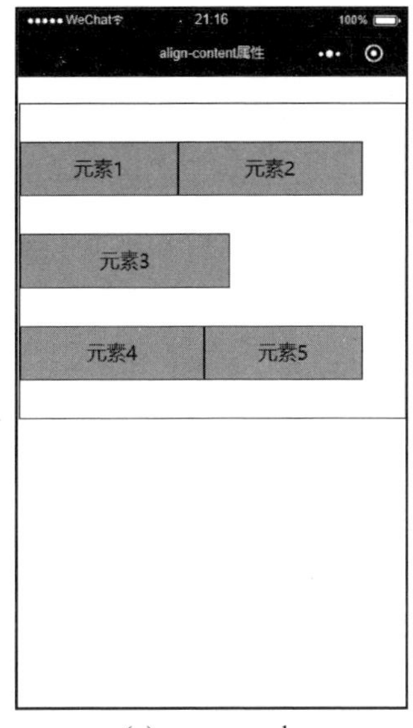

(g) space-evenly

图 4.13 (续)

4.6.3 flex 项目属性

1. order 属性

order 属性用于设置项目在主轴方向上的排列顺序，默认值为 0，容器中的项目会按照数值从小到大排列。

其语法格式如下：

```
.item {
    order: < integer >;
}
```

视频讲解

例 **4-9** order 属性小案例，运行效果如图 4.14 所示。

pages/order/order.wxml 文件代码如下：

```
< view class = "flex - container">
  < view class = "item a">元素 1 </view >
  < view class = "item b">元素 2 </view >
  < view class = "item c">元素 3 </view >
</view >
```

pages/order/order.wxss 文件代码如下：

```
.flex - container {
  margin: 50rpx auto;
  width: 740rpx;
  height: 600rpx;
  border: 1px solid #000;
  display: flex;
```

```
  flex-direction: row;
}
.item {
  width: 200rpx;
  height: 150rpx;
  line-height: 150rpx;
  border: 1px solid #000;
  text-align: center;
  background-color: skyblue;
}
.a {
  order: 1;
}
.b {
  order: 2;
}
.c {
  order: 3;
}
```

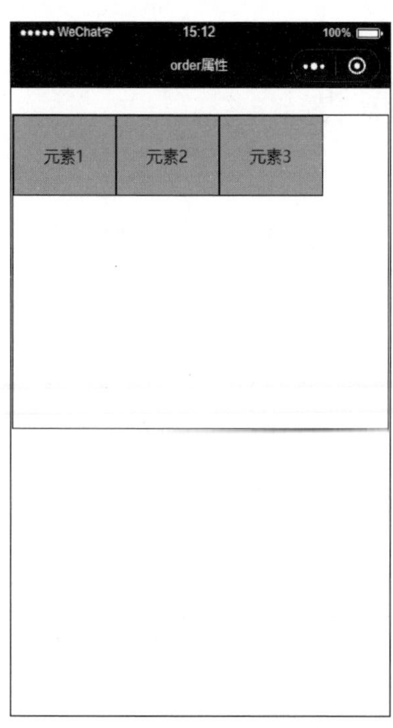

图 4.14　order 属性效果图

将容器中主轴方向设置为水平从左到右，3 个项目的 order 值分别为 1、2、3，按照从小到大的顺序依次排列。

2. flex-shrink 属性

flex-shrink 属性用于设置项目的收缩比率，当超出主轴方向上父容器的宽度，按照比例对项目进行压缩。

其语法格式如下：

```
.item {
  flex - shrink: 1(默认值)| <number>;
}
```

flex-shrink 的默认值为 1，如果没有定义该属性，将会自动按照默认值进行压缩。

压缩总权重的计算公式如下：

压缩总权重＝长度 1×收缩因子 1＋长度 2×收缩因子 2＋…＋长度 N×收缩因子 N

被移除溢出量的计算公式如下：

被移除溢出量＝原长度×溢出长度×收缩因子/压缩总权重

被压缩后的长度的计算公式如下：

被压缩后的长度＝原长度－被移除溢出量

以水平方向为例，假设有 3 个项目 a、b、c 宽度均为 300rpx，项目的收缩因子分别为 1、2 和 3。

WXSS 文件中：

```
.a {
  width: 300rpx; flex - shrink: 1;
}
.b {
  width: 300rpx; flex - shrink: 2;
}
.c {
  width: 300rpx; flex - shrink: 3;
}
```

假设父容器宽度为 600rpx，主轴水平从左向右，3 个项目的宽度均为 300rpx，会导致溢出 300rpx，现计算每个容器被压缩后的实际长度。

首先计算压缩总权重：

压缩总权重：300×1＋300×2＋300×3＝1800rpx

然后计算每个容器被移除的溢出量：

a 被移除溢出量：300×(300×1/1800)≈50rpx

b 被移除溢出量：300×(300×2/1800)≈100rpx

c 被移除溢出量：300×(300×3/1800)≈150rpx

最后计算每个容器被压缩后的宽度：

a 被压缩后的宽度：300－50＝250rpx

b 被压缩后的宽度：300－100＝200rpx

c 被压缩后的宽度：300－150＝150rpx

由上例可以看出：收缩因子不同，每个容器被压缩后的宽度也不相同，收缩因子数值越大，被移除溢出量越大。

视频讲解

例 4-10 flex-shrink 属性小案例，运行效果如图 4.15 所示。

pages/flex-shrink/flex-shrink. wxml 文件代码如下：

```
<view class = "flex - container">
```

```
  <view class = "item a">元素 1 </view>
  <view class = "item b">元素 2 </view>
  <view class = "item c">元素 3 </view>
</view>
```

图 4.15　flex-shrink 属性示例图

pages/flex-shrink/flex-shrink. wxss 文件代码如下：

```
.flex - container {
  margin: 50rpx auto;
  width: 600rpx;
  height: 600rpx; border: 1rpx solid #000000;
  display: flex; flex - direction: row; flex - wrap: nowrap;
}
.item {
  height: 100rpx;
  line - height: 100rpx;
  text - align: center;
  border: 1rpx solid #000000;
  background - color: skyblue;
}
.a {
  width: 300rpx; flex - shrink: 1;
}
.b {
  width: 300rpx; flex - shrink: 2;
}
.c {
```

```
  width: 300rpx; flex - shrink: 3;
}
```

容器被压缩前后的对比如图 4.16 所示。

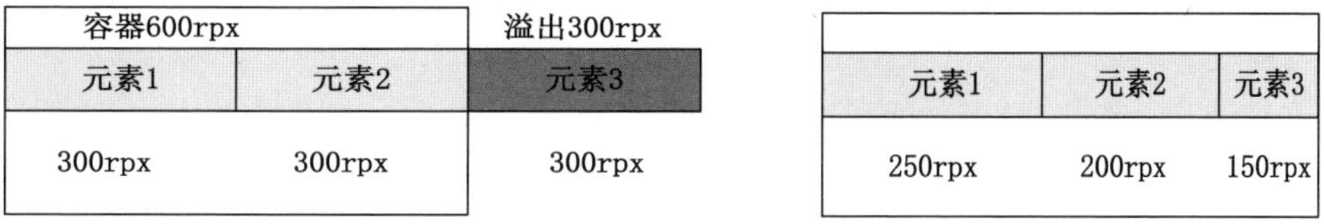

容器600rpx　溢出300rpx
元素1　元素2　元素3
300rpx　300rpx　300rpx

(a) 容器被压缩前

元素1　元素2　元素3
250rpx　200rpx　150rpx

(b) 容器被压缩后

图 4.16　容器被压缩前后的对比图

3. flex-grow 属性

flex-grow 属性用于设置项目的扩张比率。当项目在主轴方向上留有剩余空间时，通过对项目按照比例扩张来覆盖剩余空间。

其语法格式如下：

```
item {
  flex - grow: 0(默认值)| < number >;
}
```

flex-grow 的默认值为 0，如果没有定义该属性，项目不扩张。

扩张量的计算公式如下：

扩张量＝剩余空间/(扩张因子 1＋扩张因子 2＋…＋扩张因子 N)

被扩张后长度的计算公式如下：

被扩张后长度＝原长度＋扩张单元×扩张因子 N

以水平方向为例，假设有 3 个项目 a、b、c 宽度均为 200px，项目的扩张因子分别为 1、2、3。

WXSS 文件中：

```
.a {
  width: 200px; flex - grow: 1;
}
.b {
  width: 200px; flex - grow: 2;
}
.c {
  width: 200px; flex - grow: 3;
}
```

假设父容器宽度为 600rpx，3 个项目的宽度均为 100rpx，会导致剩余空间为 300rpx，现计算每个容器被扩张后的实际长度。

首先计算扩张量：300/(1＋2＋3)＝50rpx

然后计算每个容器的被扩张量：

a 被扩张量：50×1＝50rpx

b 被扩张量：50×2＝100rpx

c 被扩张量：50×3＝150rpx

最后计算每个容器被扩张后的宽度：

a 被扩张后的宽度：100＋50＝150rpx

b 被扩张后的宽度：100＋100＝200rpx

c 被扩张后的宽度：100＋150＝250rpx

由上例可以看出：扩张因子不同，每个容器被扩张后的宽度也不相同，扩张因子数值越大，被扩张量越大。

例 **4-11** flex-grow 属性小案例，运行效果如图 4.17 所示。

视频讲解

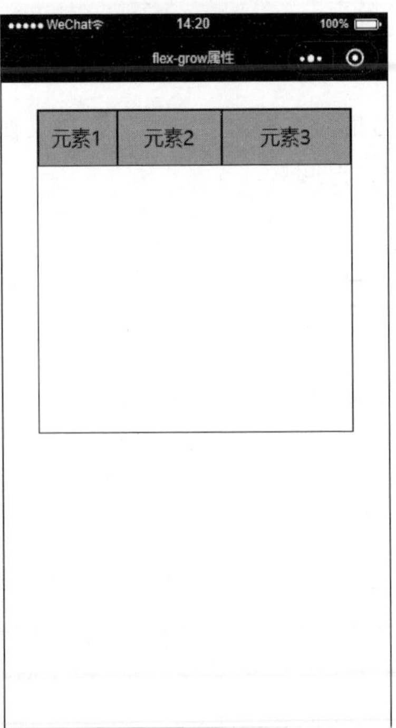

图 4.17 flex-grow 属性示例图

pages/flex-grow/flex-grow.wxml 文件代码如下：

```
<view class = "flex - container">
  <view class = "item a">元素 1 </view>
  <view class = "item b">元素 2 </view>
  <view class = "item c">元素 3 </view>
</view>
```

pages/flex-grow/flex-grow.wxss 文件代码如下：

```
.flex - container {
  margin: 50rpx auto; width: 600rpx;
  height: 600rpx; border: 1rpx solid #000000;
  display: flex; flex - direction: row;
}
.item {
  height: 100rpx; line - height: 100rpx;
```

```
    text-align: center; border: 1rpx solid #000000;
    background-color: skyblue;
}
.a {
    width: 100rpx; flex-grow: 1;
}
.b {
    width: 100rpx; flex-grow: 2;
}
.c {
    width: 100rpx; flex-grow: 3;
}
```

容器被扩张前后的对比如图 4.18 所示。

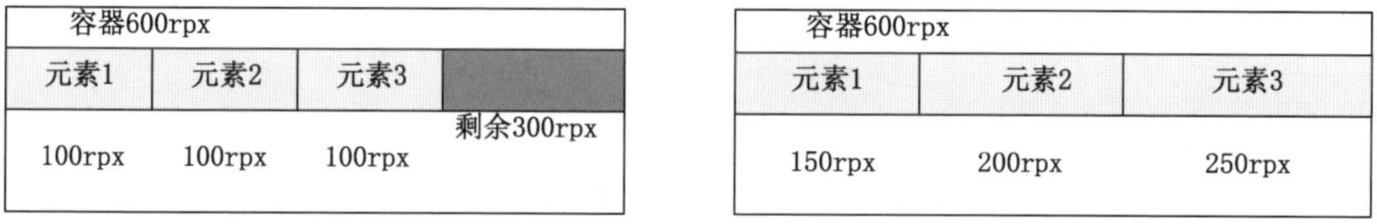

(a) 容器被扩张前　　(b) 容器被扩张后

图 4.18　容器被扩张前后的对比图

4. flex-basis 属性

flex-basis 属性用于代替主轴方向上项目的宽或高。

其语法格式如下：

```
.item {
    flex-basis: auto(默认值) | <number>
}
```

对应的属性值如下。

- 当容器属性设置为 flex-direction：row 或 flex-direction：row-reverse 时，如果 flex-basis 和 width 属性同时存在数值时，flex-basis 会代替 width 属性。
- 当容器属性设置为 flex-direction：column 或 flex-direction：column-reverse 时，如果 flex-basis 和 height 属性同时存在数值时，flex-basis 会代替 height 属性。
- 数值的优先级高于 auto，如果 flex-basis 属性值为 auto，当项目设置了高度或宽度时会取代 auto。

例 4-12　flex-basis 属性小案例，运行效果如图 4.19 所示。3 个项目的宽度均设置为 300rpx，通过对第一个项目设置属性 flex-basis：200rpx 使其宽度变为 200rpx，如图 4.20 所示。

视频讲解

pages/flex-basis/flex-basis.wxml 文件代码如下：

```
<view class="flex-container">
  <view class="item a">元素 1</view>
  <view class="item b">元素 2</view>
  <view class="item c">元素 3</view>
</view>
```

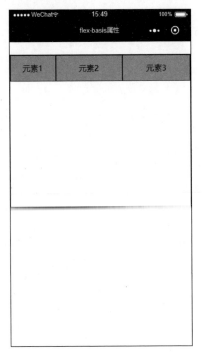

图 4.19　flex-basis 属性示例图

项目1	项目2	项目3
200rpx	300rpx	300rpx

图 4.20　项目宽度示例图

pages/flex-basis/flex-basis/wxss 文件代码如下：

```
.flex-container {
  margin: 50rpx auto; width: 740rpx;
  height: 600rpx; border: 1px solid #000;
  display: flex; flex-direction: row;
}
.item {
  height: 100rpx; line-height: 100rpx;
  border: 1px solid #000; text-align: center; background-color: skyblue;
}
.a {
  width: 300rpx; flex-basis: 200rpx;
}
.b {
  width: 300rpx;
}
.c {
  width: 300rpx;
}
```

5. align-self 属性

align-self 属性用于单独为 flex 弹性盒子内部的项目设置在交叉轴方向上的对齐方式，其属性会覆盖容器的 align-items。其语法格式如下：

```
.item {
```

```
  align - self: auto (默认值)| stretch | flex - start | center | flex - end | baseline;
}
```

对应的属性值如下。

- auto：默认项目继承父容器设置的 align-items 的属性，如果没有设置，auto 替换为 stretch。其他属性与 align-items 一致。

例 4-13　align-self 属性小案例，运行效果如图 4.21 所示。

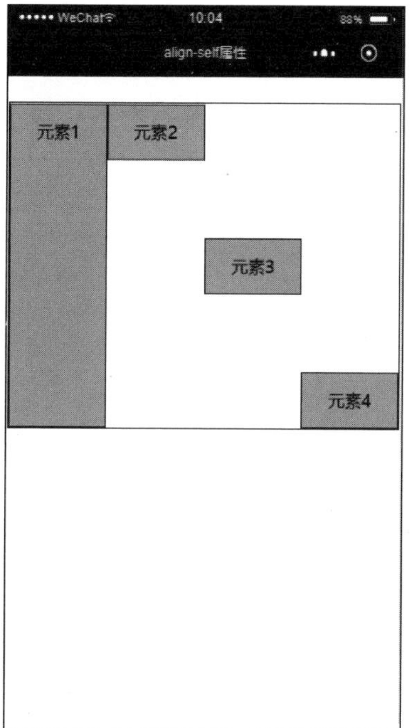

视频讲解

图 4.21　align-self 属性示例图

pages/align-self/align-self.wxml 文件代码如下：

```
< view class = "flex - container">
  < view class = "item a">元素 1 </view >
  < view class = "item b">元素 2 </view >
  < view class = "item c">元素 3 </view >
  < view class = "item d">元素 4 </view >
</view >
```

pages/align-self/align-self.wxss 文件代码如下：

```
.flex - container {
  margin: 50rpx auto;
  width: 740rpx;
  height: 600rpx;
  border: 1px solid #000;
  display: flex;
  flex - direction: row;
align - items: stretch;        /* align - items 默认取 stretch,固本行可以省略 */}
```

```
.item {
  width: 200rpx;
  line - height: 100rpx;
  border: 1px solid #000;
  text - align: center;
  background - color: skyblue;
}
.a {
  align - self: stretch;
}
.b {
  height: 100rpx;
  align - self: flex - start;
}
.c {
  height: 100rpx;
  align - self: center;
}
.d {
  height: 100rpx;
  align - self: flex - end;
}
```

【代码讲解】 如图 4.21 所示,项目 1 设置属性 align-self: stretch,并且不能设置项目的高度否则属性无效;项目 2 设置属性 align-self: flex-start 使得项目位于交叉轴顶端;项目 3 设置属性 align-self: center 使得项目位于交叉轴中间;项目 4 设置属性 align-self: flex-end 使得项目位于交叉轴底部。本项目中若去掉 4 个项目的 align-self 属性,则项目执行效果同图 4.12(a)。

align-self 属性与 align-items 属性的区别在于: align-self 属性单独作用于某一项目,可以覆盖 align-items 属性; align-items 属性作用于容器中所有的项目。

4.7 layer 布局

在 4.6 节 flex 布局中已介绍,flex 布局能够对容器中项目之间的位置关系进行设置,不同项目之间可以并列在一起,然而元素之间的位置关系没有涉及相互覆盖。在对小程序页面进行设计时,当需要实现元素之间的覆盖以及定位时,可以使用 layer 布局。

layer 布局基于定位的思想,分别有绝对定位(position: absolute)、相对定位(position: relative)以及固定定位(position: fixed)。

1. 绝对定位

将元素设置为绝对定位,首先设置元素的属性 position: absolute,该元素会相对于最近的一个具有定位属性的父元素作为参考,然后通过 top、bottom、left、right 属性设置上、下、左、右的偏移量完成绝对定位。如果不存在具有定位属性的父元素,该元素会以屏幕左上方的节点作为参考。

视频讲解

例 **4-14** 绝对定位小案例,运行效果如图 4.22 所示。

pages/absolute/absolute. wxml 文件代码如下:

```
< view >使用绝对定位</view >
```

pages/absolute/absolute. wxss 文件代码如下:

```
view {
  width: 300rpx; height: 500rpx;  position: absolute;  left: 200rpx;
  top: 200rpx; background - color:skyblue; border: 1px solid #000000;
}
```

【代码讲解】 在图 4.22 中,通过设置属性 left: 200rpx 和 top: 200rpx 使< view >容器相对于屏幕左边的距离为 200rpx,相对于屏幕上边的距离为 200rpx,实现了绝对定位。

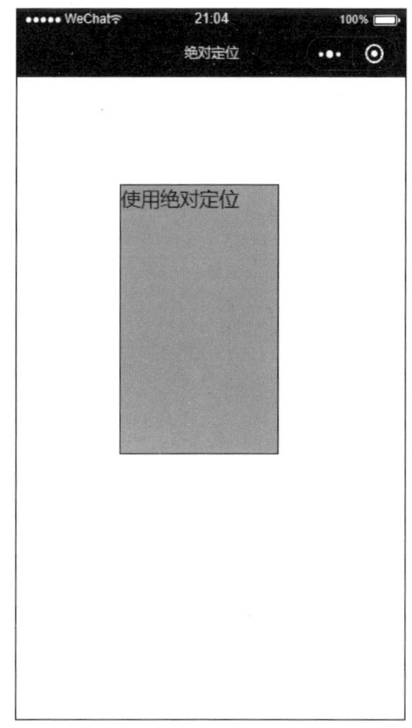

图 4.22 绝对定位

2. 相对定位

相对定位表示子元素相对于父元素作为参考的一种定位方式,子元素设置其属性 position: absolute,父元素设置其属性 position: relative,然后通过 top、bottom、left、right 属性设置上、下、左、右的偏移量完成相对定位。

例 **4-15** 相对定位小案例,运行效果如图 4.23 所示。

pages/relative/relative. wxml 文件代码如下:

视频讲解

```
< view class = "item1">
  < view class = "item2">使用相对定位</view >
</view >
```

pages/relative/relative.wxss 文件代码如下：

```
.item1 {
    position: relative; width: 700rpx;
    height: 400rpx; border: 1px solid #000; margin-top: 150rpx;
}
.item2 {
    position: absolute; width: 150rpx;
    height: 100rpx; left: 490rpx; top: 230rpx;
    background-color: skyblue; border: 1px solid #000000;
}
```

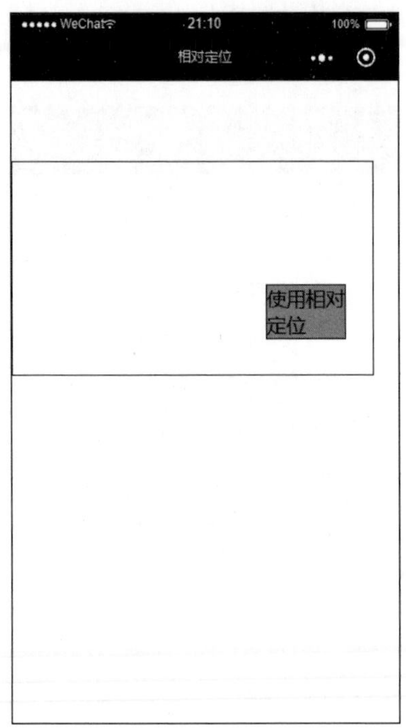

图 4.23　相对定位

【代码讲解】 在图 4.23 中，通过设置属性 left：490rpx 和 top：230rpx 使<view>容器相对于父容器左边的距离为 490rpx，相对于父容器上边的距离为 230rpx，实现了相对定位。

3. 固定定位

固定定位用于将元素固定在屏幕中，不会随着页面的滚动而发生位置变化，需要设置其属性 position：fixed，元素位置的定位与绝对定位的方式类似，不同之处在于元素会始终位于屏幕的某个位置。

例 4-16　固定定位小案例，运行效果如图 4.24 所示。

视频讲解

pages/fixed/fixed.wxml 文件代码如下：

```
<view class="item">使用固定定位</view>
```

pages/fixed/fixed.wxss 文件代码如下：

```
.item {
  position: fixed; width: 90%; height: 130rpx;
  background-color: skyblue; top: 1000rpx;
  border: 1px solid #000000;
}
```

图 4.24　固定定位

【代码讲解】 在图 4.24 中，当设置属性为 position：fixed 时，容器将固定在页面中，并且屏幕滚动时元素位置不会发生改变。

4.8 float 布局

块级元素独占一行，如果需要两个块级元素并排显示，就可以使用 float（浮动）布局，默认情况下元素是不浮动的，通过设置属性 float：left 表示元素向着屏幕左边浮动，设置属性 float：right 表示元素向着屏幕右边浮动，浮动的条件必须是子元素的总宽度小于父容器的宽度。

例 4-17　float 布局小案例，运行效果如图 4.25 所示。

pages/float/float.wxml 文件代码如下：

视频讲解

```
<view class="left">左浮动</view>
<view class="left">左浮动</view>
<view class="left">左浮动</view>
<view class="right">右浮动</view>
<view class="right">右浮动</view>
```

```
<view class = "right">右浮动</view>
```

pages/float/float. wxss 文件代码如下：

```
view {
  width: 150rpx; height: 250rpx;
  border: 1px solid #ffffff; background-color: skyblue;
  }
.left {
  float: left;
}
.right {
  float: right;
}
```

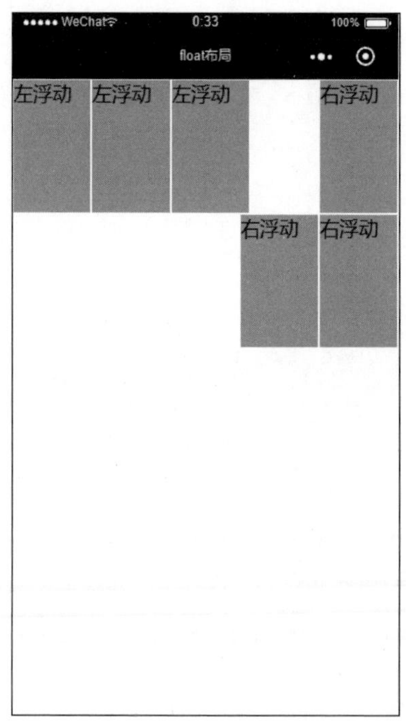

图 4.25 float 布局小案例

【代码讲解】 在图 4.25 中，设置 3 个< view >容器左浮动，并排显示在屏幕左边，每个< view >容器的宽度为 150rpx 加上 1px 的外边框后，宽度超过模拟器的总宽度 750rpx，于是后面 3 个< view >容器在设置了右浮动后，只有一个浮动在第一行的右边，后面两个并排浮动在第二行的右侧。

4.9 小程序布局实战

小程序常见的界面布局可以分为列表式、转盘式、多面板以及标签式，现分别依次进行讲解。

4.9.1 列表式

视频讲解

列表式布局是一种常见的排版方式,其排列方式为由上而下垂直排列,在每个列表元素当中存放内容,示例如图 4.26 所示。

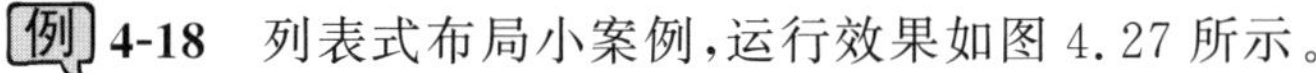
例 4-18 列表式布局小案例,运行效果如图 4.27 所示。

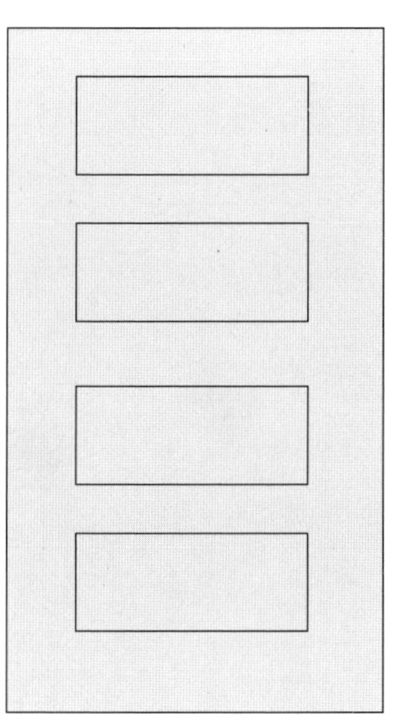
图 4.26 列表式示例图

图 4.27 电影列表小案例

pages/movie/movie.wxml 文件代码如下:

```
<! -- 导航标题 -- >
< view class = "title">
  < view class = "select">正在热映</view >
  < view class = "default">即将上映</view >
</view >
<! -- 轮播图 -- >
< view class = "haibao">
  < swiper indicator - dots = "{{indicatorDots}}" autoplay = "{{autoplay}}"
  interval = "{{interval}}">
    < block wx:for = "{{imgUrls}}">
      < swiper - item >
        < image src = "{{item}}"></image >
      </swiper - item >
    </block >
  </swiper >
</view >
```

```
<!--列表元素-->
<block wx:for="{{movies}}">
  <view class="list">
    <view class="pic">
      <image src="{{item.image}}"></image>
    </view>
    <view class="movie">
      <view class="name">{{item.name}}</view>
      <view>{{item.type}}</view>
      <view>{{item.director}}</view>
      <view>{{item.actor}}</view>
      <view>{{item.showTime}}</view>
    </view>
    <view class="btn">
      <button>观看</button>
    </view>
  </view>
  <view class="hr"></view>
</block>
```

pages/movie/movie.js 文件代码如下：

```
Page({
  data: {
    indicatorDots: true,
    autoplay: true,
    interval: 3000,
    imgUrls: [
      "/images/haibao/haibao1.jpg",
      "/images/haibao/haibao2.jpg",
    ],
    movies: [{
      image: "/images/list/movie1.jpg",
      name: "流浪地球",
      type: "类型：科幻/动作片",
      director: "导演：郭帆",
      actor: "主演：屈楚萧 李光洁",
      showTime: "上映：2019 年",
    }, {
      image: "/images/list/movie2.jpg",
      name: "复仇者联盟 4",
      type: "类型：悬疑/科幻片",
      director: "导演：安东尼·罗素",
      actor: "主演：小罗伯特·唐尼 克里斯·埃文斯",
      showTime: "上映：2019 年",
    }, {
      image: "/images/list/movie3.jpg",
      name: "波西米亚狂想曲",
      type: "类型：爱情/喜剧片",
      director: "导演：布莱恩·辛格",
      actor: "拉米·马雷克 露西·宝通",
```

```
      showTime: "上映: 2019 年",
    }]
  }
})
```

pages/movie/movie.wxss 文件代码如下：

```
/* 导航样式 */
.title {
  display: flex; flex-direction: row;
}
/* 导航选中 */
.select {
  font-size: 30rpx; width: 50%; color: green; text-align: center;
  height: 100rpx; line-height: 100rpx;
  border-bottom: 2px solid green;
}
/* 导航未选中 */
.default {
  width: 40%; font-size: 30rpx; text-align: center;
  height: 100rpx; line-height: 100rpx;
}
/* 轮播图样式 */
swiper {
  height: 280rpx;
}
.haibao image {
  width: 100%; height: 280rpx;
}
/* 列表内部样式 */
.list {
  display: flex; flex-direction: row;
}
/* 图片样式 */
.pic image {
  width: 180rpx; height: 200rpx; padding: 10rpx;
}
/* 电影信息样式 */
.movie {
  font-size: 24rpx; padding-top: 20rpx; line-height: 40rpx; color: #000;
}
/* 电影名称样式 */
.name {
  font-size: 26rpx; font-weight: bold;
}
/* 按钮位置 */
.btn {
  position: absolute; right: 30rpx; margin-top: 140rpx;
}
/* 按钮样式 */
.btn button {
```

```
  width: 100rpx; height: 50rpx; font-size: 20rpx;
  color: green; border: 1px solid green;
}
/* 分隔线样式 */
.hr {
  height: 1px; width: 100%; background-color: #ccc;
  opacity: 0.5;
}
```

【代码讲解】 页面布局思路：首先设计页面上部的标题，利用< swiper >组件实现图片的切换效果；然后在轮播图下方放置< view >容器，设置为 flex 布局，主轴在水平方向上从左向右，使得电影图片、影片介绍信息和“观看”按钮可以在水平方向上排列，再分别设置内部的电影图片、电影信息以及按钮的样式；最后通过列表渲染实现 3 个电影的显示。

列表式布局是一种常见的布局方式，读者要体会到整体与局部的关系，整体内容呈现垂直向下排布，列表中的每行内容都通过样式的设置来完成布局。

4.9.2 转盘式

对于转盘式中的内容，用户可以通过左右滑动预览每个元素，示例如图 4.28 所示。

例 **4-19** 转盘式布局小案例，运行效果如图 4.29 所示。

视频讲解

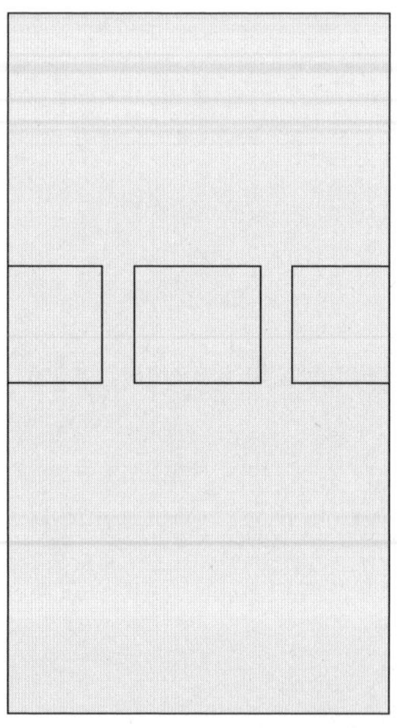

图 4.28 转盘式示例图

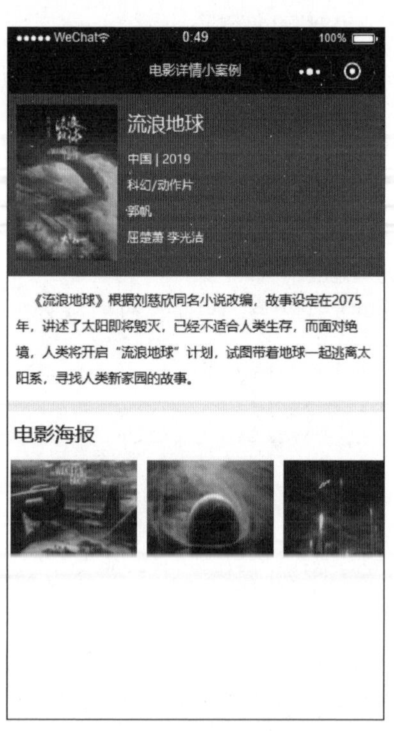

图 4.29 电影详情小案例

pages/movie/movieDetail.wxml 文件代码如下：

```
<view class="bg">
  <view class="movie">
    <view class="pic">
      <!-- 电影图片 -->
      <image src="/images/one.jpg"></image>
         </view>
    <view class="outer">
      <!-- 电影标题 -->
      <view class="name">流浪地球</view>
      <view class="item">中国 | 2019 </view>
      <view class="item">科幻/动作片</view>
      <view class="item">郭帆</view>
      <view class="item">屈楚萧 李光洁</view>
    </view>
  </view>
</view>
<view class="detail">
  <!-- 电影详情 -->
《流浪地球》根据刘慈欣同名小说改编，故事设定在 2075 年，讲述了太阳即将毁灭，已经不适合人类生存，而面对绝境，人类将开启“流浪地球”计划，试图带着地球一起逃离太阳系，寻找人类新家园的故事。
</view>
<!-- 分隔线 -->
<view class="hr"></view>
<!-- 电影海报 -->
<view class="zhinan">电影海报</view>
<scroll-view scroll-x="true">
  <view class="haibaoitem">
    <image src="/images/haibao1.jpg" style="height:180rpx;width:250rpx"></image>
  </view>
  <view class="haibaoitem">
    <image src="/images/haibao2.jpg" style="height:180rpx;width:250rpx"></image>
  </view>
  <view class="haibaoitem">
    <image src="/images/haibao3.jpg" style="height:180rpx;width:250rpx"></image>
  </view>
</scroll-view>
```

pages/movie/movieDetail.wxss 文件代码如下：

```
/* 背景颜色 */
.bg {
  width: 100%; background-color: #36648b;
}
/* 头部布局 */
.movie {
  display: flex; flex-direction: row; padding: 20rpx;
}
.pic image {
```

```
  width: 200rpx; height: 300rpx;
}
/* 电影信息样式 */
.outer {
  margin: 10rpx 20rpx;
}
.name {
  color: #fff; margin-bottom: 20rpx;
}
.item {
  font-size: 24rpx; color: #fff; line-height: 50rpx;
}
/* 电影简介样式 */
.detail {
  font-size: 26rpx; line-height: 50rpx;
  margin: 20rpx; text-indent: 1em;
}
/* 分隔线样式 */
.hr {
  height: 20rpx; width: 100%;
  background-color: #ccc; opacity: 0.2;
}
.zhinan {
  font-size: 40rpx; padding: 16rpx;
}
scroll-view {
  height: 350rpx; width: 100%; white-space: nowrap;
}
/* 电影海报样式 */
.haibaoitem {
  padding: 10rpx; display: inline-block;
}
```

【代码讲解】 本例中的电影海报部分包含多幅图片，<scroll-view>组件设置属性scroll-x="true"实现水平方向的滑动区域；<image>组件通过父组件<view>设置属性display取值inline-block，把默认是块级元素的<view>组件修改成内联块级元素，最终电影海报图片能够实现左右滑动的效果。

4.9.3 多面板

在微信小程序的界面中，多面板常用于信息的分类，示例如图4.30所示。

例4 20 多面板布局小案例，运行效果如图4.31所示。

pages/tab/tab.wxml文件代码如下：

视频讲解

```
<!-- 分隔线 -->
<view class="hr"></view>
<!-- 输入框 -->
<input placeholder="请输入商品名称"></input>
```

```
<view class="hr"></view>
<view class="content">
  <view class="left">
    <!-- 左侧部分 -->
    <scroll-view scroll-y="true">
      <block wx:for="{{list}}">
        <view>{{item}}</view>
      </block>
    </scroll-view>
  </view>
  <view class="right">
    <!-- 右侧部分 -->
    <view class="order">
      <!-- 分类部分 -->
      <view>热门推荐</view>
      <view>生活热搜</view>
      <view>专场推荐</view>
    </view>
  </view>
</view>
```

图 4.30　多面板示例图

图 4.31　多面板小案例

pages/tab/tab.js 文件代码如下：

```
Page({
  data: {
```

```
    list: ["手机数码", "男装推荐", "女装推荐", "优选水果", "家用电器", "运动户外",
      "电脑办公", "体育用品", "美妆护肤"]
  },
})
```

pages/tab/tab.wxss 文件代码如下：

```
/* 分隔线样式 */
.hr {
  border: 1px solid #EEE9E9;
  width: 100%; opacity: 0.6;
}
/* 输入框样式 */
input {
  margin: 15rpx 30rpx; border: 1px solid #ccc;
  border-radius: 50rpx;
  text-align: center; font-size: 32rpx;
}
/* 布局样式 */
.content {
  display: flex;
  flex-direction: row;
}
/* 左边样式 */
.left {
  width: 25%; font-size: 30rpx;
}
scroll-view {
  height: 90%;
}
/* 左边元素样式 */
.left view {
  text-align: center;
  height: 100rpx; line-height: 100rpx;
}
/* 右边样式 */
.right {
  width: 75%;
}
/* 分类样式 */
.order {
  display: flex; flex-direction: row;
  text-align: center;
  padding: 20 rpx;
}
.order view {
  width: 33%;
  font-size: 32rpx;
}
```

【代码讲解】 页面布局思路：页面上部放置< input >组件，为其设置相关属性完成搜索框的设计；下方分别为左右两个部分，放置< view >容器，设置为 flex 布局，主轴在水平方向上从左向右，左边部分宽度设置为 25%，利用< scroll-view >组件及列表渲染完成左边的布局，右边部分宽度设置为 75%，内部放置 3 个< view >容器，宽度均设置为 33%；最后完成样式的设置。

4.9.4 标签式

视频讲解

在微信小程序的界面中，通常在搜索框下方会有相关的标签，标签示例如图 4.32 所示。

例 4-21 标签式布局小案例，运行效果如图 4.33 所示。

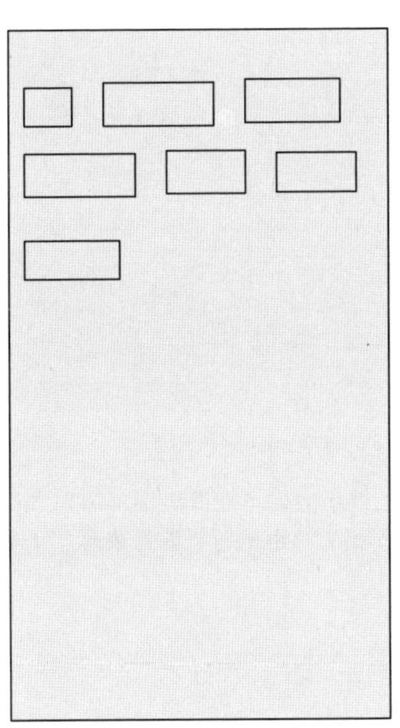

图 4.32 标签示例图

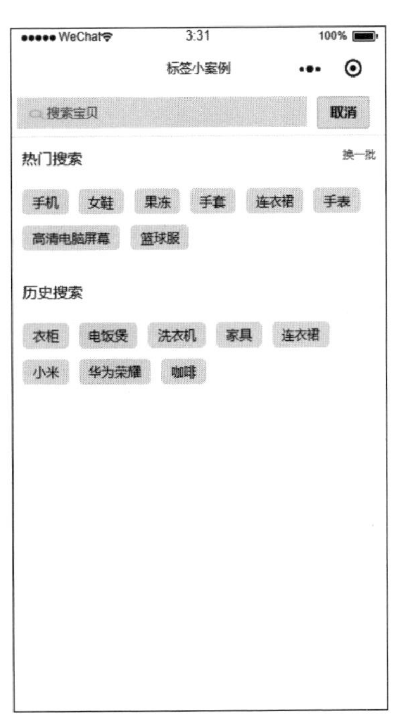

图 4.33 标签小案例

pages/tag/tag.wxml 文件代码如下：

```
<! -- 搜索框 -- >
< view class = "search">
  < view class = "searchBg">
    <! -- 搜索图标 -- >
    < image src = "/images/search.jpg" style = "width:50rpx;height:50rpx"></image >
    < input placeholder = "搜索宝贝"> </input >
  </view >
  <! -- 取消按钮 -- >
  < view class = "btn">取消</view >
</view >
```

```
<!-- 分隔线 -->
<view class="hr"></view>
<view class="title">
  <view class="left">热门搜索</view>
  <view class="right">换一批</view>
</view>
<!-- 标签内容 -->
<view class="tag">
  <block wx:for="{{label1}}">
    <view>{{item}}</view>
  </block>
</view>
<!-- 标题内容 -->
<view>
  <view class="title">
    <view class="left">历史搜索</view>
  </view>
</view>
<!-- 标签内容 -->
<view class="tag">
  <block wx:for="{{label2}}">
    <view>{{item}}</view>
  </block>
</view>
```

pages/tag/tag.js 文件代码如下：

```
Page({
  data: {
    label1: ["手机", "女鞋", "果冻", "手套", "连衣裙", "手表", "高清电脑屏幕", "篮球服"],
    label2: ["衣柜", "电饭煲", "洗衣机", "家具", "连衣裙", "小米", "华为荣耀", "咖啡"],
  },
})
```

pages/tag/tag.wxss 文件代码如下：

```
/* 搜索框样式 */
.search {
  display: flex; flex-direction: row; padding: 10rpx;
}
/* 搜索框背景样式 */
.searchBg {
  background-color: #e8e8ed; width: 80%;
  display: flex; flex-direction: row; height: 60rpx;
}
/* 搜索图标样式 */
.searchBg image {
  margin: 10rpx;
}
/* 输入框样式 */
```

```
.searchBg input {
  height: 60rpx; font-size: 30rpx;
}
/* 取消按钮样式 */
.btn {
  font-size: 26rpx; font-weight: bold;
  line-height: 60rpx; margin-left: 24rpx;
  border: 1px solid #ccc; width: 100rpx;
  text-align: center; background-color: #e8e8ed; border-radius: 6rpx;
}
/* 分隔线样式 */
.hr {
  border: 1px solid #eee9e9; opacity: 0.6;
}
/* 标题样式 */
.title {
  display: flex; flex-direction: row; padding: 20rpx;
}
/* 热门搜索样式 */
.left {
  width: 80%; font-size: 30rpx;
}
/* 换一批样式 */
.right {
  width: 20%; font-size: 25rpx;
  color: #ec3131; text-align: right;
}
/* 标签布局样式 */
.tag {
  padding-left: 20rpx; display: flex;
  flex-direction: row; flex-wrap: wrap;
}
/* 标签样式 */
.tag view {
  background-color: #e8e8ed; padding-left: 24rpx;
  padding-right: 24rpx; height: 50rpx;
  line-height: 50rpx; border-radius: 10rpx;
  text-align: center; font-size: 26rpx;
  margin-right: 20rpx; margin-bottom: 20rpx;
}
```

【代码讲解】 页面布局思路：页面的上部放置< view >容器，设置为 flex 布局主轴在水平方向上从左向右，左边放置< view >容器，用于增加背景色以及设置布局样式，将搜索图标、< input >组件以及“取消”按钮放置其中；搜索框下方的“热门搜索”和“换一批”也采用 flex 布局方式，标题下方的标签每个元素水平向右排列并且换行，因此在外面的< view >设置为 flex 布局，主轴在水平方向上从左向右，并且设置允许换行显示，再利用列表渲染完成数据的显示，最终根据不同元素之间的位置关系，编写标签内部的样式代码。

4.10 实训项目——仿京东首页小案例

视频讲解

为了加深读者对样式与布局的认识，本实训项目设计实现仿京东首页小案例，项目主要使用 flex 布局实现各个栏目的排列显示，同时使用了第 2 章关于循环渲染的知识点。项目执行效果如图 4.34 所示。

图 4.34　仿京东购物小程序首页

具体实现步骤如下。

(1) 打开微信 Web 开发者工具创建项目。

(2) 新建 images 文件夹，用 Adobe Fireworks 软件将界面的素材图片截取保存。

(3) 在 app.js 中配置项目属性和底部 tabBar。

(4) 新建 jd-index 文件，进行样式与布局的设计。

app.json 文件代码如下：

```
{
  "pages": [
    "pages/jd - index/jd - index"
  ],
  "window": {
    "backgroundTextStyle": "light",
    "navigationBarBackgroundColor": "#ffffff",
```

```
    "navigationBarTitleText": "仿京东购物",
    "navigationBarTextStyle": "black"
  },
  "tabBar": {
    "color": "#4D4D4D",
    "selectedColor": "#FF0000",
    "borderStyle": "black",
    "list": [
      {
        "selectedIconPath": "icon/index0.png",
        "iconPath": "icon/index.png",
        "pagePath": "pages/jd-index/jd-index",
        "text": "首页"
      },
      {
        "selectedIconPath": "icon/sort0.png",
        "iconPath": "icon/sort.png",
        "pagePath": "pages/jd-index/jd-index",
        "text": "分类"
      },
      {
        "selectedIconPath": "icon/shop0.png",
        "iconPath": "icon/shop.png",
        "pagePath": "pages/jd-index/jd-index",
        "text": "购物圈"
      },
      {
        "selectedIconPath": "icon/cart0.png",
        "iconPath": "icon/cart.png",
        "pagePath": "pages/jd-index/jd-index",
        "text": "购物车"
      },
      {
        "selectedIconPath": "icon/me0.png",
        "iconPath": "icon/me.png",
        "pagePath": "pages/jd-index/jd-index",
        "text": "我的"
      }
    ]
  },
  "sitemapLocation": "sitemap.json"
}
```

pages/jd-index/jd-index.wxml 文件代码如下：

```
<!-- 顶部轮播图 -->
<swiper autoplay="{{autoplay}}" interval="{{interval}}">
  <block wx:for="{{imgUrls}}">
    <swiper-item>
      <image src="{{item}}"></image>
    </swiper-item>
```

```
    </block>
  </swiper>
  <!-- 10 个京东图标 -->
  <view class = "content">
    <block wx:for = "{{elements}}">
      <view class = "content - item">
        <view>
          <image src = "{{item.image}}" style = "width:86rpx;height:78rpx;"></image>
        </view>
        <view>
          {{item.name}}
        </view>
      </view>
    </block>
  </view>
  <!-- 商品展示 -->
  <view class = "mid">
    <image src = "/images/index/11.png" style = "width:116rpx;height:120rpx"></image>
    <image src = "/images/index/12.png" style = "width:600rpx;height:120rpx"></image>
  </view>
  <view>
    <image src = "/images/index/13.png" style = "width:100 %;height:200rpx"></image>
  </view>
  <view class = "hr"></view>
  <view class = "footer">
    <text>京东拼购</text>
    <image src = "/images/index/14.png" style = "width:300rpx;height:80rpx"></image>
  </view>
```

pages/jd-index/jd-index.js 文件代码如下：

```
Page({
  data: {
    autoplay: true,
    interval: 5000,
    imgUrls: [
      "/images/haibao/haibao - 1.png",
      "/images/haibao/haibao - 2.png"
    ],
    elements: [{
        image: "/images/index/1.png",
        name: "领优惠券",
      }, {
        image: "/images/index/2.png",
        name: "9.9 元拼",
      },
      {
        image: "/images/index/3.png",
        name: " 找折扣",
      },
      {
```

```
        image: "/images/index/4.png",
        name: "闪购",
      },
      {
        image: "/images/index/5.png",
        name: "领京豆",
      },
      {
        image: "/images/index/6.png",
        name: "打卡有奖",
      }, {
        image: "/images/index/7.png",
        name: "京东服饰",
      }, {
        image: "/images/index/8.png",
        name: "京东生鲜",
      }, {
        image: "/images/index/9.png",
        name: "京东手机",
      }, {
        image: "/images/index/10.png",
        name: "全部频道",
      },
    ],
  }
})
```

pages/jd-index/jd-index.wxss 文件代码如下：

```
/* 顶部图片大小 */
swiper image {
  width: 100%; height: 300rpx;
}
/* 10 个京东图标外部容器布局 */
.content {
  display: flex; flex-direction: row; flex-wrap: wrap;
}
/* 10 个京东图标内部容器样式 */
.content-item {
  width: 20%; text-align: center;
  font-size: 24rpx; margin: 8rpx 0;
}
/* 分隔线样式 */
.hr {
  width: 100%; height: 30rpx; background-color: #f4f5f6;
}
/* 页面中间外部容器布局 */
.mid {
  display: flex; flex-direction: row; padding: 40rpx;
}
/* 页面底部外部容器布局 */
```

```
.footer {
  display: flex; flex - direction: row;
  justify - content: space - between; padding: 20rpx 40rpx;
}
/* 京东拼购内容样式 */
text {
  font - weight: bold;
}
```

【代码讲解】 在 app.json 文件中首先对页面的标题及底部导航栏进行设置，然后对页面的内容进行布局，页面的顶部使用< scroll-view >组件，< scroll-view >组件的属性值和内部图片的地址存放在 JS 文件中；中间的 10 个京东图标使用 flex 布局进行设计，主轴设置为水平方向从左到右并且允许换行，每个图标的宽度设置为 20%，使得每行显示 5 个京东图标，最后对图标的样式进行设置，下方内容均采用 flex 布局，完成商品图片的放置。

第5章 JavaScript 基础

JavaScript 是小程序逻辑层的编程语言，JavaScript 代码大约占小程序代码量的一半，学好 JavaScript 是学好小程序开发的关键。

本章主要目标

- 熟练掌握 JavaScript 语法格式；
- 熟练掌握 JavaScript 变量、数据类型、运算符、函数等基本概念；
- 熟练掌握小程序事件函数中 this 和 that 的使用；
- 熟练掌握 JavaScript 在小程序中的交互场景应用。

5.1 JavaScript 简介

JavaScript 是一种轻量、解释型、支持面向对象编程风格的脚本语言，它是一种直译式、动态类型、弱类型以及基于原型的语言。JavaScript 语言不仅可用于 Web 前端开发，也广泛用于后端开发和智能手机开发。JavaScript 的标准是 ECMAScript。2015 年 6 月 17 日，ECMA 国际组织发布了 ECMAScript 的第六版，该版本正式名称为 ECMAScript 2015，但通常被称为 ECMAScript 6 或者 ES6。

JavaScript 具有以下特性：

- JavaScript 是一种基于对象的脚本语言，它不仅能够创建对象，而且可以使用对象；
- JavaScript 是一种轻量级的编程语言；
- JavaScript 是可插入 HTML 页面的编程代码；
- JavaScript 插入 HTML 页面后，可由现在所有的浏览器执行。

正是由于 JavaScript 易学易用的特性，使得它在最近几年应用广泛，获得了编程界的好评，同时出现了大量基于 JavaScript 的开源项目。如今 JavaScript 不仅可以实现前端的动态交互功能，而且可以运行于后端，为用户提供高性能的后端服务。在微信小程序中，

JavaScript是小程序逻辑层使用的唯一开发语言，这也客观反映出JavaScript的强大，因此学好JavaScript对实现前端的交互功能非常重要。

在iOS系统中，小程序的JavaScript代码是运行在JavaScriptCore中，然后由WKWebView来渲染。

在安卓系统中，小程序的JavaScript代码是由X5 JS core来解析，由X5内核渲染。

在微信开发者工具(IDE)中，小程序的JavaScript代码是运行在nw.js中，由Chrome WebView来渲染。其中，nw.js是基于Chromium和Node.js运行的，封装了Webkit内核和Node.js，提供了桌面应用的运行环境，让在浏览器运行的程序也可以在桌面端运行。

这3个运行环境(iOS、安卓和微信开发者工具)使用的ECMA标准是不一样的，目前ECMAScript(简称ES)有8个版本，小程序使用的是ES5和ES6标准。但截至目前，iOS 8和iOS 9并没有完全兼容到ES6的标准，即ES6中的一些语法和关键字不被兼容，所以经常会发现，在微信开发者工具里和手机真机上的代码表现不一致，对此可以用微信开发者工具里的远程调试功能，在真机上进行调试。

5.2 JavaScript基础语法

本节将介绍JavaScript的基础语法，包括变量、数据类型、运算符、逻辑控制语句等。

5.2.1 变量

变量是存储信息的容器，所有JavaScript变量必须以唯一的名称标识，标识称为变量名。定义变量名称的规则为：名称可包含字母、数字、下画线；名称必须以字母开头，对大小写敏感(x和X是不同的变量)；JavaScript的关键词不能作为变量名称。

示例代码：

```
var a = 1;                        //声明变量 a 并且赋值为数字 1
var b = "abc";                    //声明变量 a 并且赋值为字符串 abc
```

如上述代码所示，注释是指编程中用于解释说明的部分，用于提高代码的可读性。JavaScript支持单行注释和多行注释。单行注释用//标注；多行注释用/ * … * /标注。

5.2.2 数据类型

JavaScript变量能够保存多种数据类型：字符串、数字、逻辑值、数组、对象。本节将逐一介绍。

1. 字符串(String)类型

字符串用于存储一系列字符，使用单引号或双引号包裹字符串的内容。

示例代码：

```
var hello = "Hello xiaochengxu";
```

```
var hello = 'Hello xiaochengxu';
```

无论是单引号还是双引号,都需要成对出现,不能出现不一致现象,否则在编译时会报错。

2. 数字(Number)类型

JavaScript 只有一种数字类型,数值后面的小数点可省略,用科学记数法能够表示极大值和极小值。

```
var x1 = 1.00;                //带小数点
var x2 = 1;                   //不带小数点
var m = 123e5;                //12300000
var n = 123e-5;               //0.00123
```

3. 布尔(Boolean)类型

布尔(逻辑)类型只有两个值:true 和 false,在使用布尔值时,不能出现"false"或"true",否则会被解析为字符串。

示例代码:

```
var x = true;
var y = false;
```

变量 x 和变量 y 都是布尔类型,如果按照下面写法:

```
var x = "true";
var y = "false";
```

变量 x 和变量 y 都是字符串类型,字符串里面的内容分别为 true 和 false。

4. 数组类型

JavaScript 中的数组用方括号书写,数组中的元素由逗号分隔。

示例代码:

```
var list = ["a", "b", "c", "d", "e", "f"];
```

上述代码定义一个数组,数组名为 list,数组中包含 6 个元素。数组的索引 index 从 0 到数组的个数减 1,通过索引值可以取得数组中的元素。例如 list[2]可以取得数组中的第 3 个元素。数组中包含一些常用的方法,在微信小程序中常用的 Array 对象方法如表 5.1 所示。

表 5.1 Array 对象方法

方法名	说明
pop()	删除并返回数组的最后一个元素
push()	向数组的末尾添加一个或更多元素,并返回新的长度
reverse()	颠倒数组中元素的顺序
shift()	删除并返回数组的第一个元素
slice()	从某个已有的数组返回选定的元素
sort()	对数组的元素进行排序
splice()	删除元素,并向数组添加新元素
toString()	把数组转换为字符串,并返回结果
unshift()	向数组的开头添加一个或更多元素,并返回新的长度

以表 5.1 中的 push()方法为例，示例代码如下：

```
var fruits = ["Apple", "Banana", "Lemon", "Grape"];
fruits.push("Cherry");
for (var i = 1; i < fruits.length; i++) {
  console.log(fruits[i]);
}
```

上述代码执行完以后，会在 Console 控制台输出“Apple，Banana，Lemon，Grape，Cherry”。

5. 对象类型

JavaScript 对象用大括号来书写，内容放置在大括号中，对象的属性通过名称和值(name：value)来定义，属性之间用逗号分隔。示例代码如下：

```
var person = {
  firstName: "tom",
  age: 23,
  hairColor: "black"
};
```

上述代码中对象(person)有 3 个属性：firstName、age 和 hairColor。

5.2.3 运算符

JavaScript 运算符常用于执行算术运算、赋值、比较运算、逻辑运算。

1. 算术运算符

算术运算符用于变量或值之间的算术运算，如表 5.2 所示。

表 5.2 JavaScript 算术运算符

运 算 符	描 述	例 子
+	加	x=y+1
−	减	x=y−1
*	乘	x=y * 1
/	除	x=y/1
%	求余数(保留整数)	x=y%1
++	累加	x=++y
−−	递减	x=−−y

对于两个数字型的变量，变量之间使用“+”运算符表示相加；当两个变量都为字符型，“+”运算符表示字符串的连接；当一个变量为字符型，另一个变量为数值型，数值型会被解析为字符型进行字符串的连接。

2. 赋值运算符

对变量进行赋值使用赋值运算符，如表 5.3 所示。

表 5.3 JavaScript 赋值运算符

运 算 符	例 子	等 同 于
=	x=y	
+=	x+=y	x=x+y
-=	x-=y	x=x-y
*=	x*=y	x=x*y
/=	x/=y	x=x/y
%=	x%=y	x=x%y

3. 比较运算符

比较运算符表示变量之间的逻辑关系,如表 5.4 所示。

表 5.4 JavaScript 比较运算符

运 算 符	描 述	比较及结果
==	等于	5=6 为 false
===	全等(值和类型)	5===5 为 true,5===''5''为 false
!=	不等于	5!=6 为 true
>	大于	5>6 为 false
<	小于	5<6 为 true
>=	大于或等于	5>=6 为 false
<=	小于或等于	5<=6 为 true

4. 逻辑运算符

逻辑运算符用于确定变量或值之间的逻辑关系,假设 x=5 and y=2,逻辑运算符的使用如表 5.5 所示。

表 5.5 JavaScript 逻辑运算符

运 算 符	描 述	例 子
&&	与	(x < 10 && y > 1)为 true
||	或	(x == 3 || y == 5)为 false
!	非	!(x == y)为 true

5.2.4 逻辑控制语句

逻辑控制语句分为条件判断语句和循环语句。

1. 条件判断语句

条件判断语句基于分支的思想,面对不同的情况或条件执行相应的选择,通用于某些代码的判断和重复执行。

1) if 语句

当小括号内的条件为 true 时,大括号内部的代码才会执行。语法如下:

```
if (条件) {
```

```
  当条件为 true 时才执行该代码
}
```

示例代码：

```
if (data > 100) {
  console.log(data)
}
```

上述代码中当变量 data 值大于 100 时，会在 Console 控制台输出这个值。

2）if…else 语句

当条件为 true 时执行 if 之后的代码，当条件为 false 时执行 else 之后的代码。语法如下：

```
if (条件) {
  当条件为 true 时执行该代码
} else {
  当条件为 false 时执行该代码
}
```

示例代码：

```
if (data > 100) {
  console.log("输入的数字大于 100")
} else {
  console.log("输入的数字小于或等于 100")
}
```

上述代码中当变量 data 值大于 100 时，会在 Console 控制台输出“输入的数字大于 100”，否则，会在 Console 控制台输出“输入的数字小于或等于 100”。

3）if…else if…else 语句

在多个代码块中选择满足条件的一个去执行，先判断 if 条件，如果条件不成立，返回 false 后判断 else if 条件，如果条件成立，返回 true 并执行该条 else if 后的代码，如果仍不成立，继续判断下一个 else if 的条件。如果直到 else 时都没有条件成立，则执行 else 中的代码。语法如下：

```
if (条件 1) {
  当条件 1 为 true 时执行该代码
} else if (条件 2) {
  当条件 2 为 true 时执行该代码
} else {
  当条件 1 和条件 2 都为 false 时执行该代码
}
```

示例代码：

```
if (data < 60) {
  console.log("输入的数字小于 60")
} else if (data >= 60 && data <= 80) {
  console.log("输入的数字在 60～80 中")
```

```
} else {
  console.log("输入的数字大于 80")
}
```

当变量 data 值小于 60 时，执行 if 之后的代码，会在 Console 控制台输出“输入的数字小于 60”；当变量 data 值为 60～80 时，执行 else if 之后的代码，输出“输入的数字在 60～80 中”；当前面的条件都不满足时，执行 else 后面的语句，输出“输入的数字大于 80”。

4）switch 语句

在多个代码块中选择满足条件的一个去执行，判断表达式通常为变量，表达式的值会与 case 中的值做匹配，如果匹配正确，会执行该 case 之后的代码块，使用 break 可以阻止代码继续执行，当条件都不满足时执行 default 之后的代码。

示例代码：

```
switch (n) {
  case 1:
    console.log("今天是星期一")
    break;
  case 2:
    console.log("今天是星期二")
    break;
  case 3:
    console.log("今天是星期三")
    break;
  case 4:
    console.log("今天是星期四")
    break;
  case 5:
    console.log("今天是星期五")
    break;
  case 6:
    console.log("今天是星期六")
    break;
  case 7:
    console.log("今天是星期日")
    break;
  default:
    console.log("输入有误请重新输入")
}
```

当判断条件符合其中某一条件时，在 Console 控制台输入对应的星期数，然后程序执行 break 语句跳出循环；当条件不满足所有 case 时，会执行 default 之后的代码，输出“输入有误请重新输入”。

2. 循环语句

循环语句通过判断条件来控制循环的次数，如果需要重复执行相关的动作就需要使用循环语句，循环语句可以提高代码的精简性。

例如,需要输出一个数组的全部元素,不使用循环语句代码如下:

```
console.log(array[0]);
console.log(array[1]);
console.log(array[2]);
console.log(array[3]);
```

使用循环语句代码如下:

```
for (var i = 0; i < array.length; i++) {
  console.log(array[i]);
}
```

上述代码通过循环将数组 array 中的所有元素在 Console 控制台输出。

1) for 循环

for 循环是经常使用的循环语句,语法如下:

```
for (语句 1; 语句 2; 语句 3)
{
  被执行的代码
}
```

语句 1: 在循环开始前执行初始化操作; 语句 2: 循环的条件,当返回值为 true 时,执行循环体,当返回值为 false 时,跳出该循环; 语句 3: 当语句 2 返回值为 true 时执行。

示例代码:

```
for (i = 1; i < 10; i++) {
  console.log(i)
}
```

上述代码会在 Console 控制台依次输出 1 到 9。

2) for/in 循环

for/in 循环用于循环对象属性,当循环体中的代码每执行一次,就会对数组的元素或者对象的属性进行一次操作。

示例代码:

```
var person = {
  firstName: "tom",
  age: 23,
  hairColor: "black"
};
var string = "";
var x;
for (x in person) {
  string += person[x];
}
```

程序执行结束后 string 的值是: tom23black。

3）while 循环

当条件满足时执行代码块，在每次执行完后会进行条件判断，条件为真会再次执行，直到条件不为真时跳出循环。语法如下：

```
while (条件) {
  执行其中的代码
}
```

示例代码：

```
var i = 1;
while (i < 10) {
  console.log("这是代码执行的第" + i + "次");
  i++
}
```

变量 i 从初始值 1，到不满足循环条件 i＜10，共执行了 9 次循环，程序在 Console 控制台执行了 9 次语句的输出。

4）do…while 循环

首先执行一次 do 里面的代码块，如果条件为真会重复执行，当条件不满足时跳出循环。语法如下：

```
do {
  执行其中的代码
}while (条件);
```

示例代码：

```
var i = 1;
do {
  console.log("这是代码执行的第" + i + "次");
i++;
}while (i < 1);
```

首先执行 do 里面的代码，然后变量 i 变为 2，不满足 while 里面的条件直接跳出循环。

5.2.5 定义和调用函数

在小程序中函数的定义如下：

```
add:function(a,b){
return a + b;
}
```

函数的调用是指使用事先定义好的函数，示例代码：

```
add(1,2);                    //返回值为 3
```

5.2.6 小程序中 this 和 that 的使用

this 是 JavaScript 语言的一个关键字,可以调用函数。当函数运行时,this 可以在函数内部使用,当函数使用场合发生变化时,this 的值也会发生变化。在小程序开发中,小程序提供的 API 接口经常会有 success、fail 等回调函数来处理后续逻辑。当需要获取当前页面对象来对视图层进行渲染时,this 只会指向调用函数的对象,如果想要获取页面的初始数据,在回调函数里面就不能使用 this.data 来获取,同时也不能使用 this.setData()函数来更新数据,而是通过语句 var this=that 将 this 指向的对象复制到 that 中才可以执行后续操作。

关于 this 和 that 的用法可参看 6.4.2 节例 6-3 案例加以了解。

5.3 JavaScript 在小程序中常见的交互场景

JavaScript 是小程序编程中的基础语言,从本书附带案例的代码来看,JavaScript 代码大约占整个小程序项目一半的代码量。全局文件 app.js 和所有的页面的 JS 文件都是由 JavaScript 来编写的,JavaScript 代码主要实现业务逻辑处理和用户交互两方面的作用。5.2 节主要从语法的角度介绍了 JavaScript 的变量、数据类型、控制语句和函数定义与调用,本节将以 JavaScript 在小程序中常见的交互场景出发,以案例教学的方式带领读者更深层次地理解 JavaScript 在小程序编程中的使用。

5.3.1 购物车场景

尽管张小龙在 2018 微信公开课上指出,"小程序不是专门为电商准备的",但由于强社交属性和微信支付的便捷性,电商成为小程序的重要应用场景。

例 5-1 电商小程序中经常要用到购物车,购物车是 JavaScript 在小程序交互场景中的经典应用。本例实现一个简单的购物车,购物车初始状态和用户加购商品之后的购物车状态如图 5.1 和图 5.2 所示。

pages/addCaricon/addCaricon.wxml 文件代码如下:

视频讲解

```
<view class = "title">1.显示图标小案例</view>
<view class = "goods - box">
  <! -- 商品展示 -->
  <image src = "/images/goods.png"></image>
  <view class = "carts - icon">
    <! -- 购物车图标 -->
    <image src = "/images/cart.png"></image>
    <! -- 加入购物车的数量 -->
    <text class = "carts - num" wx:if = "{{hasCart}}">{{totalNum}}</text>
  </view>
```

```
  <!-- 加入购物车导航 -->
  <view class="operation">
    <text class="operation-num">数量 {{num}}</text>
    <text class="operation-add" bindtap="addCount">+</text>
    <text bindtap="addCart">加入购物车</text>
  </view>
</view>
```

图 5.1 购物车初始状态

图 5.2 加购之后的购物车

pages/addCaricon/addCaricon.js 文件代码如下：

```
Page({
  data: {
    num: 1,                         //数量初始化为 1
    totalNum: 0,                    //加入购物车的商品数量初始化为 0
    hasCart: false,                 //初始化默认不加入购物车
  },
  //增加数量
  addCount: function() {
    var num = this.data.num;
    num++;                          //数量增加 1
    this.setData({
      num: num                      //更新数量
    })
```

```
  },
  //加入购物车
  addCart: function() {
    const num = this.data.num;              //获取数量
    var total = this.data.totalNum;         //获取总数
    this.setData({
      hasCart: true,                        //用于控制加入购物车时的显示图标
      totalNum: num + total                 //累计数量
    })
    wx.showToast({
      title: "加入购物车成功",              //弹出消息提示框
      duration: 3000,
    })
  },
})
```

pages/addCaricon/addCaricon.wxss 文件代码如下：

```
/*页面盒子*/
.goods-box {
  position: relative;
  padding: 0 30rpx;
  text-align: center;
}
/*购物车位置*/
.carts-icon {
  position: absolute;
  right: 600rpx;
  top: 490rpx;
  width: 70rpx;
  height: 70rpx;
}
/*购物车图标*/
.carts-icon image {
  width: 100%;
  height: 100%;
}
/*显示加入购物车图标*/
.carts-num {
  position: absolute; right: 36rpx; width: 40rpx;
  height: 40rpx; color:white;
  font-size: 26rpx;
  line-height: 40rpx;
  border-radius: 50%;
  background: #e4393c;
}
/*加入购物车导航*/
.operation {
  height: 80rpx; line-height: 80rpx;
```

```
    color:white; border - radius: 50rpx;
    background: #ff9600;
    font - size: 30rpx;
}
/* 操作文本 */
.operation text {
    display: inline - block;
    height: 80rpx;
}
/* 加入购物车数量 */
.operation - num {
    width: 160rpx;
}
/* 加入购物车符号 */
.operation - add {
    width: 80rpx;
    margin - right: 50rpx;
}
```

【代码讲解】 addCaricon.wxml 文件中"+"和"加入购物车"两个按钮绑定了点击事件。在 addcart.js 文件中为"+"按钮定义了事件函数 addCount(),用于实现当用户点击"+"按钮时商品数量加 1。为"加入购物车"按钮定义的函数 addToCart(),用于实现当用户点击"加入购物车"时,一次性向购物车添加 num 件商品。当用户有加购行为,即点击了"加入购物车"按钮时,hasCart 被赋值为 true,则在购物车图标的左上角会出现当前购物车商品数量。

5.3.2 下拉菜单场景

JavaScript 在 Web 开发中经常被用来实现动态效果,在微信小程序开发者中也存在类似的场景。下拉菜单是 JavaScript 在小程序中常见的交互场景之一,当用户单击一级菜单时,二级菜单被弹出,当再次单击一级菜单时,二级菜单又消失。

例 5-2 本例设计常见的小程序下拉菜单,图 5.3 是二级菜单没有弹出的状态,图 5.4 是二级菜单弹出的状态。

pages/subMenu/subMenu.wxml 文件代码如下:

视频讲解

```
<view class = "title"> 2.下拉菜单场景小案例</view>
<view class = "page">
  <! -- 导航内容 -->
  <view class = "nav">
    <view class = "nav - item {{shownavindex == 1?'active':''}}"
      bindtap = "listmenu" data - nav = "1">
      <view class = "content">排序</view>
      <view class = "icon"></view>
    </view>
    <view class = "nav - item">
      <view class = "content">时间</view>
      <view class = "icon"></view>
```

```
        </view>
        <view class="nav-item">
          <view class="content">价格</view>
          <view class="icon"></view>
        </view>
      </view>
      <!-- 下拉内容 -->
      <view class="list {{openif?'down':'up'}} ">
        <view wx:for="{{content}}">
          {{item}}
        </view>
      </view>
    </view>
```

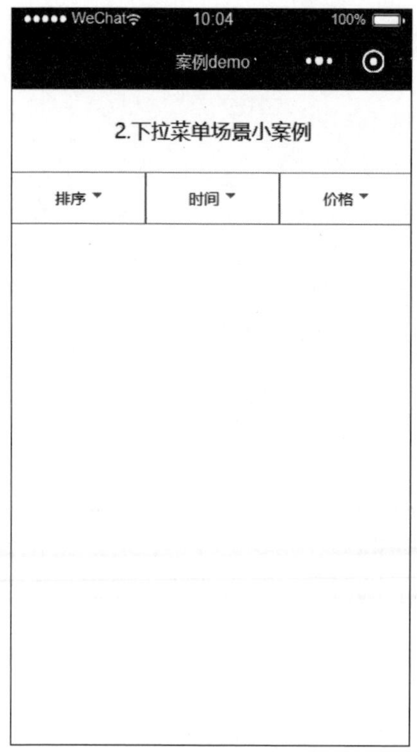

图 5.3　下拉菜单弹出之前

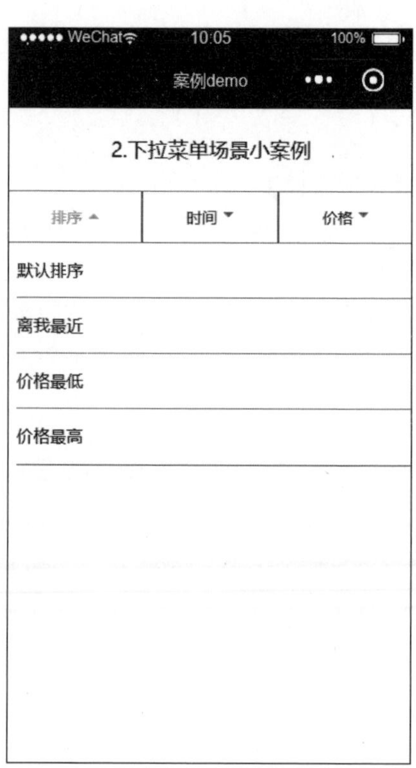

图 5.4　下拉菜单弹出之后

pages/subMenu/subMenu.js 文件代码如下：

```
Page({
  data: {
    li: ['默认排序', '离我最近', '价格最低', '价格最高'],
    shownavindex: 0                                    //数据初始化
  },
  //下拉事件
  listmenu: function(e) {
    if (this.data.openif) {
      this.setData({
```

```
        openif: false,                          //当前菜单没有下拉
        shownavindex: 0                         //控制图标样式
      })
    } else {
      this.setData({
        content: this.data.li,                  //获取数组数据
        openif: true,                           //当前菜单下拉
        shownavindex: e.currentTarget.dataset.nav //控制图标样式
      })
    }
  }
})
```

pages/subMenu/subMenu.wxss 文件代码如下：

```
/* 页面溢出隐藏 */
.page {
  overflow: hidden;
}
/* 导航外部样式 */
.nav {
  position: relative;
  z-index: 1;
  display: flex;
  flex-direction: row;
  background: white;
}
/* 导航内部样式 */
.nav-item {
  display: flex;
  flex: 1;
  text-align: center;
  height: 90rpx;
  align-items: center;
  justify-content: center;
  font-size: 30rpx;
  border: 1px solid gray;
}
/* 导航下拉图标 */
.icon {
  border: 10rpx solid transparent;
  border-top: 10rpx solid gray;
  margin-left: 12rpx;
}
/* 下拉内部样式 */
.list {
  display: none;
  width: 100%;
  overflow-y: scroll;
  padding: 0 0 0 20rpx;
```

```
  line - height: 100rpx;
  background: white;
}
/* 下拉内容样式 */
.list view {
  border - bottom: 1px solid gray;
  font - size: 32rpx;
}
/* 点击导航内容文字后的样式 */
.nav - item.active .content {
  color: skyblue;
}
/* 点击导航下拉图标后的样式 */
.nav - item.active .icon {
  border - bottom: 10rpx solid skyblue;
  border - top: 0;
}
/* 下拉动画样式 */
.down {
  display: block;
  animation: slidown 0.5s ease - in both;
}
@keyframes slidown {
  from {
    transform: translateY( - 100 % );
  }
  to {
    transform: translateY(0 % );
  }
}
/* 收起动画样式 */
.up {
  display: block;
  animation: slidup 0.5s ease - in both;
}
@keyframes slidup {
  from {
    transform: translateY(0 % );
  }
  to {
    transform: translateY( - 100 % );
  }
}
```

【代码讲解】 本例是 JavaScript 和样式布局结合的案例，在 subMenu.js 文件中设置初始值 shownavindex：0，使得导航的字体初始值为黑色。当用户点击菜单的时候激活事件函数 listmenu()，此时 content 被赋值使得下拉菜单被弹出，shownavindex 从 0 到 1 的变化使得一级菜单和向下箭头的样式也发生变化。

5.3.3 栏目切换场景

在文章管理或者商品管理小程序中，往往有分类或者栏目的切换，当点击某一栏（某一区域），显示区域的文章和商品被重置为被点击的栏目下的文章和商品。

例 5-3 本例设计商品栏目的切换效果，右侧显示区域根据左侧栏目的选择来显示对应的内容，运行效果如图 5.5 和图 5.6 所示。

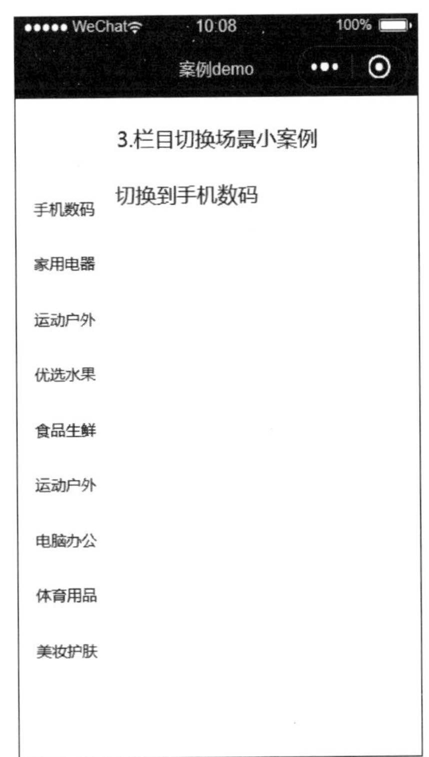

图 5.5 “手机数码”栏目

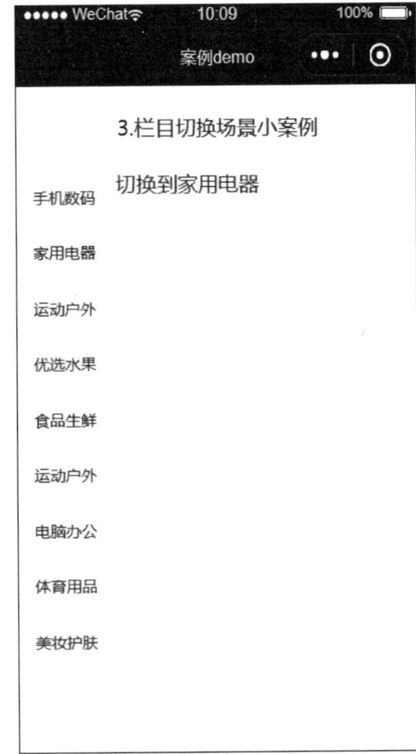

图 5.6 “家用电器”栏目

视频讲解

pages/switchTab/switchTab.wxml 文件代码如下：

```
<view class = "title">3.栏目切换场景小案例</view>
<view class = "content">
  <view class = "left">
    <! -- 左侧部分 -->
    <scroll - view scroll - y>
      <block wx:for = "{{list1}}" wx:for - index = "id">
        <view id = "{{id}}" bindtap = "switchTab">{{item}}</view>
      </block>
    </scroll - view>
  </view>
  <view class = "right">
    <! -- 右侧部分 -->
    <swiper current = "{{currentTab}}" >
```

```
      <block wx:for = "{{list2}}">
        <swiper - item>
          {{item}}
        </swiper - item>
      </block>
    </swiper>
  </view>
</view>
```

pages/switchTab/switchTab.js 文件代码如下：

```
Page({
  data: {
    currentTab: 0,                                 //初始化数据
    list1: ["手机数码", "家用电器", "运动户外", "优选水果", "食品生鲜", "运动户外", "电脑办公", "体育用品", "美妆护肤"],
    list2: ["切换到手机数码", "切换到家用电器", "切换到运动户外", "切换到优选水果", "切换到食品生鲜", "切换到运动户外", "切换到电脑办公", "切换到体育用品", "切换到美妆护肤"],
  },
  switchTab: function(e) {
    var that = this;
    var id = e.target.id;                          //获取 id
    if (this.data.currentTab == id) {
      return                                       //与 currentTab 值一致返回
    } else {
      that.setData({
        currentTab: id                             //设置 currentTab 值为 id
      });
    }
  }
})
```

pages/switchTab/switchTab.wxss 文件代码如下：

```
/* 布局样式 */
.content {
  display: flex;
  flex - direction: row;
}
/* 左边样式 */
.left {
  width: 25 %;
  font - size: 30rpx;
}
scroll - view {
  height: 90 %;
}
/* 左边元素样式 */
.left view {
  text - align: center;
  height: 100rpx;
```

```
  line - height: 100rpx;
}
/* 右边样式 */
.right {
  width: 75%;
}
```

【代码讲解】 本例 switchTab.wxml 使用 flex 布局实现栏目和内容的左右分布，左侧是< scroll-view >组件，右侧是< swiper >组件，当用户点击左侧栏目的时候 switchTab.js 获取栏目的 id，并把 id 赋值给 current，此时右侧的< swiper >组件的第 id 项被显示，从而实现了左右的同步。本例巧妙地使用了< swiper >组件的 current 属性实现内容的切换。

5.3.4 系统设置场景

大部分的 App 和小程序都有系统设置功能，系统设置功能主要是用户输入或者选中数据，JavaScript 获取用户设置的数据来修改系统配置参数。系统设计场景是 JavaScript 在微信小程序中常见的应用场景。

视频讲解

例 5-4 本例以辩论赛小程序为背景，系统设置功能可以对立论、驳立论、质辩、自由辩论和总结陈词等环节的辩论时间和倒计时时间进行设置，系统设置页效果如图 5.7 所示，项目首页简单设计如图 5.8 所示。

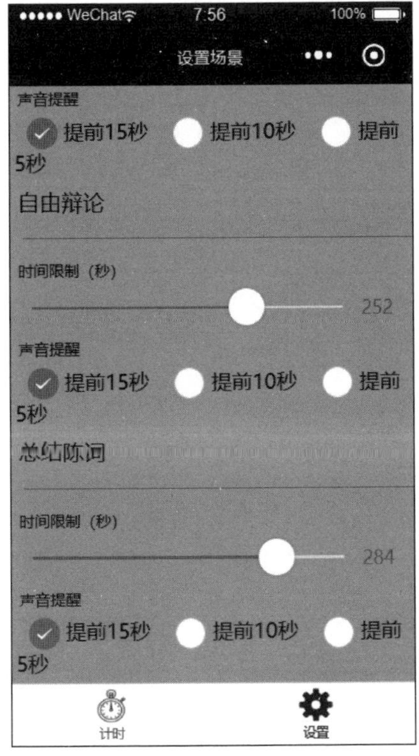

图 5.7　设置页面

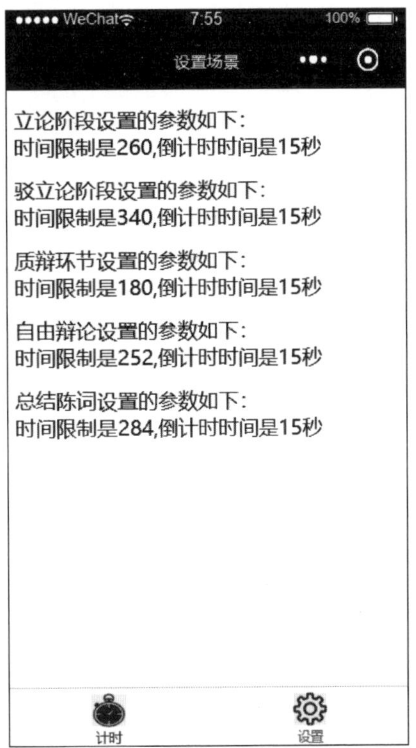

图 5.8　项目首页

app.js 文件代码如下：

```
App({
  data: { configs: [{ name: "立论阶段", time: 180, voice: 15 }, { name: "驳立论阶段", time: 180, voice: 15 }, { name: "质辩环节", time: 180, voice: 15 }, { name: "自由辩论", time: 180, voice: 15 }, { name: "总结陈词", time: 180, voice: 15 }]},
  onLaunch: function() {
    wx.setStorageSync('configs',this.data.configs);
  }
})
```

pages/setting/setting.wxml 文件代码如下：

```
<!-- 立论阶段 -->
<view>
  <view class="title">立论阶段</view>
  <view class="hr"></view>
  <view>
    <text>时间限制(秒)</text>
    <slider id="0" bindchange="sliderChange" show-value min="10" max="360" value="{{configs[0].time}}" />
  </view>
  <view>
    <text>声音提醒</text>
    <radio-group id="0" class="item" bindchange="radioChange">
      <label>
        <radio value="15" checked/>提前 15 秒</label>
      <label>
        <radio value="10" />提前 10 秒</label>
      <label>
        <radio value="5" />提前 5 秒</label>
    </radio-group>
  </view>
  <!-- 驳立论阶段 -->
  <view class="title">驳立论阶段</view>
  <view class="hr"></view>
  <view>
    <text>时间限制(秒)</text>
    <slider id="1" bindchange="sliderChange" show-value min="10" max="360"
    value="{{configs[1].time}}" />
  </view>
  <view>
    <text>声音提醒</text>
    <radio-group id="1" class="item" bindchange="radioChange">
      <label>
        <radio value="15" checked/>提前 15 秒</label>
      <label>
        <radio value="10" />提前 10 秒</label>
      <label>
        <radio value="5" />提前 5 秒</label>
    </radio-group>
```

```
</view>
  <! -- 质辩环节 -- >
<view class = "title">质辩环节</view>
<view class = "hr"></view>
<view>
  <text>时间限制(秒)</text>
  <slider id = "2" bindchange = "sliderChange" show - value min = "10" max = "360"
  value = "{{configs[2].time}}" />
</view>
<view>
  <text>声音提醒</text>
  <radio - group id = "2" class = "item" bindchange = "radioChange">
    <label>
      <radio value = "15" checked/>提前 15 秒</label>
    <label>
      <radio value = "10" />提前 10 秒</label>
    <label>
      <radio value = "5" />提前 5 秒</label>
  </radio - group>
</view>
 <! -- 自由辩论 -- >
<view class = "title">自由辩论</view>
<view class = "hr"></view>
<view>
  <text>时间限制(秒)</text>
  <slider id = "3" bindchange = "sliderChange" show - value min = "10" max = "360"
  value = "{{configs[3].time}}" />
</view>
<view>
  <text>声音提醒</text>
  <radio - group id = "3" class = "item" bindchange = "radioChange">
    <label>
      <radio value = "15" checked/>提前 15 秒</label>
    <label>
      <radio value = "10" />提前 10 秒</label>
    <label>
      <radio value = "5" />提前 5 秒</label>
  </radio - group>
</view>
<! -- 总结陈词 -- >
<view class = "title">总结陈词</view>
<view class = "hr"></view>
<view>
  <text>时间限制(秒)</text>
  <slider id = "4" bindchange = "sliderChange" show - value min = "10" max = "360"
  value = "{{configs[4].time}}" />
</view>
<view>
  <text>声音提醒</text>
  <radio - group id = "4" class = "item" bindchange = "radioChange">
    <label>
```

```
        <radio value="15" checked/>提前 15 秒</label>
      <label>
        <radio value="10" />提前 10 秒</label>
      <label>
        <radio value="5" />提前 5 秒</label>
    </radio-group>
  </view>
</view>
```

pages/setting/setting.js 文件代码如下：

```
Page({
  data:{
    configs:[]
  },
  onLoad:function(options){
    var configs = wx.getStorageSync('configs');
    this.setData({configs:configs});
  },
  sliderChange:function(e){
    var id = e.target.id;
    this.data.configs[id].time = e.detail.value
    wx.setStorageSync('configs', this.data.configs);
    console.log(this.data.configs)
  },
  radioChange:function(e){
    var id = e.target.id;
    this.data.configs[id].voice = e.detail.value
    wx.setStorageSync('configs', this.data.configs);
    console.log(this.data.configs)
  }
})
```

pages/setting/setting.wxss 文件代码如下：

```
/*页面样式*/
page {
  background-color: skyblue; color: black;
}
/*标题样式*/
.title {
  font-size: 40rpx; height: 72rpx;
  line-height: 72rpx; padding-left: 20rpx;
}
/*分隔线样式*/
.hr {
  margin: 12px; width: 100%;
  height: 1px; background-color: grey;
}
/*文字样式*/
text {
  padding-left: 20rpx; font-size: 30rpx;
}
```

```
/* 选项样式 */
.item {
  margin: 12rpx;
}
/* 标签样式 */
.item label {
  margin: 10px;
}
```

pages/index/index.wxml 文件代码如下：

```
<view  wx:for="{{configs}}">
<view class="setitem"></view>
<view>{{item.name}}设置的参数如下：</view><view>时间限制是{{item.time}},倒计时时间是
{{item.voice}}秒</view>
</view>
```

pages/index/index.js 文件代码如下：

```
Page({
  data: {
    configs: []
  },
  onShow: function (options) {
    var configs = wx.getStorageSync('configs');
    this.setData({ configs: configs });
  },
})
```

pages/index/index.wxss 文件代码如下：

```
.setitem{
  height: 30rpx;
}
view {
  padding-left: 10rpx;
}
```

【代码讲解】 本例中辩论赛数据 configs 存放在 app.js 文件中，通过 wx.setStorageSync() 和 wx.getStorageSync()接口实现多个页面之间数据的共享和维护，setting 页面的<slider>和<radio>组件输入的数据被事件函数 sliderChange()和 radioChange()用来修改全局数据 configs。项目需注意修改全局数据 configs 时的下标问题。

5.4 实训项目——计算器小案例

视频讲解

本实训项目设计一个计算器，实现了加、减、乘、除运算。程序运行效果如图 5.9 所示。本项目没有涉及小程序复杂的接口，使用的是第 4 章样式与布局和第 5 章 JavaScript 基础的知识点。项目的设计思路是先编写 WXML 文件，然后编写 JS 接收 WXML 的数据的代

码，再编写加、减、乘、除运算的逻辑代码，最后编写 WXSS 文件。程序执行效果如图 5.9 所示。

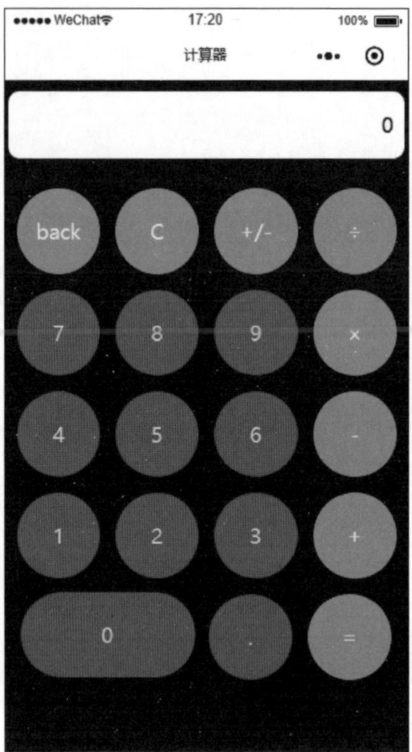

图 5.9　计算器示例图

pages/calculator/addList.wxml 文件代码如下：

```
<view class="demo-box">
  <!-- 输入框 -->
  <view class="input-screen">{{screenData}}</view>
  <!-- 按钮 -->
  <view class="btn-group">
    <view class="item green" bindtap="onclickButton" id="{{id1}}">back</view>
    <view class="item green" bindtap="onclickButton" id="{{id2}}">C</view>
    <view class="item green" bindtap="onclickButton" id="{{id3}}">+/-</view>
    <view class="item green" bindtap="onclickButton" id="{{id4}}">÷</view>
  </view>
  <view class="btn-group">
    <view class="item blue" bindtap="onclickButton" id="{{id5}}">7</view>
    <view class="item blue" bindtap="onclickButton" id="{{id6}}">8</view>
    <view class="item blue" bindtap="onclickButton" id="{{id7}}">9</view>
    <view class="item green" bindtap="onclickButton" id="{{id8}}">×</view>
  </view>
  <view class="btn-group">
    <view class="item blue" bindtap="onclickButton" id="{{id9}}">4</view>
    <view class="item blue" bindtap="onclickButton" id="{{id10}}">5</view>
    <view class="item blue" bindtap="onclickButton" id="{{id11}}">6</view>
```

```
    <view class = "item green" bindtap = "onclickButton" id = "{{id12}}"> - </view>
  </view>
  <view class = "btn - group">
    <view class = "item blue" bindtap = "onclickButton" id = "{{id13}}"> 1 </view>
    <view class = "item blue" bindtap = "onclickButton" id = "{{id14}}"> 2 </view>
    <view class = "item blue" bindtap = "onclickButton" id = "{{id15}}"> 3 </view>
    <view class = "item green" bindtap = "onclickButton" id = "{{id16}}"> + </view>
  </view>
  <view class = "btn - group">
    <view class = "item1 blue" bindtap = "onclickButton" id = "{{id17}}"> 0 </view>
    <view class = "item blue" bindtap = "onclickButton" id = "{{id18}}">. </view>
    <view class = "item green" bindtap = "onclickButton" id = "{{id19}}"> = </view>
  </view>
</view>
```

pages/calculator/addList.js 文件代码如下：

```
Page({
  data: {
    id1: "back",
    id2: "clear",
    id3: "negative",
    id4: " ÷ ",
    id5: "7",
    id6: "8",
    id7: "9",
    id8: " × ",
    id9: "4",
    id10: "5",
    id11: "6",
    id12: " - ",
    id13: "1",
    id14: "2",
    id15: "3",
    id16: " + ",
    id17: "0",
    id18: ".",
    id19: " = ",
    screenData: "0",                                  //屏幕数字初始化
    lastInput: false,                                 //控制输入
    array: [],                                        //定义一个空数组
  },
  onclickButton: function(e) {
    var id = e.target.id;
    if (id == this.data.id1) {                        //退格
      var data = this.data.screenData;
      if (data == 0) {
        return;
      }
      data = data.substring(0, data.length - 1);      //删除最后一位
      if (data == "" || data == " - ") {              //如果为空或者符号置零
```

```
      data = 0;
    }
    this.setData({
      screenData: data
    });
    this.data.array.pop();
    this.data.array.push(data);
  } else if (id == this.data.id2) {                       //清屏
    this.setData({
      screenData: "0"
    });
    this.data.array.length = 0;
  } else if (id == this.data.id3) {                       //正负号
    var data = this.data.screenData;
    if (data == 0) {
      return;
    }
    var firstWord = data.substring(0, 1);
    if (firstWord == "-") {
      data = data.substring(1, data.length);              //去掉负号
      this.data.array.shift();
    } else {
      data = "-" + data;
      this.data.array.unshift("-");                       //增加负号
    }
    this.setData({
      screenData: data
    });
  } else if (id == this.data.id19) {                      //=
    var data = this.data.screenData;
    if (data == 0) {
      return;
    }
    var lastWord = data.substring(data.length - 1, data.length);
    if (isNaN(lastWord)) {
      return;                                             //最后一位不是数字返回
    }
    var num = "";                                         //定义一个字符串
    var lastInput;
    var array = this.data.array;
    var optaration = [];
    for (var i in array) {
      if (isNaN(array[i]) == false || array[i] == this.data.id18 || array[i]
        == this.data.id3) {
        num += array[i];                                  //对小数点或者正负号进行合并处理
      } else {                                            //加减乘除单独处理
        lastInput = array[i];
        optaration.push(num);
        optaration.push(array[i]);
        num = "";
      }
```

```
        }
        optaration.push(Number(num));
        var result = Number(optaration[0]) * 1.0;
        for (var i = 1; i < optaration.length; i++) {
          if (isNaN(optaration[i])) {
            if (optaration[1] == this.data.id4) {          //除运算
              result /= Number(optaration[i + 1]);
            } else if (optaration[1] == this.data.id8) {  //乘运算
              result *= Number(optaration[i + 1]);
            } else if (optaration[1] == this.data.id12) { //减运算
              result -= Number(optaration[i + 1]);
            } else if (optaration[1] == this.data.id16) { //加运算
              result += Number(optaration[i + 1]);
            }
          }
        }
        this.data.array.length = 0;
        this.data.array.push(result);
        this.setData({
          screenData: result + ""                         //将计算结果赋值用于在屏幕上显示
        });
      } else {
        if (id == this.data.id4 || id == this.data.id8 || id == this.data.id12 || id ==
this.data.id16) {
          if (this.data.lastInput == true || this.data.screenData == 0) {
            return;                                       //开始不允许输入加减乘除
          }
        }
        var vd = this.data.screenData;
        var data;
        if (vd == 0) {
          data = id;
        } else {
          data = vd + id;                                 //用于连续输入数字
        }
        this.setData({
          screenData: data
        });
        this.data.array.push(id);
        if (id == this.data.id4 || id == this.data.id8 || id == this.data.id12 || id ==
this.data.id16) {
          this.setData({
            lastInput: true
          });
        } else {
          this.setData({
            lastInput: false
          });
        }
      }
    }
```

```
})
```

pages/calculator/addList.wxss 文件代码如下：

```
/* 外部布局 */
.demo-box {
  background-color: black; padding: 10rpx;
}
/* 输入框样式 */
.input-screen {
  background-color: white; border-radius: 20rpx; text-align: right;
  width: 710rpx; height: 120rpx; line-height: 120rpx;
  padding-right: 20rpx; margin-bottom: 40rpx;
}
/* 按钮布局方式 */
.btn-group {
  display: flex; flex-direction: row;
}
/* 按钮样式 */
.item {
  width: 150rpx; height: 150rpx;
  margin: 15rpx; border-radius: 100rpx;
  text-align: center; line-height: 150rpx;
}
/* 按钮 0 样式 */
.item1 {
  width: 320rpx; height: 150rpx;
  margin: 10rpx; border-radius: 100rpx;
  text-align: center; line-height: 150rpx;
}
/* 颜色样式 */
.green {
  color: #f7f7f7; background: #7ccd7c;
}
/* 颜色样式 */
.blue {
  color: #f7f7f7; background: #0095cd;
}
```

【代码讲解】 本项目在 addList.wxml 文件和 addList.wxss 文件中完成页面的布局，为每个按钮绑定点击事件；在 JS 文件中首先设置允许连续输入操作数并且不允许连续输入操作符，然后分别为每个按钮设置交互功能，包括退格、清零和正负号，接着定义等号的交互，数字在连续输入时将中间结果保存在数组中，最后实现加减乘除运算。由于考虑到篇幅问题，设计功能完整的计算器代码较多，故没有涉及运算的优先级处理，有兴趣的读者可以自行修改本项目。

数据库操作

随着数据量的增加，程序开发对数据库的依赖也就随之而来。小程序的数据除了在WXML和JS文件中可以声明之外，开发者也可以把有规律的数据存放在后端数据库中，再通过通信连接接口对数据库中的数据进行读写操作。本章将以MySQL数据库为蓝本，介绍微信小程序如何利用wx.request()接口读写数据库数据。本章结尾还引入了一个开源项目html2wxml，它可以将HTML页面顺利地转化成小程序的WXML文件，从而为快速开发小程序页面提供了另一种解决方案。

本章主要目标

- 了解小程序前端与后端开发的分工；
- 熟练掌握MySQL数据库及Navicat的操作；
- 熟练掌握Servlet读取数据库数据、JSON格式处理和wx.request()网络请求操作；
- 熟练掌握数据缓存Storage和富文本插件html2wxml的使用。

6.1 MySQL数据库

MySQL是一个小型关系型数据库管理系统，被广泛地应用在Internet上的中小型网站中。由于其体积小、速度快、成本低，尤其是开放源码这一特点，许多中小型网站为了降低网站总体开发和运营成本而选择它作为网站数据库。下面将介绍MySQL数据库的安装、管理和可视化管理工具Navicat，有相关方面学习经历的读者可以跳过此节内容。

6.1.1 MySQL数据库介绍

MySQL由瑞典MySQL AB公司开发，目前属于Oracle旗下产品。MySQL是最流行的关系型数据库管理系统之一，在Web应用方面，MySQL是最好的RDBMS (Relational

Database Management System,关系数据库管理系统)应用软件之一。关系型数据库将数据保存在不同的表中,而不是将所有数据放在一个大仓库内,从而提升了速度并提高了灵活性。相比其他数据库,MySQL 具有以下特点和优势:

- MySQL 是开源的,所以不需要支付额外的费用;
- MySQL 支持大型的数据库,可以处理拥有上千万条记录的大型数据;
- MySQL 使用标准的 SQL 数据语言形式;
- MySQL 可以运行于多个系统上,并且支持多种语言。这些编程语言包括 C、C++、Python、Java、Perl 和 PHP 等;
- MySQL 对 PHP 有很好的支持,PHP 是目前较常用的 Web 开发语言;
- MySQL 支持大型数据库,支持 5000 万条记录的数据仓库,32 位系统支持的最大表文件为 4GB,64 位系统支持的最大表文件为 8TB;
- MySQL 是可以定制的,采用了 GPL 协议,用户可以修改源码来开发自己的 MySQL 系统。

6.1.2 MySQL 数据库下载和安装

目前针对不同用户,MySQL 提供了两个不同的版本。

- MySQL Community Server:社区版,该版本完全免费,但是官方不提供技术支持。
- MySQL Enterprise Server:企业版,它能够高性价比地为企业提供数据仓库应用,支持 ACID 事务处理,提供完整的提交、回滚、崩溃恢复和行级锁定功能。但是该版本需付费使用,官方提供电话及文档等技术支持。

目前最新的 MySQL 版本为 MySQL 8,可以在官方网站下载该软件,如图 6.1 所示。首先单击右下角的 Download 超链接,然后按照提示步骤操作就可以将 MySQL 软件下载到本地计算机中了。

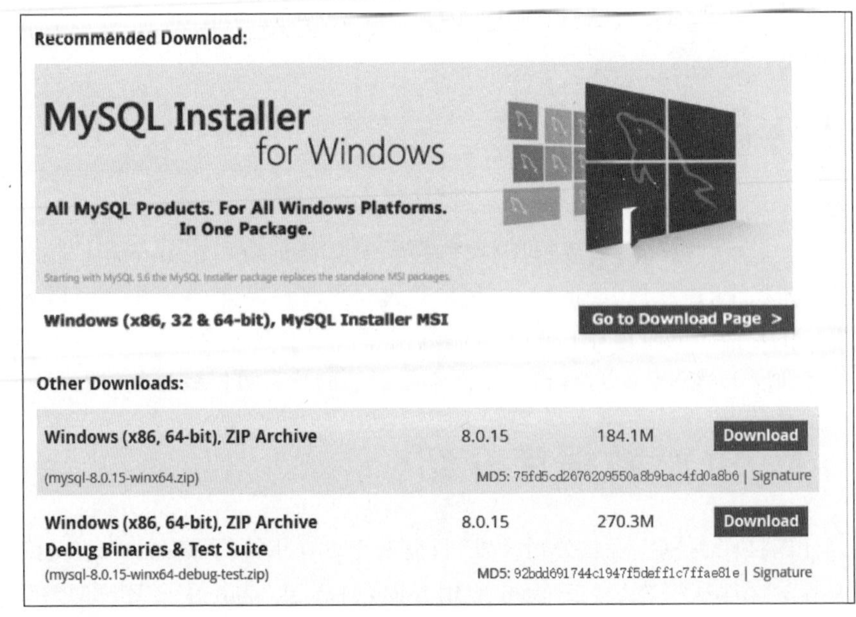

图 6.1 MySQL 官方网站

MySQL下载完成后，找到下载到本地的文件，按照相关安装图文说明进行安装即可。因为安装过程篇幅比较长，这里就不做介绍了。

如图6.2所示，在安装的过程中安装程序会需要用户设置一个MySQL Root Password，这个密码是数据库超级管理员root的密码，用户需要记住自己设置的root管理员的密码，在后续的数据库操作中需要用到这个密码，在图中用户还可以新增其他用户。

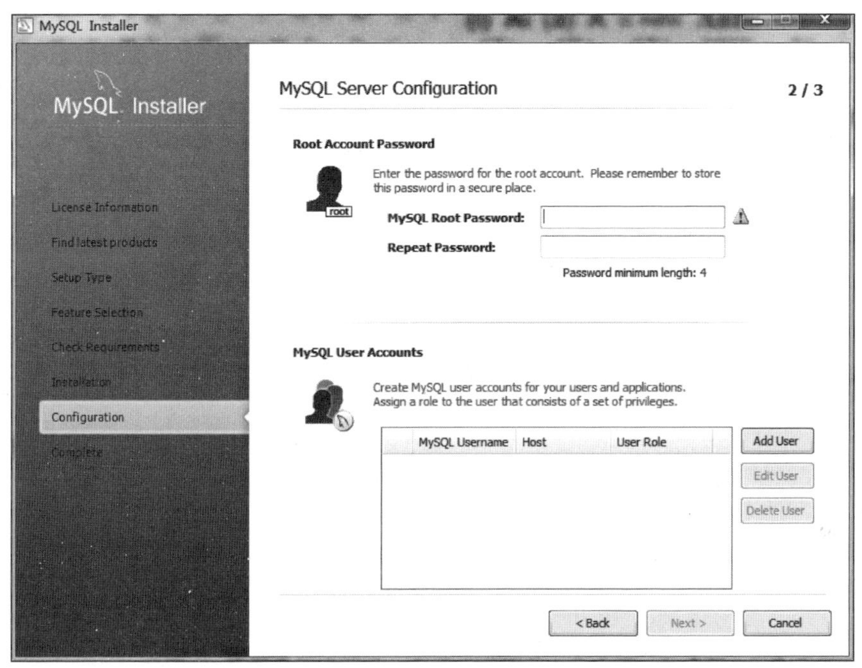

图6.2 MySQL配置密码和新增用户界面

完成MySQL数据库的整个安装配置过程后，用户打开任务管理器，可以在其中看到MySQL服务进程mysqld.exe已经启动了，这说明MySQL服务器正常在后台运行，如图6.3所示。至此，说明用户在Windows上顺利地安装了MySQL。接下来就可以对数据库进行操作了。

安装好MySQL数据库后，还需要在CMD命令环境下对MySQL进行启动、登录、操作数据库、操作表和字段、录入数据和关闭MySQL数据库等操作，详情可以参考MySQL数据库手册和相关书籍。为了方便操作，笔者建议读者使用Navicat for MySQL可视化工具，详见6.2节的讲解，该工具可以极大地提高数据库操作效率。

6.1.3 使用phpStudy安装MySQL

使用6.1.2节中下载MySQL数据库并安装的方式，过程比较复杂，而且需要一些配置操作，很多读者安装到最后会发现无法启动，为此笔者推荐大家使用更加快捷地安装MySQL的方法，就是使用phpStudy安装MySQL数据库。

phpStudy是一个PHP调试环境的程序集成包。该程序包集成最新的Apache+PHP+MySQL+phpMyAdmin+ZendOptimizer，可实现一次性安装，无须配置即可使用，是非

任务管理器

文件(F) 选项(O) 查看(V)

进程 性能 应用历史记录 启动 用户 详细信息 服务

名称	状态	29% CPU	47% 内存	0% 磁盘	0% 网络
Microsoft Windows Search P...		0%	1.7 MB	0 MB/秒	0 Mbps
Microsoft Windows Search P...		0%	1.3 MB	0 MB/秒	0 Mbps
Microsoft Windows Search ...		0%	10.3 MB	0 MB/秒	0 Mbps
MidwareUIMfc (32 位)		0%	0.8 MB	0 MB/秒	0 Mbps
mysqld.exe (32 位)		0%	16.8 MB	0 MB/秒	0 Mbps
NVIDIA Container		0%	3.4 MB	0 MB/秒	0 Mbps
NVIDIA Container		0%	1.3 MB	0 MB/秒	0 Mbps
NVIDIA Container		0%	1.6 MB	0 MB/秒	0 Mbps
NVIDIA Container (32 位)		0%	6.5 MB	0 MB/秒	0 Mbps
NVIDIA Wireless Controller S...		0%	0.4 MB	0 MB/秒	0 Mbps
pcas service (32 位)		0.2%	2.3 MB	0 MB/秒	0 Mbps
phpStudy (32 位)		0.3%	14.7 MB	0 MB/秒	0 Mbps
PresentationFontCache.exe		0%	0.7 MB	0 MB/秒	0 Mbps
QQ安全防护进程 (Q盾) (32 位)		0%	3.1 MB	0.1 MB/秒	0 Mbps

简略信息(D) 结束任务(E)

图 6.3 任务管理器中的 Mysqld.exe

常方便、好用的 PHP 调试环境。该程序不仅包括 PHP 调试环境，还包括开发工具、开发手册等。总之学习 PHP 只需一个包，因为它集成了 MySQL 数据库，而且不需要配置，这就给广大开发者安装 MySQL 提供了一种更加快捷的方式。

先到 phpStudy 的官方网站 http://phpStudy.php.cn/下载最新的版本，下载到本地后直接安装即可，安装过程如同安装 QQ 等通用软件一样简单，所以这里不做介绍。安装完成后用户可以打开程序，以 phpStudy 2018 为例，如图 6.4 所示。

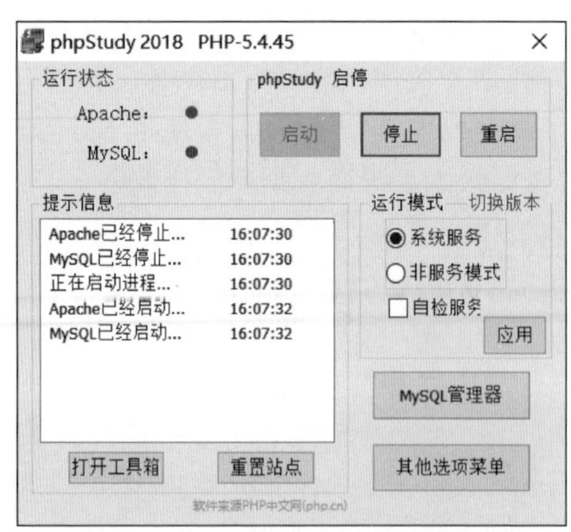

图 6.4 phpStudy 2018 启动界面

开发者可以单击图 6.4 中的“启动”按钮，这时 phpStudy 帮开发者启动了 MySQL 数据库和 Apache 服务器，而 Apache 服务器又是做小程序开发时不需要的，为了减轻计算机的负担，读者可以按照图 6.5 来操作。

图 6.5 只启动 MySQL

在 Windows 系统的任务栏中找到 phpStudy 的图标，如图 6.5 中的①所示，右击 phpStudy 图标，然后鼠标依次移动到②服务器管理、③MySQL 和④启动位置，即可只启动 MySQL 而不必启动 Apache。

6.2 可视化工具 Navicat for MySQL

6.2.1 Navicat 介绍与安装

通过 6.1.2 节和 6.1.3 节的学习，MySQL 数据库已经可以被顺利安装好并启动了。但是在 CMD 命令环境中操作 MySQL 数据库还是比较复杂的，为此读者可以使用可视化的操作工具 Navicat。

Navicat for MySQL 是管理和开发 MySQL 或 MariaDB 的理想解决方案。它是一套单一的应用程序，能同时连接 MySQL 和 MariaDB 数据库，并与 Amazon RDS、Amazon Aurora、Oracle Cloud、Microsoft Azure、阿里云、腾讯云和华为云等云数据库兼容。这套全面的前端工具为数据库管理、开发和维护提供了一款直观而强大的图形界面。读者可以到官方网站付费下载。

双击桌面上的快捷图标运行 Navicat for MySQL，然后选择菜单“文件”|“新建连接”选项来新建一个连接，如图 6.6 所示。

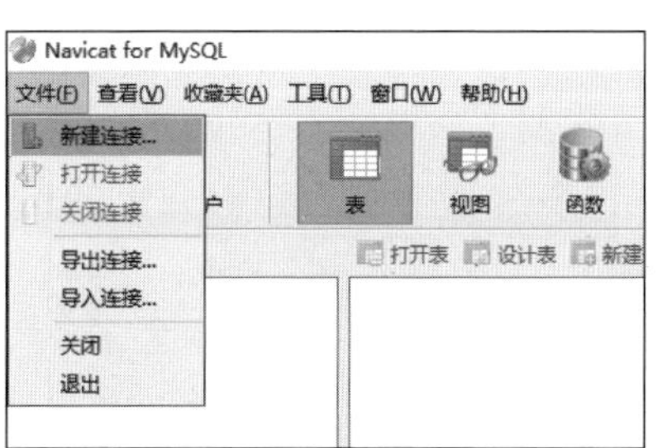

图 6.6 新建连接

在图 6.7“连接”对话框中,连接名是自己定义的,主机名 localhost、端口号 3306 和用户名 root 是软件默认填上的,密码是在安装 MySQL 数据库的时候自己设置的。如果采用 phpStudy 来安装的 MySQL,则密码默认也是 root。

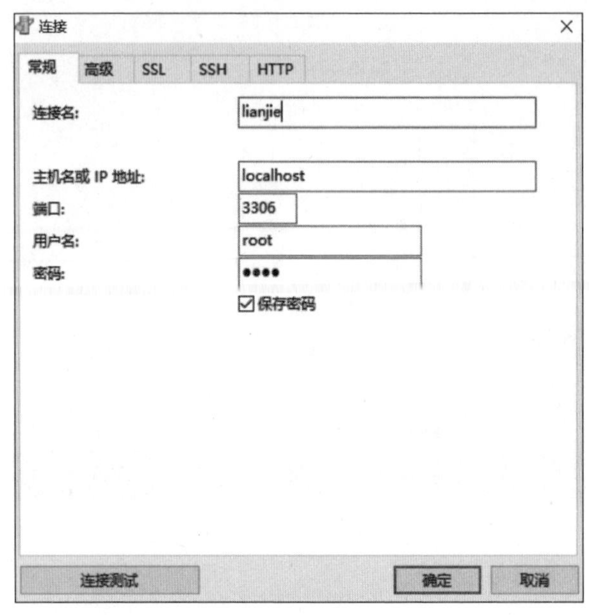

图 6.7 新建连接

单击“确定”按钮即完成了连接的建立,当然在创建连接之前需要先开启 MySQL 数据库,否则创建连接会失败。

6.2.2 在 Navicat 中创建数据库

双击 6.2.1 节中创建的连接 lianjie,读者会发现有几个 MySQL 预先创建好的数据库,可暂且不管它们。右击创建好的连接 lianjie,选择“新建数据库”命令,如图 6.8 所示。

此时弹出“创建新数据库”对话框,如图 6.9 所示,数据库名自定义为 xinwen,“字符集”选择 uft8--UTF-8 Unicode,“校对”选择 utf8_general_ci。单击“确定”按钮即可成功创建数据库。

在图 6.10 中双击创建的数据库 xinwen,右击“表”,选择“新建表”选项,弹出的窗口如图 6.11 所示。在图 6.11 中输入 id、title、img、content 和 cTime 字段分别表示一条新闻的 id、标题、图片、文章内容和发表时间,设置好它们的类型、长度,其中把 id 设置为关键字,把表命名为 wenzhang。

图 6.8 新建数据库

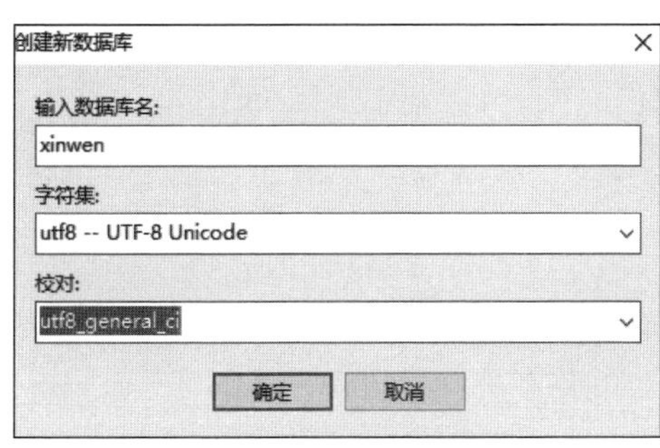

图 6.9 填写数据库编码方式

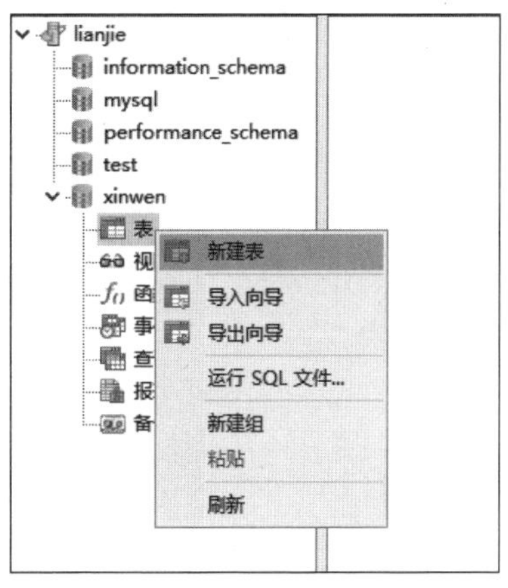

图 6.10 新建表

名	类型	长度	小数点	允许空值 (	
id	int	10	0	☐	1
title	varchar	255	0	☑	
img	varchar	50	0	☑	
content	text	0	0	☑	
cTime	varchar	10	0	☑	

图 6.11 新建字段

完成表的创建之后，可以向表中录入数据，如图 6.12 所示。至此，数据库方面的准备工作就完成了。

id	title	img	content	cTime
1	携号转网来临前，那些"狼来了"的日子	/img/01.jpg	本网讯（党群工作部）10月24日上午，.	2019-03-15
2	任正非：要敢于与美国争夺人才 我们待遇	/img/02.jpg	本网讯（教务处）10月20-21日，2018	2019-03-15
3	高通苹果专利侵权案庭审结束，高通赔偿	/img/03.jpg	根据国外媒体报道，当地时间周三高通	2019-03-15
4	3.15后多家电商APP下架"电子烟"关键词	/img/04.jpg	钛媒体3月16日消息，央视315曝光"电子	2019-03-15
5	还有什么力量能把大盘推上3000点?	/img/05.jpg	本周，沪指在3000点一带展开拉锯，各	2019-03-15
6	市场监管总局及时布置央视3·15晚会曝光	/img/06.jpg	3月15日中央电视台3·15晚会结束后，市	2019-03-15
7	纽约：大批武装警察持枪守卫清真寺	/img/07.jpg	3月16日报道，当地时间3月15日，新西	2019-03-15
8	与人民在一起——习近平的两会时间	/img/08.jpg	2019年全国两会期间，习近平总书记六	2019-03-15
9	习近平就新西兰枪击事件向新西兰总督雷	/img/09.jpg	新华社北京3月15日电国家主席习近平3	2019-03-15
10	十三届全国人大二次会议在京闭幕	/img/10.jpg	新华社北京3月15日电 第十三届全国人	2019-03-15

图 6.12 录入信息

在 Navicat 中导入和导出数据库数据也是常用的操作。如图 6.13 所示，"转储 SQL 文件"是把建立好的数据导出到本地硬盘形成 SQL 文件；"运行 SQL 文件"则是将硬盘上的

SQL 文件导入到 MySQL 数据库中来使用。

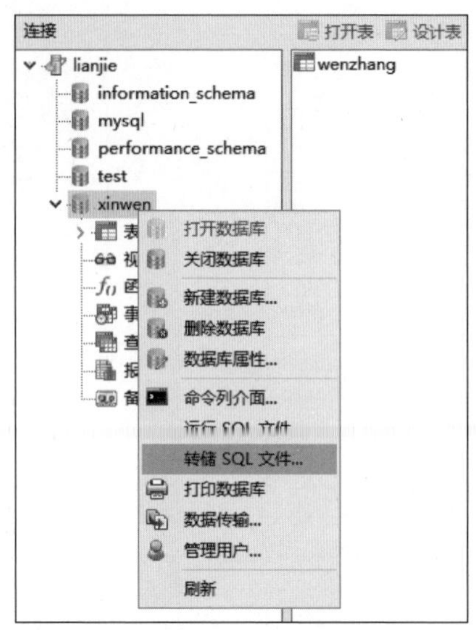

图 6.13　导入/导出数据库数据

6.3　基于 Java 的后端 JSON 接口

微信小程序没有提供直接和 MySQL 数据库连接的接口，而是采用前端小程序连接后端程序，再通过后端程序连接数据库的方式来间接连接数据库。前端小程序连接后端程序是通过小程序网络接口 wx.request()来实现的，因为 wx.request()只能读取 JSON 格式的数据，所以后端程序要形成 JSON 数据才可以供前端小程序使用。而后端程序可以选择自己熟悉的语言来开发，例如 PHP 或者 Java，本书后续的后端代码将统一采用 Java 来编写，图 6.14 描述了前端、后端和数据库之间的关系。

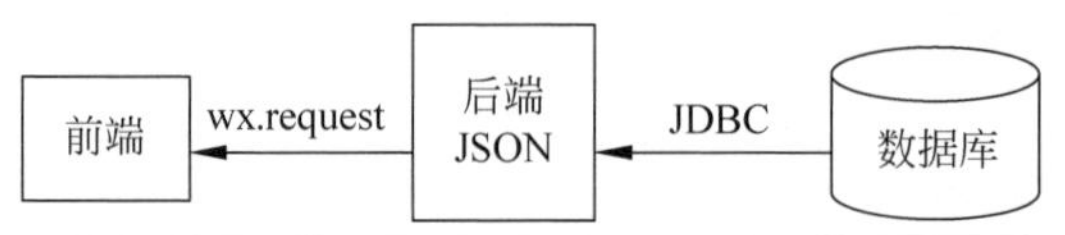

图 6.14　小程序和数据库的交互示意图

6.3.1　JDBC

JDBC 英文全称为 Java Data Base Connectivity (Java 数据库连接)，可以为多种关系数据库提供统一的访问。JDBC 是 Sun 开发的一套数据库访问编程接口，是一种 SQL 级的 API。它由 Java 语言编写完成，具有很好的跨平台特性，使用 JDBC 编写的数据库应用程序可以在任何支持 Java 的平台上运行，而不必在不同的平台上编写不同的应用程序。

JDBC 的主要功能如下：

(1) 建立与数据库或者其他数据源的连接。

(2) 向数据库发送 SQL 命令。

(3) 处理数据库的返回结果。

JDBC 常用接口有连接到数据库(Connection)、建立操作指令(Statement)、执行查询指令(executeQuery)和获得查询结果(ResultSet)等。

例 6-1 JDBC 连接 MySQL 数据库代码演示。

视频讲解

```
package request;
import java.sql.DriverManager;
import java.sql.ResultSet;
import java.sql.SQLException;
import java.sql.Statement;
import java.sql.Connection;
import java.sql.DriverManager;
import java.sql.SQLException;
public class jdbc {
public static void main(String[ ] args) throws ClassNotFoundException, SQLException
    {
    String uri = "jdbc:mysql://127.0.0.1:3306/xinwen?useUnicode = true&amp;
    characterEncoding = utf - 8";
    String user = "root";
    String password = "root";
    Class.forName("com.mysql.jdbc.Driver");          //加载 JDBC 驱动程序
    //创建第一个 Connection 类型的 con 对象来建立连接
    Connection con = DriverManager.getConnection(uri, user, password);
    //创建第二个 Statement 类型的对象 st 作为 sql 语句的载体实现增删改查
    Statement st = con.createStatement();
    ResultSet rs = st.executeQuery("select * from wenzhang");
    //处理数据库的返回结果(使用 ResultSet 类)
     while(rs.next()){
  System.out.println(rs.getString("title") + "   " + rs.getString("cTime"));}
     //关闭资源
    rs.close();
    st.close();
    con.close();
    }
}
```

【代码讲解】 本例连接 6.2.2 节中创建的 xinwen 数据库下面的 wenzhang 表中的内容,并读取了表中的标题和时间两列字段。本例代码主要实现了加载 JDBC 驱动程序、3 个对象的建立和释放,其中 3 个对象指的是 Connection 类型的 con、Statement 类型的 st 和 ResultSet 类型的 rs。在读到数据之后使用 while 循环来输出数据,因为代码是带有 main 函数的 Java 函数,所以输出结果在 Console 中显示。JDBC 是 Java 和 JSP 课程的内容,读者可以参考相关书籍。需要说明的是,本书中后端 Java 部分的开发工具是 MyEclipse 2015 版本,当然其他的开发工具不影响结果的正确性。而连接数据库是需要有 JDBC 驱动程序的,

开发者计算机上的 MySQL 版本不一样,JDBC 驱动程序的版本也需要不一样,笔者采用的是 mysql-connector-java-5.1.41-bin.jar,读者也可以拥有适合自己的 JDBC 驱动程序。程序运行结果如图 6.15 所示。

图 6.15　运行结果

6.3.2　JSON 接口

有了 JDBC,只是 Java 程序可以读取到数据库的数据,但是小程序只能读取 JSON 格式的数据,也就是说开发者在读取数据库数据的时候需要进行 JSON 格式处理。考虑到 wx.request()需要以 Web 的方式来访问数据,开发者需要找到一种合适的方法来改造例 6-1 中的呈现方式,笔者选择了 Servlet 的方式。对于 Servlet 的理解,读者可参考 JSP 的相关书籍,可以简单地理解为 Servlet 是可以 Web 访问的 Java 程序,即例 6-1 的 Java 程序呈现结果都在控制台中显示,而 Servlet 是把结果呈现在 Web 浏览器上。

视频讲解

例 6-2　使用 Servlet 读取数据库数据,并以 JSON 格式来呈现。

如图 6.16 所示,先创建一个 Web 项目 mini,然后在项目的 src 目录下创建一个包 request,右击 request 包,选择 New|Servlet 选项,弹出 Create Servlet 对话框,如图 6.17 所示。

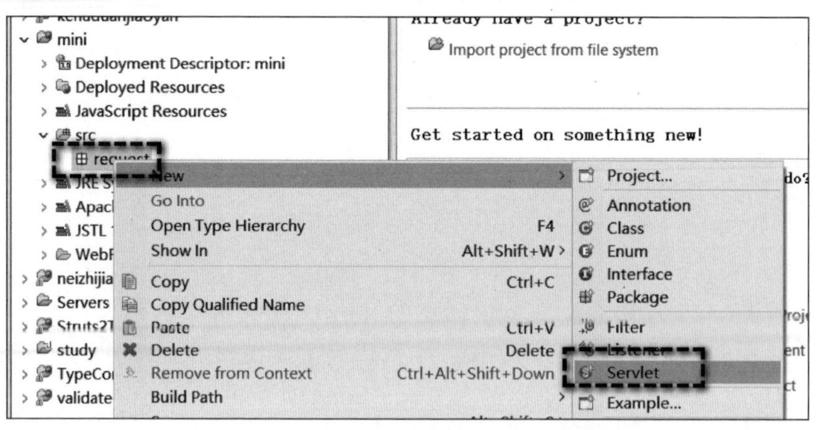

图 6.16　新建 Servlet

在图 6.17 中,只有 Class name 一项需填写,其余都是 MyEclipse 软件自动填上的。接下来编辑代码如下:

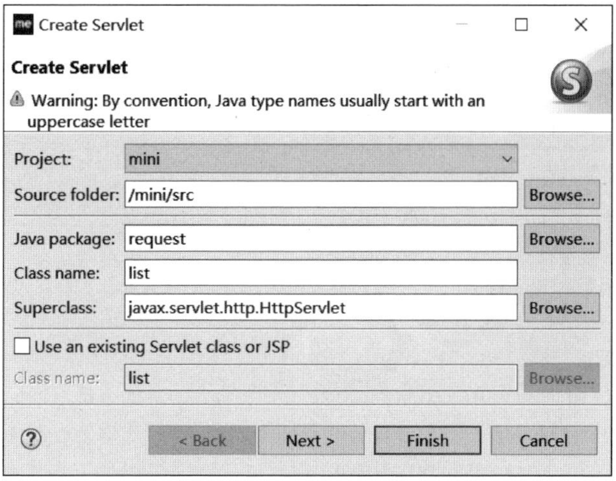

Create Servlet

Create Servlet

Warning: By convention, Java type names usually start with an uppercase letter

Project:	mini	
Source folder:	/mini/src	Browse...
Java package:	request	Browse...
Class name:	list	
Superclass:	javax.servlet.http.HttpServlet	Browse...

☐ Use an existing Servlet class or JSP

Class name: list　Browse...

< Back　Next >　Finish　Cancel

图 6.17　填写 Servlet 的类名字

```
package request;
import java.io.IOException;
import javax.servlet.ServletException;
import javax.servlet.annotation.WebServlet;
import javax.servlet.http.HttpServlet;
import javax.servlet.http.HttpServletRequest;
import javax.servlet.http.HttpServletResponse;
import java.io.IOException;
import java.io.PrintWriter;
import java.sql.*;
import net.sf.json.JSONArray;
import net.sf.json.JSONObject;
/*** Servlet implementation class list */
@WebServlet("/list")
public class list extends HttpServlet {
    private static final long serialVersionUID = 1L;

    /*** @see HttpServlet#HttpServlet() */
    public list() {
        super();
        //TODO Auto-generated constructor stub
    }
    /*** @see
HttpServlet#doGet(HttpServletRequest request, HttpServletResponse response)
     */
    protected void doGet(HttpServletRequest request, HttpServletResponse response) throws
ServletException, IOException {
    response.setContentType("text/json; charset=utf-8");
    PrintWriter out = response.getWriter();
        try {
    Class.forName("com.mysql.jdbc.Driver");
     String url = "jdbc:mysql://localhost/xinwen?user=root&password
```

```
    = root&useUnicode = true&characterEncoding = UTF - 8";
    Connection con = DriverManager.getConnection(url);
    Statement stmt = con.createStatement();          //创建 Statement 对象
    String sql;
    sql = "SELECT * FROM wenzhang";
    ResultSet rs = stmt.executeQuery(sql);
    JSONArray jsonarray = new JSONArray();
    JSONObject jsonobj = new JSONObject();
    //展开结果集数据库
    while(rs.next()){
    //通过字段检索
    jsonobj.put("id", rs.getString("id"));           //这个是返回的内容 1
    jsonobj.put("title", rs.getString("title"));     //这个是返回的内容 2
    jsonobj.put("img", rs.getString("img"));         //这个是返回的内容 3
    jsonobj.put("cTime", rs.getString("cTime"));     //这个是返回的内容 4
    jsonarray.add(jsonobj);
    }
    //输出数据
    out = response.getWriter();
    out.println(jsonarray);
    //完成后关闭
    rs.close();
    stmt.close();
    con.close();
    }catch (Exception e)
{ out.print("get data error!");
    e.printStackTrace(); }
    }
    /*** @see HttpServlet#doPost(HttpServletRequest request, HttpServletResponse response) */
    protected void doPost(HttpServletRequest request, HttpServletResponse response) throws
ServletException, IOException {
        this.doGet(request, response);
    }
}
```

【代码讲解】 本例中连接的仍是 6.2.2 节中创建的 xinwen 数据库下面的 wenzhang 表中的内容,并读取了表中的 id、title、img 和 cTime 4 个字段。从功能代码上看,和例 6-1 中的功能差不多,但是多了以下两行代码:

```
JSONArray jsonarray = new JSONArray();
JSONObject jsonobj = new JSONObject();
```

其中,jsonobj 用来存放从数据库中读取的一行数据,它是 JSON 格式,即程序已经把数据处理成了 JSON 格式;jsonarray 是 JSONArray 类型,是把一行一行的 JSON 格式的数据封装成一个数组。

关于 Servlet 代码结构,需要说明的是:开发者只需要关注 doGet 和 doPost 方法,核心代码写在 doGet 和 doPost 方法中即可。Get 请求，调用 doGet 方法；Post 请求，则调用 doPost 方法。为了保险起见，如果核心代码写在了 doGet 里，就应在 doPost 方法中加上“this.doGet(request, response);”这条语句，这相当于在 doGet 方法中把代码复制了一份到 doPost 方法中。

另外，因为每一个 Tomcat 服务器包中含有的 JAR 包是不一样的，本例中需要用到 mysql-connector-java-5.1.41-bin.jar、commons-beanutils-1.8.0.jar、commons-lang-2.5.jar、commons-collections-3.2.1.jar、commons-logging-1.1.1.jar、ezmorph-1.0.6.jar、json-lib-2.3.jar 和 servlet-api.jar，缺少上述 JAR 文件的读者可以自行下载。

如果程序调试正确的话，运行结果将如图 6.18 所示，数据的最外层是一对中括号，表示是一个 JSON 数组，而数组的里面是一个一个的大括号，大括号中的数据是 JSON 对象，对应于数据库中的一行数据。有了 JSON 数据，小程序前端就可以访问了。

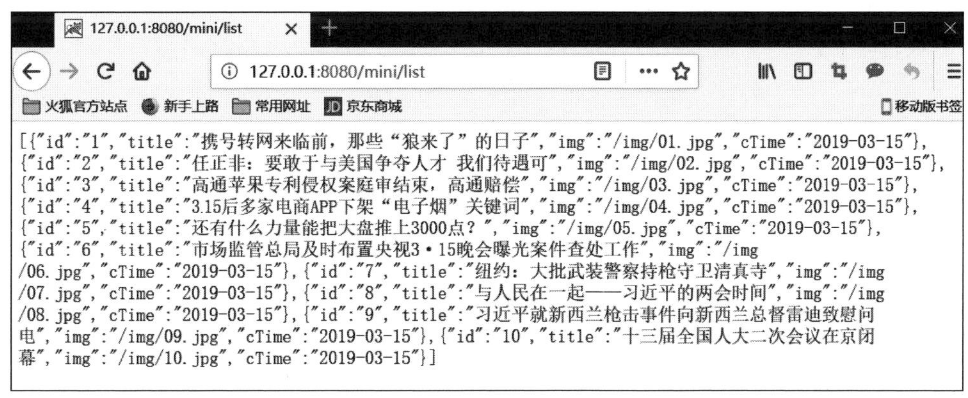

图 6.18 Servlet 读出来的 JSON 格式数据

6.4 数据库操作

在 Web 环境中发起 HTTPS 请求是很常见的，但是微信小程序是腾讯内部的产品，不能直接打开一个外部的链接。例如，在微信小程序里面无法直接打开 www.taobao.com 网站，但是，在小程序开发的时候，如果需要请求一个网站的内容或者服务，如何实现？虽然微信小程序里面不能直接访问外部链接，但是腾讯为开发者封装好了一个 wx.request(object)的 API。在 6.2 节和 6.3 节的基础上，开发者已经准备好了数据库和 JSON 格式的数据读取接口，开发者可以使用 wx.request()接口来获取后端的数据了。

6.4.1 wx.request()接口

wx.request()是腾讯公司封装好的一个 request 请求的函数，类似于其他程序语言的自带函数，开发者只需把这些内置函数复制过来使用即可，无须注意函数底层代码实现部分。wx.request()接口属性如表 6.1 所示。

表 6.1　wx.request()属性表

属　　性	类　　型	默认值	必填	说　　明
url	string		是	开发者服务器接口地址
data	string|object|ArrayBuffer		否	请求的参数
header	Object		否	设置请求的 header，header 中不能设置 Referer。content-type 默认为 application/json
method	string	GET	否	HTTP 请求方法
dataType	string	json	否	返回的数据格式
responseType	string	text	否	响应的数据类型
success	function		否	接口调用成功的回调函数
fail	function		否	接口调用失败的回调函数

微信公众平台官方给出的 wx.request()示例代码如下：

```
wx.request({
  url: 'test.php',                                   //仅为示例，并非真实的接口地址
  data: {
    x: '',
    y: ''
  },
  header: {
    'content-type': 'application/json'               //默认值
  },
  success(res) {
    console.log(res.data)
  }
})
```

1）接口地址 url

微信小程序里面的数据是由接口地址 url 来获取的，它非常重要。接口地址 url 就是 6.3.2 节中的 Servlet 地址，其返回结果是 JSON 格式数据。因为 JSON 格式数据不仅处理起来方便，而且传输安全稳定，容易保存。所以，一般的第三方服务商提供的接口返回的数据都是以 JSON 格式返回的。

url 是提供 JSON 格式数据的接口地址，一般是开发者专门开发的或第三方服务商提供的接口地址。例如快递查询和天气预报等功能在网络上都有相应 JSON 接口的调用地址，其中一些接口是商业收费的。

2）请求参数 data

当小程序前端对 url 发起 HTTPS 请求时，实际上跟在浏览器打开一个网址是一个道理，在浏览器打开网址 http://127.0.01:8080/mini/detail? id=5，实际上是向这个域名所在的服务器发送了一个 HTTPS 请求，在这个请求里面使用了参数 id=5，这里的 id=5 是在请求 url 时需要传递过去的参数。

```
url:'http://127.0.01:8080/mini/detail',
data: {
   id: '5'
}
```

等价于：

```
url:'http://127.0.01:8080/mini/detail?id=5'
```

3）请求方法 method 和请求头 header

wx. request()本质上是 HTTP 请求，header 是请求的消息头，method 是请求的方法。示例代码如下：

```
method: 'GET',
header: {
        'content - type': 'application/json'
    }
method: 'POST',
header: {
        'content - type': 'application/x - www - form - urlencoded'
        },
```

method 取值'GET'或者'POST'（较少用'PUT'和'DELETE'）。HTTP 请求的 header 包含 Accept、Accept-Charset 和 Content-Type 等多项协议头，wx. request()接口中需要填写的是 header 的 Content-Type 选项。Content-Type 取值为'application/x-www-form-urlencoded'和'application/json'。那么，method 和 header 的 Content-Type 就存在如下 4 种组合：

method：'GET'　'content-type'：'application/json'　(1)

method：'GET'　'content-type'：'application/x-www-form-urlencoded'　(2)

method：'POST'　'content-type'：'application/json'　(3)

method：'POST'　'content-type'：'application/x-www-form-urlencoded'　(4)

需要说明的是，前端最终发送给服务器的数据是 String 类型，如果传入的 data 不是 String 类型，会被转换成 String。转换规则如下：

- 对于 GET 方法的数据，会将数据转换成 query string。(encodeURIComponent(k)=encodeURIComponent(v)&encodeURIComponent(k)=encodeURIComponent(v)...)
- 对于 POST 方法且 header['content-type'] 为 application/json 的数据，会对数据进行 JSON 序列化。
- 对于 POST 方法且 header['content-type'] 为 application/x-www-form-urlencoded 的数据，会将数据转换成 query string。(encodeURIComponent(k)=encodeURIComponent(v)&encodeURIComponent(k)=encodeURIComponent(v)...)

因为后端开发者默认前端传递过来的数据是 query string 类型，所以对组合(1)、(2)和(4)有效的后端程序会对组合(3)无效，开发者可避免使用组合(3)。

4）success 函数

当一个 HTTPS 请求成功时，小程序就会自动触发这个返回成功信息的函数，这个函数是腾讯公司封装好的函数，无须开发者自己编写。开发者获取的 JSON 数组在 res. data 中。

6.4.2　基于数据库的新闻列表页案例

有了 6.2 节和 6.3 节的基础，笔者已经写好了小程序读数据库数据的后端部分，现在只需要编写小程序前端部分就可以读 6.3 节中准备好的 JSON 数据了。微信小程序实现前端和后端的网络交互的功能封装在 wx. request()接口中。

视频讲解

例 6-3　基于 MySQL 数据库的新闻列表页的实现。

pages/list/list. wxml 文件代码如下：

```
<view class="body">
<!--文章列表模板 begin-->
<template name="itmes">
  <navigator url="../../pages/detail/detail?id={{id}}"
    hover-class="navigator-hover">
    <view class="imgs"><image src="{{img}}" class="in-img"
    background-size="cover" model="scaleToFill"></image></view>
    <view class="infos">
      <view class="title">{{title}}</view>
      <view class="date">{{cTime}}</view>
    </view>
  </navigator>
</template>
<!--文章列表模板 end-->
<!--循环输出列表 begin-->
<view wx:for="{{shuzu}}" class="list">
  <template is="itmes" data="{{...item}}" />
</view>
</view>
```

【代码讲解】 list.wxml 使用了 wx:for 渲染和模板组件 template，每一次循环就从JSON 数组中读一个对象，并把该 JSON 对象的 img、title 和 cTime 3 个属性显示出来。关于后端代码，可以参看 6.3 节内容。

pages/list/list.js 文件代码如下：

```
Page({
  /*** 页面的初始数据 ***/
  data: {
   id1:1,
    shuzu: []
  },
  /*** 生命周期函数--监听页面加载 ****/
  onLoad: function(options) {
    var that = this
    wx.request({
      url: 'http://127.0.0.1:8080/mini/list',        //仅为示例，并非真实的接口地址
      data: {
        x: '',
        y: ''
      },
      header: {
        'content-type': 'application/json'           //默认值
      },
      success(res) {
        console.log(res)
        that.setData({
          shuzu:res.data
        })
      }
    })
  },
  dian:function(e)
  {
    var a =  e.target.id
    console.log(a)
    wx.navigateTo({
```

```
            url:"/pages/detail/detail?id = " + a,
        })
    },
    /*** 用户点击右上角分享 */
    onShareAppMessage: function() {
    }
})
```

【代码讲解】 list.js 中的 onLoad 函数使用 wx.request()接口向 http://127.0.0.1:8080/mini/list 发起网络请求，success 返回 JSON 数组，并把它以数据绑定的方式传给 list.wxml 文件。dian:function()是 list.wxml 中定义的点击事件，负责跳转到详情页 detail.wxml。

pages/list/list.wxss 文件代码如下：

```
.body{
    height: 100%;display: flex;
    flex-direction: column; padding: 20rpx;
}
navigator { overflow: hidden;}
.list {margin-bottom: 20rpx;height: 200rpx;position: relative;}
.imgs {float: left;}
.imgs image {display: block; width: 200rpx;height: 200rpx;}
.infos {float: left; width: 480rpx; height: 200rpx;padding: 20rpx 0 0 20rpx;}
.title {font-size: 20px;}
.date {font-size: 16px;color: #aaa; position: absolute;bottom: 0;}
.loadMore {text-align: center;margin: 30px;color: #aaa;font-size: 16px}
page{
  background-color: #d1d3d4;
}
```

【代码讲解】 list.wxss 使用 page{}标签选择器对页面的背景色进行设置。

程序运行结果如图 6.19 所示。

图 6.19 新闻列表页运行结果

6.4.3 基于数据库的新闻详情页案例

例 **6-4** 6.4.2 节已经实现了基于 MySQL 数据库的新闻列表页，现在要做的是进一步实现详情页。

视频讲解

src/servlet/detail.java 代码如下：

```
package request;
import java.io.IOException;
import java.io.PrintWriter;
import java.sql.Connection;
import java.sql.DriverManager;
import java.sql.ResultSet;
import java.sql.Statement;
import javax.servlet.ServletException;
import javax.servlet.annotation.WebServlet;
import javax.servlet.http.HttpServlet;
import javax.servlet.http.HttpServletRequest;
import javax.servlet.http.HttpServletResponse;
import net.sf.json.JSONArray;
import net.sf.json.JSONObject;
/*** Servlet implementation class detail */
@WebServlet("/detail")
public class detail extends HttpServlet {
    private static final long serialVersionUID = 1L;
    /*** @see HttpServlet#HttpServlet() */
    public detail() {
       super();
       //TODO Auto-generated constructor stub
    }
    /*** @see HttpServlet#doGet(HttpServletRequest request,
HttpServletResponse response) */
    protected void doGet(HttpServletRequest request, HttpServletResponse response) throws
ServletException, IOException {
    response.setContentType("text/json; charset=utf-8");
    PrintWriter out = response.getWriter();
    response.setHeader("Access-Control-Allow-Origin", "*");
    response.setHeader("Access-Control-Allow-Methods", "GET,POST");
    String id = request.getParameter("id");
    int id1 = Integer.parseInt(id);
    System.out.println("id=" + id);                    //验证数据传递情况
    try {
         Class.forName("com.mysql.jdbc.Driver");
         String url = "jdbc:mysql://localhost/xinwen?user=root&password=
         root&useUnicode=true&characterEncoding=UTF-8";
         Connection con = DriverManager.getConnection(url);
         Statement stmt = con.createStatement();
         String sql = "SELECT * FROM wenzhang where id=" + id1;
         ResultSet rs = stmt.executeQuery(sql);
         JSONArray jsonarray = new JSONArray();
         JSONObject jsonobj = new JSONObject();
         while(rs.next()){
         jsonobj.put("title", rs.getString("title"));
         jsonobj.put("cTime", rs.getString("cTime"));
```

```
            jsonobj.put("img", rs.getString("img"));
            jsonobj.put("content", rs.getString("content"));
             jsonarray.add(jsonobj);
            }
            out = response.getWriter();
            out.println(jsonarray);
            rs.close();
            stmt.close();
            con.close();
            }catch (Exception e) {
            out.print("get data error!");
            e.printStackTrace(); }
        }
    /*** @see HttpServlet#doPost(HttpServletRequest request, HttpServletResponse response) */
        protected void doPost(HttpServletRequest request, HttpServletResponse response) throws
    ServletException, IOException {
        this.doGet(request, response);
        }
    }
```

【代码讲解】 本例和列表页的 JSON 接口最大的区别是 doGet 函数的开头部分多了两行代码：

```
response.setHeader("Access-Control-Allow-Origin", "*");
response.setHeader("Access-Control-Allow-Methods", "GET,POST");
```

第 1 行代码是跨域访问控制语句，同域指的是“协议＋域名＋端口号”均相同，而小程序和自行搭建的服务器一般不在一台服务器，后端程序加上跨域访问控制语句才合理。第 2 行代码控制允许的前端请求方法，可以结合 6.4.1 节的 method 方法来合理编写。“String sql = "SELECT * FROM wenzhang where id="+id1;”则是去查找指定 id 的行。

如果程序调试正确，访问网址 http://127.0.0.1:8080/mini/detail?id=5 就可以读出数据库中的一行记录，并且以 JSON 格式呈现，如图 6.20 所示。

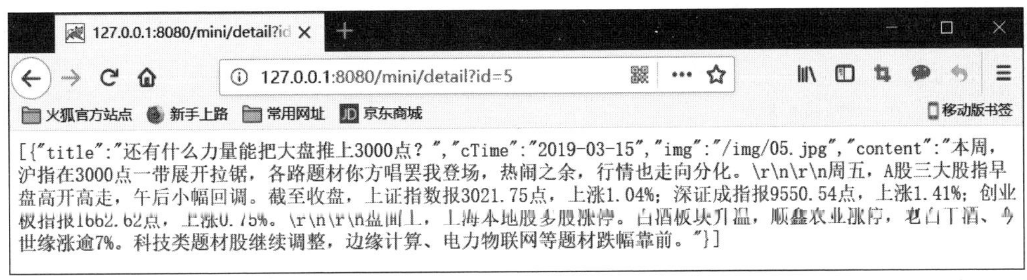

图 6.20　详情页的 JSON 结果

pages/detail/detail.wxml 的代码如下：

```
<!-- detail.wxml -->
<view class = "warp">
<view wx:for = "{{shuzu}}">
  <view class = "title">{{item.title}}</view>
  <view class = "cTime">{{item.cTime}}</view>
  <view class = "img"><image src = "{{item.img}}" class = "in-img"
```

```
    background - size = "cover" model = "scaleToFill"></image></view>
  <view class = "content">{{item.content}}</view>
  <view class = "close" bindtap = "closepage"> 返回 </view>
</view></view>
```

【代码讲解】 detail.wxml 和 6.4.2 节中的 list.wxml 的功能近似，不过本例的 shuzu 数组只是一条数据库记录的 JSON 数据。

pages/detail/detail.js 的代码如下：

```
//detail.js
var app = getApp()
Page({
  data: {
    id1: 1,
    shuzu: []
  },
  onLoad: function(options) {
    var that = this
    wx.request({
      url: 'http://127.0.0.1:8080/mini/detail',         //仅为示例，并非真实的接口地址
      data: { id: decodeURIComponent(options.id) },
      method: 'POST',
      header: {
        'content - type': 'application/x - www - form - urlencoded'  //默认值
      },
      success(res) {
        console.log(res.data)
        that.setData({
          shuzu: res.data
        })
      }
    })
  },
  closepage: function(){
      wx.navigateBack()
  },
  toastChange: function(){
    this.setData({toastHidden:true})
    wx.navigateBack()
  },
})
```

【代码讲解】 本例中的 decodeURIComponent(options.id)是 list.wxml 中带参数的 url 被点击之后得到的，例如在 list.wxml 中点击了 id 为 5 的那篇文章，则 decodeURIComponent(options.id)就等于 5，这样就可以去查看指定文章的详情页了。本例中要注意 header 的写法。

pages/detail/detail.wxss 的代码如下：

```
.warp {
    height: 100 %; display: flex;
```

```
    flex-direction: column; padding: 20rpx;
    font-size: 16px;
}
.title {
    text-align: center; padding: 20rpx;
    font-size: 20px;
}
.cTime {
    color: #aaa;
}
.img {
    text-align: center; padding: 20rpx;
}
.img image {
    width: 120px; height: 120px;
}
.content {
    text-indent: 2em;
}
.close {
    text-align: center; margin: 30px;
    font-size: 20px; color: #aaa
}
page{
  background-color: #d1d3d4;
}
```

程序运行结果如图 6.21 所示。

图 6.21　详情页的执行效果

6.5 数据缓存 Storage

在 Web 开发中，服务器可以为每个用户创建一个会话对象(session 对象)，注意：一个浏览器独占一个 session 对象(默认情况下)。因此，在需要保存用户数据时，服务器程序可以把用户数据写到用户浏览器独占的 session 中，当用户使用浏览器访问其他程序时，其他程序可以从用户的 session 中取出该用户的数据。

很多人学了 JSP 和 PHP 这些 Web 程序语言之后对 session 仍然一知半解，其实 session 是容器为访问者开辟的一个内存空间，在通常情况下这块内存空间将为用户保留 30 分钟。同一个用户访问 Web 系统使用的是同一块内存空间也即同一个 session，不同用户拥有不同的 session。微信小程序提供了 Storage 接口实现类似 Web 开发中的 session 的机制。

小程序提供了 wx. setStorage()、wx. getStorage()、wx. removeStorage() 和 wx. clearStorage()4 个接口以实现 Storage 数据的设置、获取、删除和清空操作。同时，小程序还提供了 wx. setStorageSync()、wx. getStorageSync()、wx. removeStorageSync() 和 wx. clearStorageSync()4 个同步版本接口。

wx. setStorage(Object object)将数据存储在本地缓存指定的 key 中，会覆盖原来该 key 对应的内容。数据存储生命周期跟小程序本身一致，即除用户主动删除或超过一定时间被自动清理，数据都一直可用。单个 key 允许存储的最大数据长度为 1MB，所有数据存储上限为 10MB。wx. setStorage()属性如表 6.2 所示。

表 6.2　wx. setStorage()属性表

属　　性	类　　型	必填	说　　明
key	string	是	本地缓存中指定的 key
data	any	是	需要存储的内容。只支持原生类型、Date 及能够通过 JSON. stringify 序列化的对象
success	function	否	接口调用成功的回调函数
fail	function	否	接口调用失败的回调函数
complete	function	否	接口调用结束的回调函数(调用成功、失败都会执行)

wx. getStorage(Object object)从本地缓存中异步获取指定 key 的内容，它的属性如表 6.3 所示。

表 6.3　wx. getStorage()属性表

属　　性	类　　型	必填	说　　明
key	string	是	本地缓存中指定的 key
success	function	否	接口调用成功的回调函数
fail	function	否	接口调用失败的回调函数
complete	function	否	接口调用结束的回调函数(调用成功、失败都会执行)

例 6-5 本例在第一个页面使用 wx. setStorage()接口向数据缓存区设置 name 和 password 两个变量，在第二个页面使用 wx. getStorage()接口获取 name 和 password 两个变量，程序运行效果如图 6.22 所示。

视频讲解

Console Sources Network Security AppData
top Filter
wx.setStorage传过来的name=张三
wx.setStorage传过来的password=123456

图 6.22 使用 wx. getStorage()获取缓存区数据

pages/setStorage/setStorage. wxml 的代码如下：

```
<view class = "container" bindtap = 'test'>
传数据过去下一页面
</view>
```

pages/setStorage/setStorage. js 的代码如下：

```
Page({
  data: {
    password:"123456"
  },
  test: function() {
    wx.setStorage({
      key: "name",
      data:"张三"
    })
    wx.setStorage({
      key: "password",
      data: "123456"
    })
   wx.wx.navigateTo({
     url: '../getStorage/getStorage',
   })
  }
})
```

pages/getStorage/getStorage. wxml 的代码如下：

```
<view class = "container log - list">
  传过来的数据是{{name}}和{{password}}
</view>
```

pages/getStorage/getStorage. js 的代码如下：

```
Page({
  data: {
    name: "",
    password:''
  },
  onLoad: function(option) {
    var that = this
```

```
    wx.getStorage({
      key: 'name',
      success(res) {
        console.log("wx.setStorage 传过来的 name = " + res.data)
        that.setData({
          name: res.data
        })
      }
    })
    wx.getStorage({
      key: 'password',
      success(res) {
        console.log("wx.setStorage 传过来的 password = " + res.data)
        that.setData({
          password: res.data
        })
      }
    })
  }
})
```

例 6-6 wx.setStorage()和 wx.getStorage()接口有一对同步接口 wx.setStorageSync()和 wx.getStorageSync(),异步接口和同步接口实现的功能是一样的,但是使用方法却差别很大,本例用数据缓存的同步接口改写例 6-5,程序运行效果如图 6.23 所示。

视频讲解

图 6.23 使用 wx.getStorageSync()获取缓存区数据

pages/setStorageSync/setStorageSync.wxml 的代码如下:

```
<view class = "container" bindtap = 'test'>
传数据过去下一页面
</view>
```

pages/setStorageSync/setStorageSync.js 的代码如下:

```
Page({
  data: {
  },
  test: function() {
  wx.setStorageSync("name", "李四")
    wx.setStorageSync("password", "123456")
   wx.redirectTo({
     url: '../getStorageSync/getStorageSync',
   })
  }
})
```

pages/getStorageSync/getStorageSync. wxml 的代码如下：

```
<view class = "container log - list">
  传过来的数据是{{name}}和{{password}}
</view>
```

pages/getStorageSync/getStorageSync. js 的代码如下：

```
Page({
  data: {
    name: "",
    password:''
  },
  onLoad: function(option) {
    this. setData({
     name:wx. getStorageSync("name"),
    password:wx. getStorageSync("password")
    })
    console. log("setStorageSync 传过来的 name = " + this. data. name)
    console. log("setStorageSync 传过来的 password = "  +  this. data. password)
  }
})
```

【代码讲解】 认真对比例 6-5 和例 6-6，同步接口 wx. setStorageSync()、wx . getStorageSync()和 JSP 里面的 session 更加相像，笔者推荐使用同步接口。

6.6 html2wxml 富文本插件

6.6.1 html2wxml 插件介绍

html2wxml 是 JFinal 学院开发的富文本插件，它是微信小程序 Java 版富文本渲染解决方案，实现了 HTML 向 WXML 格式的转化，项目基于 JFinal＋jsoup＋FastJSON 开发，html2wxml 的实现功能如下：

(1) 目前仅实现解析 HTML 并转换成 JSON 格式数据，将生成的 JSON 通过接口访问返回给微信小程序端。

(2) 微信小程序端使用 html2wxml 组件版，按照 html2wxml 教程配置即可。

(3) HTML 大部分标签已经支持，如 Video、Audio 标签。

(4) Pre 标签里的代码支持代码着色高亮显示。

6.6.2 html2wxml 插件使用

此插件提供了后端的源码，读者可以到 https://gitee. com/909854136/html2wxml4J 下载后端 Java 源码，然后搭建到自己的服务器或者本地计算机上，这个搭建操作实际上是开发者自己用JFinal学院开发的接口源码搭建自己的HTML转WXML的接口服务。但是因为搭建自己的接口服务过于复杂，建议读者下载小程序端插件，然后调用JFinal学院对外公布的调用接口，这样就省去了后端接口的搭建环节。下面介绍如何下载和使用html2wxml插件。

开发者需要到网址https://gitee.com/qwqoffice/html2wxml/tree/master/demo下载插件，插件实际上是一个微信小程序项目。当开发者把插件导入微信开发者工具中时，能看到如图6.24所示的项目结构，项目多了一个html2wxml文件夹。开发者只需把html2wxml文件夹复制到小程序根目录下即可使用。在新建要使用html2wxml插件的页面时需要按照相应的引用规则来编写。

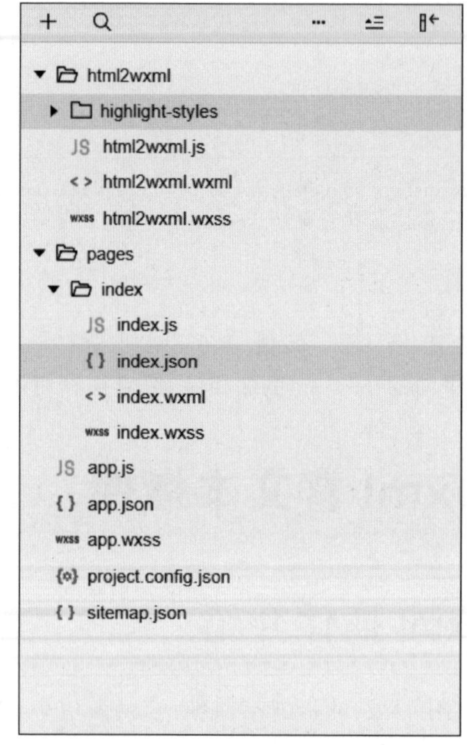

图6.24　html2wxml插件项目结构

视频讲解

例6-7　在index页面中引入https://www.qwqoffice.com/html2wxml/example.html这个页面，即开发者把https://www.qwqoffice.com/html2wxml/example.html这个HTML页面直接转化成微信小程序的WXML页面，而不需要自己写WXML页面了。

pages/index/index.wxml代码如下：

```
<import src="/html2wxml/html2wxml.wxml" />
<template is="html2wxml" data="{{wxmlData:article}}" />
```

【代码讲解】 在引用html2wxml插件的时候无须修改index.wxml文件中的代码，只需修改JS文件的内容即可。

pages/index/index.js 代码如下：

```
var html2wxml = require('../../html2wxml/html2wxml.js');
Page({
  data: {
  },
  onLoad: function() {
    wx.request({
    url: 'https://www.qwqoffice.com/html2wxml/example.html',
      success: res => {
        wx.request({
          url: 'https://www.qwqoffice.com/api/',
          data: {
            text: res.data,
            type: 'html',
            linenums: true,
            highlight: true
          },
          method: 'POST',
          header: {
            'content-type': 'application/x-www-form-urlencoded'
          },
          success: res => {
            wx.stopPullDownRefresh();
            html2wxml.html2wxml('article', res.data, this, 5);
          }
        })
      }
    })
  },
  wxmlTagATap(e) {
    console.log(e);
  },
  onPullDownRefresh() {
    this.onLoad();
  }
})
```

【代码讲解】 代码 url：'https://www.qwqoffice.com/html2wxml/example.html'是要被转化的 HTML 地址，url：'https://www.qwqoffice.com/api/'这条语句是 html2wxml 的官方转化接口，即如果想把一个 HTML 网址转成 WXML 直接显示，只需要替换第一个 url 即可。程序运行结果如图 6.25 所示。

图 6.25　html 转化的效果

html2wxml 插件可以顺利地把 HTML 转化成 WXML，但是对于 JSP 和 PHP 等动态网页的转化还不能实现。有人可能会觉得这个插件作用不大，但实际上现在大量使用的微信公众号网页都是 HTML 页面，使用 html2wxml 插件可以顺利地将这些微信公众号页面转化成小程序的 WXML 页面。

6.7 实训项目——基于数据库的注册与登录案例

视频讲解

在 6.3 节和 6.4 节通过小程序与 MySQL 数据库的连接，实现了新闻列表页和详情页的功能，本节继续以注册和登录为案例来熟悉小程序和后台数据库的交互。与新闻列表和详情页案例不同的是，注册的时候还需要将小程序提交的数据插入后台的 MySQL 数据库中。基本的步骤还是先建立 MySQL 的数据库和表格，然后写好 Java 的后端 JSON 接口文件，再在小程序端编写前端部分。程序执行效果如图 6.26～图 6.28 所示。

图 6.26　注册界面

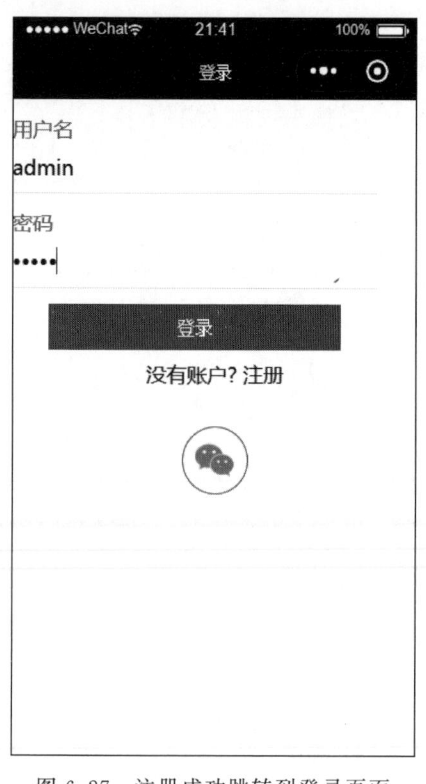

图 6.27　注册成功跳转到登录页面

(1) 创建数据库。

如图 6.29 所示，新建数据库 useradmin，并设置字符集为 utf8 -- UTF-8 Unicode，校对编码为 utf8_general_ci。单击“确定”按钮完成数据库的创建。

(2) 创建表格。

创建 user 表格，为了简单起见，user 表格中仅设计最简单的 name 和 password 两个字段，字段的定义如表 6.4 所示。在表格创建完毕之后，在 user 表格中插入 name 取值为 admin，password 取值也为 admin 的一行记录。

图 6.28 登录成功跳转到个人页面

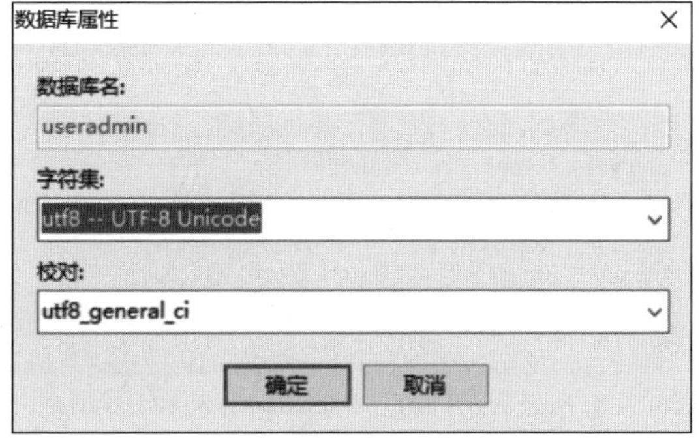

图 6.29 新建 useradmin 数据库

表 6.4 user 表

字　段	类　型	长　度	是否为空	是否为关键字
name	Varchar	10	否	是
password	Varchar	10	否	否

(3) 新建 Servlet 接口文件。

src/request/register.java 的代码如下：

```
package request;
import java.io.IOException;
import java.io.PrintWriter;
import java.sql.Connection;
import java.sql.DriverManager;
import java.sql.ResultSet;
import java.sql.Statement;
import javax.servlet.ServletException;
import javax.servlet.annotation.WebServlet;
import javax.servlet.http.HttpServlet;
import javax.servlet.http.HttpServletRequest;
import javax.servlet.http.HttpServletResponse;
import com.mysql.jdbc.PreparedStatement;
import net.sf.json.JSONArray;
import net.sf.json.JSONObject;
/**
 * Servlet implementation class register
```

```
 */
@WebServlet("/register")
public class register extends HttpServlet {
    private static final long serialVersionUID = 1L;
    /**
     * @see HttpServlet#HttpServlet()
     */
    public register() {
        super();
        //TODO Auto-generated constructor stub
    }
    /**
     * @see HttpServlet#doGet(HttpServletRequest request, HttpServletResponse response)
     */
    protected void doGet(HttpServletRequest request, HttpServletResponse response) throws
ServletException, IOException {
    response.setContentType("text/json; charset = utf - 8");
    PrintWriter out = response.getWriter();
    response.setHeader("Access - Control - Allow - Origin", " * ");
    response.setHeader("Access - Control - Allow - Methods", "GET,POST");
    String name1 = request.getParameter("name");
    String password1 = request.getParameter("password");
    System.out.println("账号是" + name1 + "密码是" + password1);  //验证数据传递情况
    try {
          Class.forName("com.mysql.jdbc.Driver");
          String url = "jdbc:mysql://localhost/useradmin?user = root&password =
          root&useUnicode = true&characterEncoding = UTF - 8";
          Connection con = DriverManager.getConnection(url);
        String sql = "insert into user(name,password) values(?,?)";
    PreparedStatement pst = (PreparedStatement) con.prepareStatement(sql);
        pst.setObject(1,name1);
       pst.setObject(2,password1);
       int jieguo = pst.executeUpdate();
       JSONArray jsonarray = new JSONArray();
       JSONObject jsonobj = new JSONObject();
       jsonobj.put("jieguo", jieguo);
       jsonarray.add(jsonobj);
       out = response.getWriter();
       out.println(jsonarray);
       pst.close();
       con.close();
       }catch (Exception e) {
       out.print("get data error!");
       e.printStackTrace(); }
    }
    /**
     * @see HttpServlet#doPost(HttpServletRequest request, HttpServletResponse response)
     */
    protected void doPost(HttpServletRequest request, HttpServletResponse response) throws
ServletException, IOException {
        this.doGet(request, response);
    }
}
```

src/request/login.java 的代码如下：

```
package request;
import java.io.IOException;
import java.io.PrintWriter;
import java.sql.Connection;
import java.sql.DriverManager;
import java.sql.ResultSet;
import java.sql.Statement;
import javax.servlet.ServletException;
import javax.servlet.annotation.WebServlet;
import javax.servlet.http.HttpServlet;
import javax.servlet.http.HttpServletRequest;
import javax.servlet.http.HttpServletResponse;
import net.sf.json.JSONArray;
import net.sf.json.JSONObject;
import com.mysql.jdbc.PreparedStatement;
/**
 * Servlet implementation class login
 */
@WebServlet("/login")
public class login extends HttpServlet {
    private static final long serialVersionUID = 1L;
    /**
     * @see HttpServlet#HttpServlet()
     */
    public login() {
        super();
        //TODO Auto-generated constructor stub
    }
    /**
     * @see HttpServlet#doGet(HttpServletRequest request, HttpServletResponse response)
     */
    protected void doGet(HttpServletRequest request, HttpServletResponse response) throws
ServletException, IOException {
        response.setContentType("text/json; charset=utf-8");
        PrintWriter out = response.getWriter();
        response.setHeader("Access-Control-Allow-Origin", "*");
        response.setHeader("Access-Control-Allow-Methods", "GET,POST");
        String name1 = request.getParameter("name");
        String password1 = request.getParameter("password");
        System.out.println("账号是" + name1 + "密码是" + password1); //验证数据传递情况
        try {
            Class.forName("com.mysql.jdbc.Driver");
            String url = "jdbc:mysql://localhost/useradmin?user=root&password=
             root&useUnicode=true&characterEncoding=UTF-8";
             Connection con = DriverManager.getConnection(url);
             Statement stmt = con.createStatement();
             String sql = "SELECT * FROM user where name='" + name1 + "'" + "
```

```
        and password = " + "'" + password1 + "'";
       ResultSet rs = stmt.executeQuery(sql);
       JSONArray jsonarray = new JSONArray();
       SONObject jsonobj = new JSONObject();
       while(rs.next()){
        jsonobj.put("jieguo",1);
       jsonarray.add(jsonobj);
       }
       out = response.getWriter();
      out.println(jsonarray);
     stmt.close();
      con.close();
     }catch (Exception e) {
     out.print("get data error!");
     e.printStackTrace(); }
  }
  /**
   * @see HttpServlet#doPost(HttpServletRequest request, HttpServletResponse response)
   */
  protected void doPost(HttpServletRequest request, HttpServletResponse response) throws
ServletException, IOException {
      this.doGet(request, response);
  }
}
```

【代码讲解】 register 和 login 两个 Servlet 和 6.4 节中的新闻列表页、详情页的 Servlet 的代码差不多,区别在于注册的时候需要从小程序端把账号和密码传递到 Servlet 中,在 register 这个 Servlet 中需要增加接收数据的代码。

```
String name1 = request.getParameter("name");
String password1 = request.getParameter("password");
```

上面两行语句是接收小程序传递过来的数据。另外,注册时需要将数据插入数据库中,所以使用了语句:

```
String sql = "insert into user(name,password) values(?,?)";
```

如果对相关代码不清楚,可以参考 JSP 程序设计的知识。

(4) 小程序部分。

pages/register/register.wxml 的代码如下:

```
<view class="zong">
<form report-submit  bindsubmit='dian'>
    <view class="login-item">
        <view class="login-item-info">用户名</view>
        <view><input name="name" /></view>
    </view>
    <view class="login-item">
        <view class="login-item-info">密码</view>
        <view class="login-pwd">
```

```
            < input style = "flex - grow:1" password = "true"   name = "password"/>
            </view >
        </view >
        < view class = "login - item bottom">
            < button class = "login - btn" form - type = 'submit'>注册</button >
        </view >
        </form >
    < view class = "item bottom - info">
        < view class = "wechat - icon">
            < image src = "/image/wechat.png"   />
        </view >
    </view >
</view >
```

pages/register/register.js 的代码如下：

```
var app = getApp();
Page({
  data: {
    username: null,
    password: null,
    jieguo:null
  },
  dian: function(e) {
   var that = this;
    that.username =  e.detail.value.name;
    that.password =  e.detail.value.password;
    console.log(that.username);
    wx.request({
      url: 'http://localhost:8080/mini/register',
      data: {
        name: that.username,
        password: that.password
      },
      header: {
        'content - type': 'application/json'                    //默认值
      },
      success: function(res) {
        console.log(res.data[0]);
        console.log(res.data[0].jieguo);
        that.jieguo = res.data[0].jieguo;
        if (that.jieguo == 1) { wx.redirectTo({ url: "../login/login" }) }
      },
      fail: function(res) {
        console.log(".....fail.....");
      }
    })
  },
  usernameInput: function(event) {
    console.log(event.detail.value);
    this.setData({ username: event.detail.value })
```

```
  },
  passwordInput: function(event) {
    this.setData({ password: event.detail.value })
  }
})
```

pages/register/register.wxss 的代码如下：

```
.mcontainer {
  height: 100%; display: flex;
  flex-direction: column; align-items: center;
  justify-content: space-between;
}
.item{
    flex-grow:1; height: 30%;
    overflow: hidden; display: flex;
    justify-content: center;
    align-items: center;
    flex-direction: column;
    width: 96%; padding: 0 2%;
}
.login-item{
    width: 90%; margin-top: 30rpx;
    border-bottom: 1px solid #eee;
    flex-grow:1;  display: flex;
    flex-direction: column;
    justify-content: flex-end;
    padding-bottom: 20rpx;
}
 .bottom{
     border-bottom: 0px;
}
.login-item-info{
    font-size: 16px; color: #888;
     padding-bottom: 20rpx;
}
.login-pwd{
    display: flex; flex-grow: 1;
    justify-content: space-between;
    align-items: center;
}
.login-pwd text{
    height: 100%; font-size: 14px;
    color: #888; display: flex;
}
.login-btn{
    width: 80%; color: white;
    background-color: green;
    border-radius: 0; font-size: 14px;
}
.login-btn:hover{
```

```
    width: 80 %; color: white;
    border - radius: 0;
}
.bottom - info{
}
.wechat - icon{
    margin - top: 30px; width: 30px;
    height: 30px;
    border: 1px solid seagreen;
    border - radius: 50 %; padding: 10px;
}
.wechat - icon image{
    width: 30px; height: 30px;
}
```

【代码讲解】 在 register. wxml 中需要将数据提交到 JS 文件中，此时< form >标签必须写 report-submit 属性，并且< button >标签的 form-type 需要设置为 submit，故有以下两行代码：

```
< form report - submit   bindsubmit = 'dian'>
< button class = "login - btn" form - type = 'submit'>注册</button >
```

register. wxml 文件提交数据之后，register. js 通过以下两行代码：

```
that.username =  e.detail.value.name;
that.password = e.detail.value.password;
```

来接收 register. wxml 的数据，再通过 wx. request()中的 data 对象传递给后端的 Servlet 文件。Servlet 处理完之后的结果又封装在 wx. request()方法的 res. data 的数组里面，因为是插入操作，结果只以"0"或者"1"标识，所以数组只有一个元素，即数据在 res. data[0]中。当返回的 jieguo==“1”说明注册成功，则通过 wx. redirectTo({ url: "../login/login" })跳转到 login 页面。

pages/login/login. wxml 的代码如下：

```
< view class = "zong">
< form report - submit   bindsubmit = 'dian'>
      < view class = "login - item">
           < view class = "login - item - info">用户名</view >
           < view >< input name = "name" /></view >
      </view >
      < view class = "login - item">
           < view class = "login - item - info">密码</view >
           < view class = "login - pwd">
           < input style = "flex - grow:1" password = "true"   name = "password"/>
           </view >
      </view >
      < view class = "login - item bottom">
           < button class = "login - btn" form - type = 'submit'>登录</button >
      </view >
      </form >
```

```
  <view class="item bottom-info">
    <view>没有账户?<text>注册</text></view>
    <view class="wechat-icon">
        <image src="/image/wechat.png" />
    </view>
  </view>
</view>
```

pages/login/login.js 的代码如下：

```
var app = getApp();
Page({
 data: {
   username: null,
   password: null,
   jieguo: null
 },
 dian: function(e) {
   var that = this;
   that.username = e.detail.value.name;
   that.password = e.detail.value.password;
   console.log(that.username);
   wx.request({
     url: 'http://localhost:8080/mini/login',
     data: {
       name: that.username,
       password: that.password
     },
     header: {
       'content-type': 'application/json'                    //默认值
     },
     success: function(res) {
       console.log(res.data[0]);
       console.log(res.data[0].jieguo);
       that.jieguo = res.data[0].jieguo;
       if (that.jieguo == 1) {
         wx.setStorageSync("username", that.username)
         wx.redirectTo({ url: "../user/user" }) }
     },
     fail: function(res) {
       console.log(".....fail.....");
     }
   })
 }
})
```

pages/login/login.wxss 的代码如下：

```
.mcontainer {
 height: 100%; display: flex;
 flex-direction: column;
```

```
 align - items: center;
 justify - content: space - between;
}
.item{
   flex - grow:1; height: 30 %;
   overflow: hidden; display: flex;
   justify - content: center;
   align - items: center;
   flex - direction: column;
   width: 96 %; padding: 0 2 %;
}
.login - item{
   width: 90 %;
   margin - top: 30rpx;
   border - bottom: 1px solid #eee;
   flex - grow:1; display: flex;
   flex - direction: column;
   justify - content: flex - end;
   padding - bottom: 20rpx;
}
 .bottom{
    border - bottom: 0px;
}
.login - item - info{
   font - size: 16px; color: #888;
   padding - bottom: 20rpx;
}
.login - pwd{
   display: flex; flex - grow: 1;
   justify - content: space - between;
   align - items: center;
}
.login - pwd text{
   height: 100 %; font - size: 14px;
   color: #888; display: flex;
}
.login - btn{
   width: 80 %; color: white;
   background - color: green;
   border - radius: 0; font - size: 14px;
}
.login - btn:hover{
   width: 80 %;  color: white;
   border - radius: 0;
}
.bottom - info{
```

```
}
.wechat-icon{
    margin-top: 30px; width: 30px;
    height: 30px;
    border: 1px solid seagreen;
    border-radius: 50%; padding: 10px;
}
.wechat-icon image{
    width: 30px; height: 30px;
}
```

【代码讲解】 后端的Servlet代码中，while(rs.next())说明存在账号和密码都正确的记录；"jsonobj.put("jieguo",1);"语句传回jieguo=1给小程序前端；wx.redirectTo({url:"../user/user"})跳转到登录成功的user页面。

pages/user/user.wxml的代码如下：

```
<view class="index-container">
    <image  src="/image/user.jpg"></image>
    <view class="nickname">
        {{username}}
    </view>
</view>
```

pages/user/user.js的代码如下：

```
var app = getApp();
Page({
  data:{
    username:null
  },
  onLoad:function(options){
    console.log((wx.getStorageSync("username")));
    this.setData({
      username:wx.getStorageSync("username")
    })
  }
})
```

pages/user/user.wxss的代码如下：

```
.index-container{
  width: 100%;
  height: 100%;
  position: relative;
}
.index-container image{
  width: 100%;
  height: 550px;
}
.nickname{
  position: absolute;
```

```
    top: 170px;
    left: 120px;
    font - size: 25px;
}
```

【代码讲解】 user. js 页面通过 wx. getStorageSync("username")获取上一个页面 login 设置的 username 数据,并通过数据绑定的方式在 user. wxml 中显示。

第 7 章

网络通信与文件上传下载操作

微信小程序常见的网络应用有 HTTPS 请求 wx.request、WebSocket 通信、上传文件 wx.uploadFile 和下载文件 wx.downloadFile，其中第 6 章的数据库操作部分已经对 wx.request 的用法进行了讲解，本章将对余下的三组网络应用进行讲解。

本章主要目标

- 了解微信小程序网络通信的应用场景；
- 了解小程序 WebSocket 的通信机制和常用接口；
- 熟练掌握微信小程序文件上传和下载操作；
- 通过网络相册实训项目加深对文件上传下载操作的理解与运用。

7.1 WebSocket

随着互联网的发展，传统的 HTTP 协议已经很难满足 Web 应用日益复杂的需求了。近年来，随着 HTML5 的诞生，WebSocket 协议被提出，它实现了浏览器与服务器的全双工通信，扩展了浏览器与服务端的通信功能，使服务端也能主动向客户端发送数据。

传统的 HTTP 协议是无状态的，每次请求（Request）都要由客户端（如浏览器）主动发起，服务端进行处理后返回 Response 结果。传统 Web 模式中客户端是主动方，服务端是被动方，服务端很难主动向客户端发送数据，对于信息变化不频繁的 Web 应用来说传统模式是足够应付的。对于涉及实时信息交互的 Web 应用，如带有即时通信、实时数据、订阅推送等功能的应用，上述模式带来了很大的不便。在 WebSocket 规范被提出之前，开发人员若要实现这些实时性较强的功能，经常会使用折中的解决方法：轮询（Polling）和 Comet 技术。后者本质上其实也是一种轮询，但有所改进。

轮询是最原始实现实时 Web 应用的解决方案。轮询技术要求客户端以设定的时间间隔周期性地向服务端发送请求，频繁地查询是否有新的数据改动。明显地，这种方法会导致过多不必要的请求，浪费流量和服务器资源。

Comet 技术可以分为长轮询和流技术。长轮询改进了上述的轮询技术，减少了无用的请求。它为某些数据设定过期时间，当数据过期后才会向服务端发送请求，这种机制适合数据改动不是特别频繁的情况。流技术通常是指客户端使用一个隐藏的窗口与服务端建立一个 HTTP 长连接，服务端会不断更新连接状态以保持 HTTP 长连接存活，据此，服务端就可以通过这条长连接主动将数据发送给客户端。

轮询和 Comet 技术都基于请求-应答模式，都不算是真正意义上的实时技术。它们的每一次请求、应答，在相同的头部信息上都浪费了一定流量，并且复杂度也较大。

伴随着 HTML5 推出的 WebSocket，真正实现了 Web 的实时通信，使 B/S 模式具备了 C/S 模式的实时通信能力。WebSocket 的工作流程是这样的：浏览器通过 JavaScript 向服务端发出建立 WebSocket 连接的请求，在 WebSocket 连接建立成功后，客户端和服务端就可以通过 TCP 连接传输数据。因为 WebSocket 连接本质上是 TCP 连接，所以不需要每次传输都带上重复的头部数据，它的数据传输量比轮询和 Comet 技术小很多。

7.1.1 WebSocket 接口

微信小程序的 WebSocket 功能提供了 wx. connectSocket()、wx. onSocketOpen()、wx. onSocketError()、wx. sendSocketMessage()、wx. onSocketMessage()、wx. closeSocket() 和 wx. onSocketClose()接口，具体功能如表 7.1 所示。

表 7.1 WebSocket 常用函数

接　口	功能和用途
wx. connectSocket()	创建 WebSocket 连接
wx. onSocketOpen()	监听 WebSocket 打开
wx. onSocketError()	监听 WebSocket 错误
wx. sendSocketMessage()	发送 WebSocket 消息
wx. onSocketMessage()	接收 WebSocket 消息
wx. closeSocket()	关闭 WebSocket 连接
wx. onSocketClose()	监听 WebSocket 关闭

因为篇幅的问题，本书只对其中主要的接口做说明，其他接口可以参考小程序官方文档。

1. wx. connectSocket(OBJECT)

基础库 1.7.0 之前，一个微信小程序在同一时刻只能创建一个 WebSocket 连接；如果当前已存在一个 WebSocket 连接，会自动关闭该连接，并重新创建一个 WebSocket 连接。基础库 1.7.0 及以后，微信小程序支持存在多个 WebSocket 连接，每次成功调用 wx. connectSocket()都会返回一个新的 SocketTask。wx. connectSocket() 接口的属性如表 7.2 所示。

表 7.2　wx.connectSocket()属性表

参　数	类　型	必填	说　明
url	String	是	开发者服务器接口地址，必须是 wss 或者 ws 协议，且域名必须是后台配置的合法域名
header	Object	否	HTTP Header，header 中不能设置 Referer
protocols	StringArray	否	子协议数组
success	Function	否	接口调用成功的回调函数
fail	Function	否	接口调用失败的回调函数
complete	Function	否	接口调用结束的回调函数(调用成功、失败都会执行)

示例代码：

```
wx.connectSocket({
  url: 'wss://example.qq.com',
  data:{
    x: '',
    y: ''
  },
  header:{
    'content-type': 'application/json'
  },
  protocols: ['protocol1'],
  method:"GET"
})
```

从示例代码中可以发现 wx.connectSocket()接口和 wx.request()接口很类似，重点需要注意的是接口地址 url 不能写 HTTP 或者 HTTPS 协议，而应写 WS 和 WSS 协议，这是 WebSocket 规范要求的。

2. wx.sendSocketMessage(OBJECT)

通过 WebSocket 连接发送数据，需要先调用 wx.connectSocket()接口，并在 wx.onSocketOpen()回调之后才能发送。wx.sendSocketMessage()接口的属性如表 7.3 所示。

表 7.3　wx.sendSocketMessage()属性表

参　数	类　型	必填	说　明
data	String | ArrayBuffer	是	需要发送的内容
success	Function	否	接口调用成功的回调函数
fail	Function	否	接口调用失败的回调函数
complete	Function	否	接口调用结束的回调函数(调用成功、失败都会执行)

示例代码：

```
var socketOpen = false
var socketMsgQueue = []
wx.connectSocket({
  url: 'test.php'
})

wx.onSocketOpen(function(res) {
```

```
    socketOpen = true
    for (var i = 0; i < socketMsgQueue.length; i++){
      sendSocketMessage(socketMsgQueue[i])
    }
    socketMsgQueue = []
  })

  function sendSocketMessage(msg) {
    if (socketOpen) {
      wx.sendSocketMessage({
        data:msg
      })
    } else {
      socketMsgQueue.push(msg)
    }
  }
```

一般来说，开发 WebSocket 项目通过 WXML 文件的点击事件来调用 wx.sendSocketMessage()，如代码中的 sendSocketMessage(msg)就是对 wx.sendSocketMessage()的调用。从示例代码中可以看出，要使用 wx.sendSocketMessage()接口，应该先使用 wx.connectSocket()和 wx.onSocketOpen()两个接口。

3. wx.onSocketMessage(CALLBACK)

wx.onSocketMessage()接口用来监听 WebSocket 从服务器接收到的消息事件。wx.onSocketMessage()接口的属性如表 7.4 所示。

表 7.4　wx.onSocketMessage()属性表

参　数	类　型	说　明
data	String|ArrayBuffer	服务器返回的消息

示例代码：

```
wx.connectSocket({
  url: 'test.php'
})
wx.onSocketMessage(function(res) {
  console.log('收到服务器内容：' + res.data)
})
```

7.1.2　基于 Node.js 的 WebSocket 案例

WebSocket 需要有后台程序的配合，而后台程序既可以采用 Tomcat 下的 Java 程序，也可以采用 Node.js 程序，本例中采用后者作为 WebSocket 的后台程序。简单地说，Node.js 就是运行在服务器端的 JavaScript。Node.js 是一个基于 Chrome JavaScript 运行时建立的平台，是一个事件驱动 I/O 服务端 JavaScript 环境，它基于 Google 的 V8 引擎，因为 V8 引擎执行 JavaScript 的速度非常快，所以性能非常好。

Node.js 的安装配置过程可以在网络中搜索到。本书附带的代码包中也提供了一个 Node.js 的 WebSocket 后台的 demo，在开发 WebSocket 小程序的时候，开发者需要开启这个 Node. js 的 WebSocket 后台程序，启动成功之后的界面如图 7.1 所示。

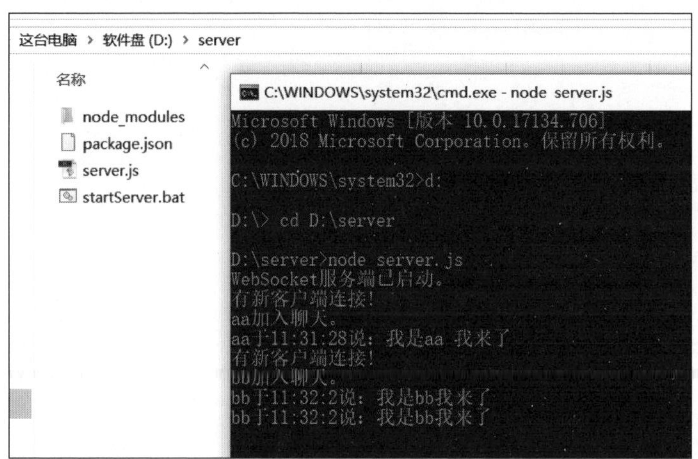

图 7.1　基于 Node. js 的 WebSocket 后台程序

视频讲解

例 7-1　WebSocket 实际上是一个多方通信接口，当小程序部署到服务器上的时候，WebSocket 允许多个手机访问后台服务器来建立通信。为了在一台计算机上演示通信过程，本例采用一个小程序和一个 HTML 页面来建立通信，具体效果如图 7.2 和图 7.3 所示。其中，图 7.2 是小程序，图 7.3 是 360 浏览器，它们访问同一个 Node. js 后端服务来建立连接。

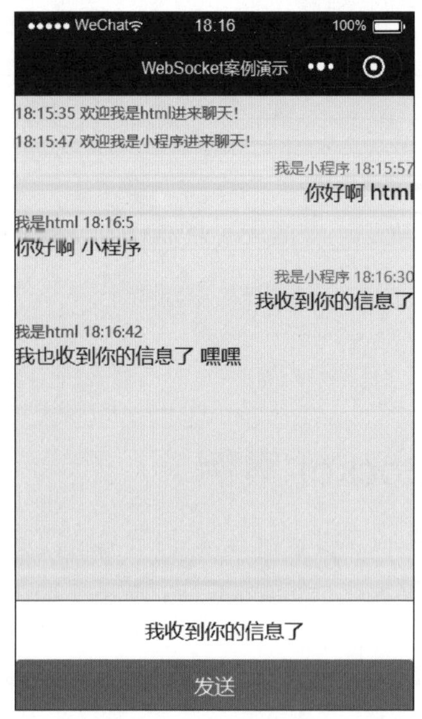

图 7.2　小程序和 HTML 建立连接的小程序端

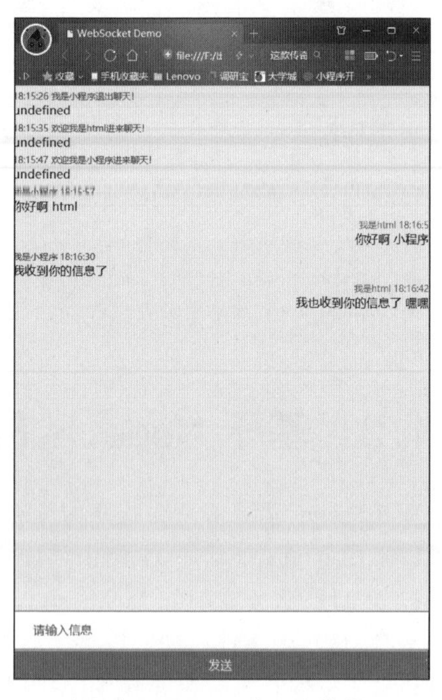

图 7.3　小程序和 HTML 建立连接的 HTML 端

pages/ websocket /websocket. wxml 的代码如下：

```
<view class="zong">
    <view class="shang">
        <view wx:for="{{messagelist}}" wx:for-index="idx" wx:for-item="itemName">
         <view class="selfMessage" wx:if="{{itemName.type == 'self'}}">
        <view class="nameInfo">{{itemName.name + " " + itemName.time}}</view>
                <view class="detailMessage">{{itemName.message}}</view>
            </view>
            <view class="otherMessage" wx:else>
         <view class="nameInfo">{{itemName.name + " " + itemName.time}}</view>
                <view class="detailMessage">{{itemName.message}}</view>
            </view>
        </view>
    </view>
    <view class="xia">
        <form report-submit  bindsubmit="fasong">
         <view class="inputArea">
    <input type="text"  name="inputValue" placeholder="{{placeholderText}}"
        class="message"/></view>
        <button size="default" type="primary" form-type="submit"
          class="sendButton">发送</button>
        </form>
    </view>
</view>
```

pages/websocket/websocket. js 的代码如下：

```
Page({
  data: {
    placeholderText: "连接后端服务器中……",
    messagelist: [],
    socketOpen: false,
  },
  onLoad: function(options) {
    var that = this;
    console.log("将要连接后端服务器。");
    wx.connectSocket({
      url: 'ws://localhost:8081'
    });
    wx.onSocketOpen(function(res) {
      console.log("连接后台服务器成功。");
      that.setData({
        placeholderText: "连接成功,请输入您的昵称。",
        socketOpen: true
      });
    });
    wx.onSocketMessage(function(res) {
      console.log('收到后端服务器内容: ' + res.data);
      var data = res.data;
      var dataArray = data.split("_");
```

```
        var newMessage = {
          type: dataArray[0],
          name: dataArray[1],
          time: dataArray[2],
          message: dataArray[3]
        };
        var newArray = that.data.messagelist.concat(newMessage);
        that.setData({
          messagelist: newArray,
          placeholderText: "请输入聊天信息"
        });
      });
    },
    onUnload: function() {
      wx.closeSocket();
    },
    fasong: function(e) {
      var that = this
      if (e.detail.value.inputValue != "") {
        if (this.data.socketOpen) {
          wx.sendSocketMessage({
            data: e.detail.value.inputValue
          })
        }
      }
    }
  });
```

【代码讲解】 WebSocket.wxml 文件通过< form report-submit bindsubmit="fasong">传递数据给 JS 文件,JS 文件通过点击函数 fasong: function(e)接收前台输入的数据 e.detail.value.inputValue,然后把 e.detail.value.inputValue 数据存到 wx.sendSocketMessage()的 data 参数中,再发送到后台的服务器端。wx.onSocketMessage()接口接收数据,接收到的数据可以是自己(聊天时候的自己)刚才发出去的,也可以是别的客户端(聊天的对方)发给服务器端的,所有的数据全部在 res.data 中。通过 console.log('收到服务器内容:' + res.data)来打印 res.data 的数据如图 7.4 所示。

图 7.4 Console 中打印显示 res.data 内容

图 7.4 在 Console 控制台打印显示了 wx.onSocketMessage()接口实时监控到的数据,接收到一条信息就自动更新 res.data 数据,再通过 self.data.messageArray.concat(newMessage)将每一个 res.data 数据连接起来形成数据数组 messageArray,然后在

WXML 通过 for 循环的方式遍历显示出来。

7.2 wx.uploadFile()文件上传

wx.uploadFile(OBJECT)接口将本地资源上传到开发者的服务器上，客户端发起一个 HTTPS 的 Post 请求，其中 content-type 为 multipart/form-data。在上传之前需要先获取本地(手机)上的资源，即使用 wx.uploadFile(OBJECT)之前应该先调用其他的接口来获取待上传的文件资源，例如先调用 wx.chooseImage()接口来获取到本地图片资源的临时文件路径，再通过 wx.uploadFile(OBJECT)接口将本地资源上传到指定服务器。wx.uploadFile()接口属性如表 7.5 所示。

表 7.5　wx.uploadFile(Object object)属性表

属　性	类　型	必填	说　　明
url	string	是	开发者服务器地址
filePath	string	是	要上传文件资源的路径
name	string	是	文件对应的 key，开发者在服务器端可以通过这个 key 获取文件的二进制内容
header	Object	否	HTTP 请求 Header，Header 中不能设置 Referer
formData	Object	否	HTTP 请求中其他额外的 form data
success	function	否	接口调用成功的回调函数
fail	function	否	接口调用失败的回调函数
complete	function	否	接口调用结束的回调函数(调用成功、失败都会执行)

示例代码：

```
wx.chooseImage({
 success(res) {
   const tempFilePaths = res.tempFilePaths
   wx.uploadFile({
    url: 'https://example.weixin.qq.com/upload',             //仅为示例，非真实的接口地址
     filePath: tempFilePaths[0],
     name: 'file',
     formData: {
       user: 'test'
     },
     success(res) {
       const data = res.data
       //do something
     }
   })
 }
})
```

7.2.1 文件上传后端

第6章中wx.request()从服务器上获取数据库的内容，服务器端需准备好读取MySQL数据库并形成小程序需要的JSON格式数据的接口。wx.uploadFile(OBJECT)接口实现小程序上传文件到服务器上，服务器端也应该有对应的接口来接收文件，本节采用Struts搭建一个后端的服务接口。Struts2是一个基于MVC设计模式的Web应用框架，它本质上相当于第6章的后端Servlet程序。在MVC设计模式中，Struts2作为控制器(Controller)来建立模型与视图的数据交互。对于不熟悉Struts的读者，可以查看Struts相关书籍。

例 7-2 文件上传操作的后端接口。图7.5是本例后端项目的项目结构图，程序运行效果如图7.6所示。

图7.5 文件上传后端项目结构图

视频讲解

图7.6 文件上传后端项目运行效果图

src/struts.xml的代码如下：

```
<!DOCTYPE struts PUBLIC
"-//Apache Software Foundation//DTD Struts Configuration 2.0//EN"
"http://struts.apache.org/dtds/struts-2.0.dtd">
<struts>
    <!-- Configuration for the default package. -->
    <constant name="struts.custom.i18n.resources" value="globalMessages" />
    <constant name="struts.i18n.encoding" value="utf-8" />
    <constant name="struts.multipart.saveDir" value="/tmp"/>
    <package name="I18N" extends="struts-default">
        <action name="upLoad" class="fileUpDown.UploadAction">
            <interceptor-ref name="fileUpload">
                <!-- 允许的 MIME 类型 -->
                <param name="allowedTypes">image/bmp,image/png,image
                  /gif,image/jpeg</param>
                <!-- 允许上传文件的最大尺寸 -->
```

```
                <param name="maximumSize">1024000</param>
            </interceptor-ref>
            <!-- 一定要写在后面 -->
            <interceptor-ref name="defaultStack"></interceptor-ref>
            <result name="input">/fileUp.jsp</result>
            <result name="success">/fileUpSuccess.jsp</result>
        </action>
    </package>
</struts>
```

src/fileUpDown/UploadAction.java 的代码如下：

```
package fileUpDown;
import com.opensymphony.xwork2.ActionSupport;
import java.io.File;
import java.io.FileInputStream;
import java.io.FileOutputStream;
import java.io.IOException;
import org.apache.commons.io.FileUtils;
import org.apache.struts2.ServletActionContext;
public class UploadAction extends ActionSupport{
private File file;
private String fileContentType;
private String fileFileName;
public File getFile() {
    return file;
}
public void setFile(File file) {
    this.file = file;
}
public String getFileContentType() {
    return fileContentType;
}
public void setFileContentType(String fileContentType) {
    this.fileContentType = fileContentType;
}
public String getFileFileName() {
    return fileFileName;
}
public void setFileFileName(String fileFileName) {
    this.fileFileName = fileFileName;
}
public String execute() throws Exception {
        //得到上传文件在服务器的路径加文件名
         String target = ServletActionContext.getServletContext().getRealPath
         ("/upload/" + fileFileName);
        //获得上传的文件
        File targetFile = new File(target);
         //通过 struts2 提供的 FileUtils 类复制
        try {
            FileUtils.copyFile(file, targetFile);
```

```
            } catch (IOException e) {
                e.printStackTrace();
            }
            return SUCCESS;
        }
    }
```

【代码讲解】 在 Struts.xml 文件中引用了 Struts 自带的 fileUpload 拦截器，对上传文件进行了限制，< param name="allowedTypes">设置了只能上传图片格式文件，< param name="maximumSize"> 1024000 </param >设置了上传文件的大小，不满足条件的上传操作将失败。因为 Struts 做了很多的功能封装，UploadAction.java 文件中的"FileUtils.copyFile(file, targetFile);"实现了把上传文件复制一份到指定的目录下的功能。

项目中有几个 JSP 文件是可以在 PC 端实现上传功能的，本例只讲解使用小程序作为客户端来上传文件。读者可以下载本书附带的此例源代码，以后用到文件上传功能就可以直接使用此项目作为后端接口。

7.2.2 文件上传前端

视频讲解

例 7-3 本例在例 7-2 的基础上进行前端小程序的编写，小程序端调用 wx.uploadFile()接口选择手机上的图片，然后上传到服务器上。小程序端运行效果如图 7.7 所示，服务器接收到的图片如图 7.8 所示。

图 7.7 小程序端运行效果

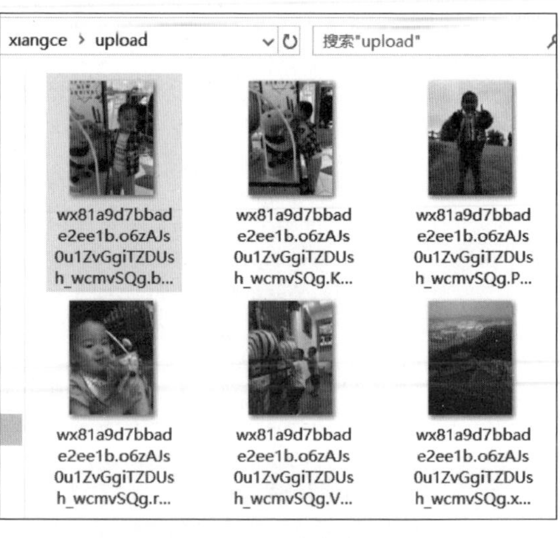

图 7.8 服务器接收到的图片

pages/uploadFile/uploadFile.wxml 的代码如下：

```
<view class = "container">
 <button bindtap = 'upfile'>选择上传文件</button>
</view>
```

pages/uploadFile/uploadFile.js 的代码如下：

```
var app  =  getApp()
Page({
  data: {
  },
  //事件处理函数
  upfile: function() {
    console.log(" -- bindViewTap -- ")
      wx.chooseImage({
        success: function(res) {
        var tempFilePaths  =  res.tempFilePaths
          wx.uploadFile({
            url: 'http://127.0.0.1:8080/upload/upLoad.action',
            header: { "Content - Type": "multipart/form - data" },
            filePath: tempFilePaths[0],
            name: 'file',
            formData:{
            }
          })
      }
    })
  },
  onLoad: function() {
  }
})
```

【代码讲解】 http://127.0.0.1:8080/upload/upLoad.action 是在例 7-2 中准备好的后端接口，因为是 Struts 项目，所以开发者可以直接访问 action。要上传的文件是通过 wx.chooseImage()在手机上选择的，关于 wx.chooseImage()接口，将在第 8 章中讲述。wx.chooseImage()把选择的图片保存在 tempFilePaths[0]中，因为图片存到 tempFilePaths[0]之后文件名已经变得不规则了，所以服务器接收到文件的名字也跟着变得不规则。图 7.8 显示的是服务器接收到的从手机上传过来的图片文件。

7.3　wx.downloadFile()文件下载

wx.downloadFile(Object object)下载文件资源到本地（手机）。客户端直接发起一个 HTTPS GET 请求，返回文件的本地临时路径。因为是临时路径，也就意味着用户不会知道真实的文件目录，所以下载到临时路径之后应该马上做后续的工作，例如把临时图片设置为头像，或者把临时文件通过别的接口真实保存到手机指定的目录下。wx.downloadFile

(Object object)参数如表 7.6 所示。

表 7.6 wx.downloadFile(Object object)参数表

属 性	类 型	必填	说 明
url	string	是	下载资源的 url
header	Object	否	HTTP 请求的 Header,Header 中不能设置 Referer
filePath	string	否	指定文件下载后存储的路径
success	function	否	接口调用成功的回调函数
fail	function	否	接口调用失败的回调函数
complete	function	否	接口调用结束的回调函数(调用成功、失败都会执行)

示例代码:

```
wx.downloadFile({
  url: 'https://example.com/audio/123',                    //仅为示例,并非真实的资源
  success(res) {
    //只要服务器有响应数据,就会把响应内容写入文件并进入 success 回调,需要自行判断是否下
    //载到了想要的内容
    if (res.statusCode == = 200) {
      wx.playVoice({
        filePath: res.tempFilePath
      })
    }
  }
})
```

视频讲解

例 7-4 本例实现两个功能:功能一实现把下载到的网络图片设置为头像,即图 7.9 中第一个头像本来是不存在的,点击了空白处之后才下载并设置了头像;功能二把图 7.9 中第二幅图(网络图片)下载到本地手机上保存,保存效果如图 7.10 所示。

pages/downloadFile/downloadFile.wxml 的代码如下:

```
<! -- index.wxml -- >
< view class = "container">
  < view   bindtap = "dian" class = "userinfo">
    < image class = "userinfo - avatar" src = "{{avatar}}" background - size = "cover"></image >
    < text class = "userinfo - nickname">{{userInfo.nickName}}</text >
  </view >
  < view class = "usermotto">
 < image src = 'https://ss3.bdstatic.com/70cFv8Sh_Q1YnxGkpoWK1HF6hhy
/it/u = 3018968254,2801372361&fm = 26&gp = 0.jpg' class = "tu"></image >
    < view bindtap = 'dian2'>下载上图</view >
  </view >
</view >
```

【代码讲解】 本例中有两个点击事件 dian 和 dian2,函数 dian 实现从网络下载图片并把该图片设置为头像(本来无头像),函数 dian2 实现从同一网络地址下载图片并把图片保存到手机相册中。

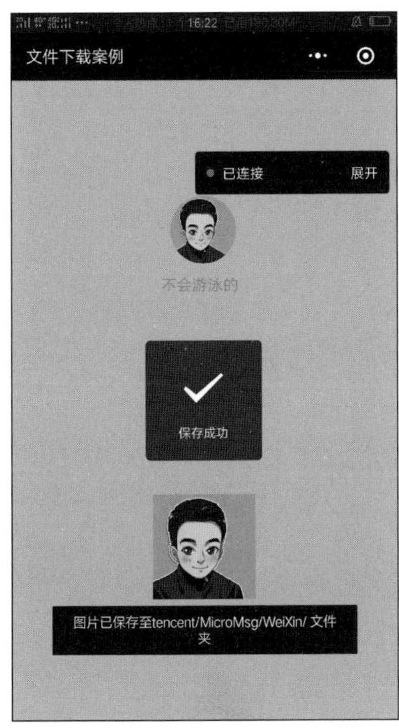

图 7.9　下载网络图片并设置头像真机调试界面

图 7.10　手机相册中已经成功保存图片

pages/downloadFile/downloadFile.js 的代码如下：

```
//downloadFile.js
var app = getApp()
Page({
  data: {
    motto: 'Hello World',
    userInfo: {},
    avatar:null
  },
  //事件处理函数
  dian: function() {
    console.log(" -- bindViewTap -- ")
    var that = this;
    wx.downloadFile({
        url: 'https://ss3.bdstatic.com/70cFv8Sh_Q1YnxGkpoWK1HF6hhy
        /it/u=3018968254,2801372361&fm=26&gp=0.jpg',
       type: 'image',
        success: function(res) {
          console.log(res)
         that.setData({avatar:res.tempFilePath})
        }
     })
  },
  onLoad: function() {
    console.log('onLoad')
```

```
        var that = this
        //调用应用实例的方法获取全局数据
        app.getUserInfo(function(userInfo){
          //更新数据
          that.setData({
            userInfo:userInfo
          })
        })
      },
      dian2: function() {
        wx.downloadFile({
          url: 'https://ss3.bdstatic.com/70cFv8Sh_Q1YnxGkpoWK1HF6hhy/it
         /u = 3018968254,2801372361&fm = 26&gp = 0.jpg',
          success: function(res) {
            console.log(res);
            var rr = res.tempFilePath;
            wx.saveImageToPhotosAlbum({
              filePath: rr,
              success(res) {
                wx.showToast({
                  title: '保存成功',
                  icon: 'success',
                  duration: 2000
                })
              }
            })
          }
        })
      }
    })
```

【代码讲解】 在函数 dian 中调用了 wx.downloadFile()接口,下载成功,图片就会保存在 res.tempFilePath 中,再把 res.tempFilePath 设置为头像。在函数 dian2 中,通过 wx.saveImageToPhotosAlbum()接口把下载成功的图片保存到手机相册。

7.4 实训项目——网络相册

视频讲解

现如今大家都喜欢把自己拍摄的照片上传到网络上,但是大部分人可能没有仔细思考过网络相册的本质是什么,网络相册其实就是图片的上传和存储,并在客户需要的时候把上传到服务器的图片引用显示。网络相册工作流程如图 7.11 所示。

问题是小程序的上传和引用没有大家想得这么简单,上传操作我们在 7.2 节已经实现了,但是引用呢?引用实际上是一个复杂的过程,它大致分为以下三个步骤:

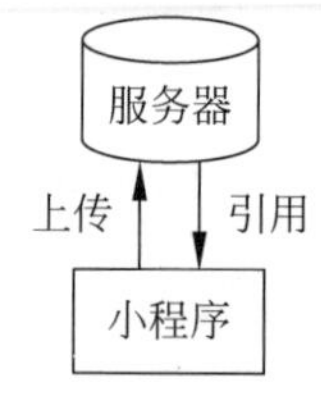

图 7.11　网络相册工作流程

(1) 读取服务器某个文件夹(每个客户一个文件夹)中所有的图片的文件名。

(2) 把读到的文件以 JSON 格式显示在一个接口文件中，如下是对图 7.8 中即上传到服务器的那些图片的 JSON 格式的返回结果，每一个大括号就是一个文件的地址。

[{"file":"http://127.0.0.1:8080/xiangce/upload/wx81a9d7bbade2ee1b.o6zAJs0u1ZvGgiTZDUsh_wcmvSQg.bYmSa6cz1Khw5b94278d919bc438c042ef13e5082cf2.jpg"},{"file":"http://127.0.0.1:8080/xiangce/upload/wx81a9d7bbade2ee1b.o6zAJs0u1ZvGgiTZDUsh_wcmvSQg.KYxfByl8SUwk029ba1250df59e9095970fb8219217fd.jpg"},{"file":"http://127.0.0.1:8080/xiangce/upload/wx81a9d7bbade2ee1b.o6zAJs0u1ZvGgiTZDUsh_wcmvSQg.Pob3GneuRu9Be852b44365c2f8c22b26e69146a3b4bc.jpg"},{"file":"http://127.0.0.1:8080/xiangce/upload/wx81a9d7bbade2ee1b.o6zAJs0u1ZvGgiTZDUsh_wcmvSQg.reYc8drJWnn4d8ed125c49a59922c59d8874a0d130b2.jpg"},{"file":"http://127.0.0.1:8080/xiangce/upload/wx81a9d7bbade2ee1b.o6zAJs0u1ZvGgiTZDUsh_wcmvSQg.V773hUbqjRgg5c835bf0c570195eab1ba4d167fa38fe.jpg"},{"file":"http://127.0.0.1:8080/xiangce/upload/wx81a9d7bbade2ee1b.o6zAJs0u1ZvGgiTZDUsh_wcmvSQg.xBenpV76xDCKb737abb3fc4b7989920fbd6a84f01c36.jpg"},{"file":"http://127.0.0.1:8080/xiangce/upload/wx81a9d7bbade2ee1b.o6zAJs0u1ZvGgiTZDUsh_wcmvSQg.zMav0IsjFtLKfc0f5b00fe67e31505fbc9398fc068b0.jpg"}]

(3) 在小程序中引用第(2)步得到的 JSON 格式的图片地址。

本项目难点在于不能假设已经知道类似 http://127.0.0.1:8080/xiangce/upload/1.jpg 的图片地址，这个地址是程序处理得到的。项目的运行效果如图 7.12 所示，后端项目结构如图 7.13 所示。

图 7.12　网络相册项目运行效果

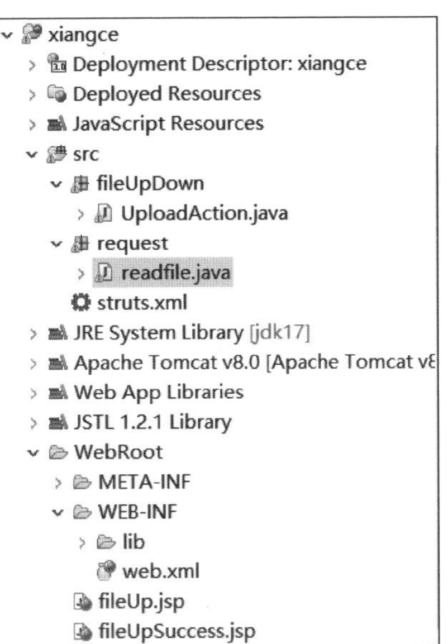

图 7.13　后端项目结构图

7.4.1 网络相册项目后端

网络后端分为上传文件的后端部分和获取文件夹下所有图片并以JSON格式返回这些文件的后端接口部分，它具体的项目结构图如图7.13所示。项目为Struts项目，UloadAction.java是Struts的Action，负责文件上传；readfile.java是一个Servlet，负责获取文件夹中的所有图片并形成JSON返回结果给小程序。

WebRoot/WEB-INF/web.xml代码如下：

```
<?xml version="1.0" encoding="UTF-8"?>
<web-app xmlns:xsi="http://www.w3.org/2001/XMLSchema-instance" xmlns="http://java.sun.com/xml/ns/javaee" xsi:schemaLocation="http://java.sun.com/xml/ns/javaee http://java.sun.com/xml/ns/javaee/web-app_3_0.xsd" id="WebApp_ID" version="3.0">
  <display-name>struts2_demo01</display-name>
  <welcome-file-list>
    <welcome-file>index.html</welcome-file>
    <welcome-file>index.htm</welcome-file>
    <welcome-file>index.jsp</welcome-file>
    <welcome-file>default.html</welcome-file>
    <welcome-file>default.htm</welcome-file>
    <welcome-file>default.jsp</welcome-file>
  </welcome-file-list>
  <filter>
    <filter-name>struts2</filter-name>
<filter-class>org.apache.struts2.dispatcher.ng.filter.StrutsPrepareAndExecuteFilter
</filter-class>
  </filter>
  <filter-mapping>
    <filter-name>struts2</filter-name>
    <url-pattern>/*</url-pattern>
  </filter-mapping>
  <servlet>
        <servlet-name>readfile</servlet-name>
        <servlet-class>request.readfile</servlet-class>
</servlet>
  <servlet-mapping>
      <servlet-name>readfile</servlet-name>
      <url-pattern>/readfile.servlet</url-pattern>
</servlet-mapping>
</web-app>
```

【代码讲解】 本例的难点之一是Struts的Action和Servlet的共存问题，Struts项目的uploadAction这个Action是用来接收文件的，readfile这个Servlet是用来返回一个文件夹中所有图片的JSON格式地址的。通常情况下Struts的Action和Servlet是不能共存的，可以通过上面web.xml的配置实现二者的共存问题。

Action和Servlet共存的核心代码：

```
<servlet>
```

```
        <servlet-name>readfile</servlet-name>
        <servlet-class>request.readfile</servlet-class>
</servlet>
<servlet-mapping>
    <servlet-name>readfile</servlet-name>
    <url-pattern>/readfile.servlet</url-pattern>
</servlet-mapping>
```

通常 Struts 项目中不能存在 Servlet 文件，要实现 Action 和 Servlet 共存，不但需要配置 Servlet，而且要注意使用< url-pattern >/readfile. servlet </url-pattern >来解决共存问题。

```
http://127.0.0.1:8080/xiangce/UploadAction                    (1)
http://127.0.0.1:8080/xiangce/UploadAction.action             (2)
http://127.0.0.1:8080/xiangce/readfile                        (3)
http://127.0.0.1:8080/xiangce/readfile.servlet                (4)
```

只要 Web 项目是 Struts 项目，就可以通过样式(1)和样式(2)的 url 来访问 Action；在非 Struts 的 Web 项目(JSP 项目)中可以通过样式(3)来访问 Servlet；在 Struts 项目中如果通过样式(3)的 url 来访问 Servlet，系统会把 Servlet 当成 Action 来处理，这时系统就找不到 Servlet 了，此时只能通过样式(4)的 url 来访问 Servlet。

src/struts. xml 代码如下：

```
<!DOCTYPE struts PUBLIC
"-//Apache Software Foundation//DTD Struts Configuration 2.0//EN"
"http://struts.apache.org/dtds/struts-2.0.dtd">
<struts>
    <!-- Configuration for the default package. -->
    <constant name="struts.custom.i18n.resources" value="globalMessages" />
    <constant name="struts.i18n.encoding" value="utf-8" />
    <constant name="struts.multipart.saveDir" value="/tmp"/>
    <package name="I18N" extends="struts-default">
        <action name="UploadAction" class="fileUpDown.UploadAction">
            <interceptor-ref name="fileUpload">
                <!-- 允许的 MIME 类型 -->
                <param name="allowedTypes">image/bmp,image/png,image/gif,
                  image/jpeg</param>
                <!-- 允许上传文件的最大尺寸 -->
                <param name="maximumSize">1024000</param>
            </interceptor-ref>
            <!-- 一定要写在后面 -->
            <interceptor-ref name="defaultStack"></interceptor-ref>
            <result name="input">/fileUp.jsp</result>
            <result name="success">/fileUpSuccess.jsp</result>
        </action>
    </package>
</struts>
```

【代码讲解】 本例中的 Struts. xml 和 7.2 节中的 Struts. xml 是同一个 Struts. xml，是对 Action 的配置，其中限制了上传文件的格式必须是图片，大小不能超过 1024KB。

src/fileUpDown/UploadAction.java 代码如下：

```
package fileUpDown;
import com.opensymphony.xwork2.ActionSupport;
import java.io.File;
import java.io.FileInputStream;
import java.io.FileOutputStream;
import java.io.IOException;
import org.apache.commons.io.FileUtils;
import org.apache.struts2.ServletActionContext;
public class UploadAction extends ActionSupport{
private File file;
private String fileContentType;
private String fileFileName;
public File getFile() {
    return file;
}
public void setFile(File file) {
    this.file = file;
}
public String getFileContentType() {
    return fileContentType;
}
public void setFileContentType(String fileContentType) {
    this.fileContentType = fileContentType;
}
public String getFileFileName() {
    return fileFileName;
}
public void setFileFileName(String fileFileName) {
    this.fileFileName = fileFileName;
}
public String execute() throws Exception {
        //得到上传文件在服务器的路径加文件名
         String target = ServletActionContext.getServletContext().getRealPath("/upload/" +
fileFileName);
        //获得上传的文件
        File targetFile = new File(target);
         //通过 struts2 提供的 FileUtils 类复制
        try {
            FileUtils.copyFile(file, targetFile);
         } catch (IOException e) {
             e.printStackTrace();
         }
         return SUCCESS;
    }

}
```

【代码讲解】 语句"ServletActionContext.getServletContext().getRealPath("/upload/");"是获取项目的路径然后在根目录下新建一个文件夹 upload，这个文件夹可以理解为，不同

的客户新建一个不同的文件夹，用来保存客户上传的文件。

src/request/readfile.java 代码如下：

```
package request;
import java.io.File;
import java.io.IOException;
import java.io.PrintWriter;
import java.sql.Connection;
import java.sql.DriverManager;
import java.sql.ResultSet;
import java.sql.Statement;
import javax.servlet.ServletException;
import javax.servlet.annotation.WebServlet;
import javax.servlet.http.HttpServlet;
import javax.servlet.http.HttpServletRequest;
import javax.servlet.http.HttpServletResponse;
import org.apache.struts2.ServletActionContext;
import net.sf.json.JSONArray;
import net.sf.json.JSONObject;
/**
 * Servlet implementation class readfile
 */
public class readfile extends HttpServlet {
    private static final long serialVersionUID = 1L;
    /**
     * @see HttpServlet#HttpServlet()
     */
    public readfile() {
        super();
        //TODO Auto-generated constructor stub
    }
    /**
     * @see HttpServlet#doGet(HttpServletRequest request, HttpServletResponse response)
     */
    protected void doGet(HttpServletRequest request,
HttpServletResponse response) throws ServletException, IOException
     { System.out.print("11111");
        response.setContentType("text/json; charset=utf-8");
        PrintWriter out = response.getWriter();
        String target = request.getSession().getServletContext().
         getRealPath("\\") + "/upload/";
        System.out.print(target);
        File f = new File(target);
        JSONArray jsonarray = new JSONArray();
        if (!f.exists())
          {
           out.println("查无文件");
           return;
          }
        File fa[] = f.listFiles();
```

```
            for(int i = 0;i < fa.length;i++)
            {
              File fs = fa[i];
              System.out.print(fs.getName());
              System.out.print("开始" + request.getContextPath() + "/"
               + fs.getName());
               JSONObject jsonobj = new JSONObject();
              jsonobj.put("file", http://127.0.0.1:8080
            + request.getContextPath() + "/upload/" + fs.getName());
                    jsonarray.add(jsonobj);
                }
                out = response.getWriter();
                out.println(jsonarray);
        }
        /**
          * @see HttpServlet#doPost(HttpServletRequest request, HttpServletResponse response)
          */
        protected void doPost(HttpServletRequest request, HttpServletResponse response) throws
ServletException, IOException {
            this.doGet(request, response);
        }
}
```

【代码讲解】 语句

```
File f = new File(target);
File fa[] = f.listFiles();
```

用来获取文件夹下所有文件，这时 fa[]中的每一个元素是类似 D:\apache-tomcat-8.0.3\webapps\xiangce\upload\1.jpg 格式的文件全路径，通过 fs.getName()获取了类似“1.jpg”格式的文件名，然后通过和 http://127.0.0.1:8080+request.getContextPath()+"/upload/"进行相加，得到类似 http://127.0.0.1:8080/xiangce/upload/1.jpg 的可以访问的网络地址，最后用以下两条语句：

```
jsonobj.put("file", filepath);
jsonarray.add(jsonobj);
```

得到小程序可以访问的 JSON 格式的图片地址库，执行效果如图 7.14 所示。readfile.java 是本项目中最难的逻辑部分，需好好理解。

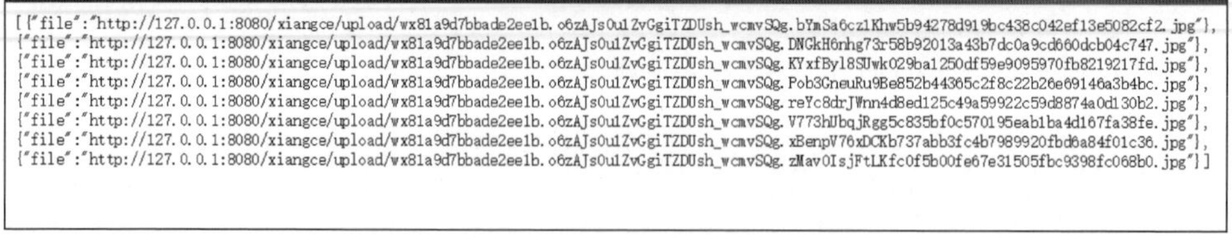

127.0.0.1:8080/xiangce/readfile
http://127.0.0.1:8080/xiangce/readfile.servlet

[{"file":"http://127.0.0.1:8080/xiangce/upload/wx81a9d7bbade2ee1b.o6zAJs0u1ZvGgiTZDUsh_wcmvSQg.bYmSa6cz1Khw5b94278d919bc438c042ef13e5082cf2.jpg"},
{"file":"http://127.0.0.1:8080/xiangce/upload/wx81a9d7bbade2ee1b.o6zAJs0u1ZvGgiTZDUsh_wcmvSQg.DNGkH6nhg73r58b92013a43b7dc0a9cd660dcb04c747.jpg"},
{"file":"http://127.0.0.1:8080/xiangce/upload/wx81a9d7bbade2ee1b.o6zAJs0u1ZvGgiTZDUsh_wcmvSQg.KYxfBy18SUwk029ba1250df59e9095970fb8219217fd.jpg"},
{"file":"http://127.0.0.1:8080/xiangce/upload/wx81a9d7bbade2ee1b.o6zAJs0u1ZvGgiTZDUsh_wcmvSQg.Pob3GneuRu9Be852b44365c2f8c22b26e69146a3b4bc.jpg"},
{"file":"http://127.0.0.1:8080/xiangce/upload/wx81a9d7bbade2ee1b.o6zAJs0u1ZvGgiTZDUsh_wcmvSQg.reYc8drJWnn4d8ed125c49a59922c59d8874a0d130b2.jpg"},
{"file":"http://127.0.0.1:8080/xiangce/upload/wx81a9d7bbade2ee1b.o6zAJs0u1ZvGgiTZDUsh_wcmvSQg.V773hUbqjRgg5c835bf0c570195eab1ba4d167fa38fe.jpg"},
{"file":"http://127.0.0.1:8080/xiangce/upload/wx81a9d7bbade2ee1b.o6zAJs0u1ZvGgiTZDUsh_wcmvSQg.xBenpV76xDCKb737abb3fc4b7989920fbd6a84f01c36.jpg"},
{"file":"http://127.0.0.1:8080/xiangce/upload/wx81a9d7bbade2ee1b.o6zAJs0u1ZvGgiTZDUsh_wcmvSQg.zMav0IsjFtLKfc0f5b00fe67e31505fbc9398fc068b0.jpg"}]

图 7.14 小程序获取到的 JSON 格式的图片地址

7.4.2 网络相册项目前端

src/photo/photo.wxml 代码如下：

```
<view class = "all">
<scroll - view scroll - y = "true" class = "body" style = "height:{{scrollHeight}}px;">
    <image bindtap = "uploadPhoto" style = "width:{{imageWidth}}rpx;height:{{imageHeight}}
rpx;" mode = "aspectFill" class = "plus - image" src = "../../image/plus.png" />
    <view class = "photo - list" wx:for = "{{photoList}}" wx:key = " * this" wx:for - item =
"imageSrc">
            < image bindtap = " previewPhoto" style = " width: {{ imageWidth}} rpx; height:
{{imageHeight}}rpx;" mode = "aspectFill" src = "{{imageSrc.file}}"/>
    </view>
</scroll - view>
<view class = "bottom">
    <button type = "default" bindtap = "two">两列</button>
    <button type = "default" bindtap = "three">三列</button>
</view>
</view>
```

【代码讲解】 语句“wx:for="{{photoList}}"”遍历图 7.14 获取到的 JSON 数值，每一次循环就读取到一个大括号，因为大括号中的 file 属性是 JS 文件需要的，imageSrc.file 就是网络图片的真实可访问地址，形如 http://127.0.0.1:8080/xiangce/upload/1.jpg。

src/photo/photo.js 代码如下：

```
var app = getApp();
Page({
    data:{
        scrollHeight:569,
        imageWidth:250,
        imageHeight:250,
        photoList:[]
    },
    uploadPhoto:function(){
    console.log("从本地选取图片");
    var that = this;
    var showFileList = that.data.photoList;
    wx.chooseImage({
    count: 9,                          //最多可以选择的图片张数，默认为 9
    sizeType: ['original','compressed'], //original: 原图,compressed: 压缩图
        sourceType: ['album', 'camera'], //album: 从相册选图,camera: 使用相机
          success: function(res){
            //后台慢慢保存文件到本地
            var tempFilePaths = res.tempFilePaths
            wx.uploadFile({
              url: 'http://127.0.0.1:8080/xiangce/UploadAction.action',
              header: { "Content - Type": "multipart/form - data" },
              filePath: tempFilePaths[0],
```

```
            name: 'file',
            formData: {
            }
          })
        }
      });
    },
    previewPhoto:function(el){
        console.log(el);
        var curTarget = el.target.dataset.src
        console.log(curTarget);
        wx.previewImage({
          current: curTarget,              //当前显示图片的链接,不填则默认为 urls 的第一张
          urls: this.data.photoList,
          success: function(res){
            //success
            console.log(res);
          }
        })
    },
    two:function(){
        var that = this;
        var length = 750/2;
        that.setData({
            imageWidth:length,
            imageHeight:length
        });
    },
    three:function(){
        var that = this;
        var length = 750/3;
        that.setData({
            imageWidth:length,
            imageHeight:length
        });
    },
    onLoad:function(){
        console.log("加载图片列表");
        var that = this;
        //获取当前窗口高度,以便设置 scrollView 的高度
        wx.getSystemInfo({
          success: function(res) {
            console.log(res.model);
            console.log(res.pixelRatio);
            console.log(res.windowWidth);
            console.log(res.windowHeight);
            console.log(res.version);
            that.setData({
                scrollHeight:res.windowHeight
            });
          }
```

```
      })
      wx.request({
        url: 'http://127.0.0.1:8080/xiangce/readfile.servlet',
          //仅为示例,并非真实的接口地址
        data: {
          x: '',
          y: ''
        },
        header: {
          'content - type': 'application/json'    //默认值
        },
        success(res) {
          console.log(res)
          that.setData({
            photoList: res.data
          })
        }
      })
  },
  onShow: function() {
  }
})
```

【代码讲解】 函数 upPhoto: function()实现图片的上传功能,它先调用 wx.chooseImage()接口去相册选择图片或者以直接拍照的方式获取本地手机图片,sourceType: ['album', 'camera']指可以选择相册图片,也可以即时拍照; wx.request()读取 Servlet 接口的 JSON 地址库,然后通过 photoList: res.data 赋值给 photoList; 函数 two 和 three 实现了相册列数的控制; previewPhoto:function(el)函数实现了具体一张图片的预览功能。

src/photo/photo.wxss 代码如下:

```
.all {
  height: 100%;
  background - color: #eee;
  padding:0;
  margin:4rpx 0 0 0;
  box - sizing: border - box;
}
.body{
  margin - bottom: 4rpx;
}
.plus - image{
  width:250rpx;
  height:250rpx;
}
.photo - list{
  display: inline - block;
}
.photo - list image {
```

```
    width:250rpx;
    height: 250rpx;
}
.bottom{
  width:100%;
  border-top:2rpx gray solid;
  display: flex;
  flex-direction: row;
  flex-wrap: nowrap;
  align-items: center;
  bottom: 0;
}
.bottom button {
  width: 100%;
  border-radius: 0;
}
```

第 8 章

媒体与设备操作

小程序提供了大量和手机硬件相关的操作接口，其中最主要的就是媒体和设备操作。本节将介绍地图、位置、图片、视频、录音、音频播放控制、背景音乐、获取系统信息、网络环境和电量等接口。微信小程序官方 2019 年对其中一些接口进行了更新，这也间接反映出这些接口在小程序开发中的重要地位。

本章主要目标

- 了解微信小程序常用的媒体和设备接口；
- 熟练掌握地图、位置、图片、视频、录音、音频播放控制、背景音乐、获取系统信息、网络环境和电量等接口操作；
- 使用 AudioContext 对象完成音乐播放器的制作。

8.1 地图与位置

小程序经常会用到地图和位置功能，这两个功能都以经度和纬度为基础，是相关性很大的操作，所以本节将它们安排在一起来讲，读者需重点理解 map 组件和 wx.openLocation() 接口的区别。

8.1.1 地图

map 是一个比较复杂的组件，它有很多参数，如表 8.1 所示。

表 8.1 map 组件参数表

属 性	类 型	必填	说 明
longitude	number	是	中心经度
latitude	number	是	中心纬度

续表

属　　性	类　　型	必填	说　　明
scale	number	否	缩放级别，取值范围为 5～18
markers	Array.<marker>	否	标记点
covers	Array.<cover>	否	即将移除，请使用 markers
polyline	Array.<polyline>	否	路线
circles	Array.<circle>	否	圆
controls	Array.<control>	否	控件(即将废弃，建议使用 cover-view 代替)
include-points	Array.<point>	否	缩放视野以包含所有给定的坐标点
show-location	boolean	否	显示带有方向的当前定位点
polygons	Array.<polygon>	否	多边形

map 组件的两个属性 longitude 和 latitude 表示当前地图中心的经度和纬度，和当前用户所在位置的经度和纬度是不同的概念，无直接关系。例如，某人在广东省东莞市，但是可以打开以北京天安门为中心的一幅地图，map 的 longitude 和 latitude 是用来控制地图中心的参数，并不是用户实时的地理位置。

例 8-1　map 组件显示地图，程序运行效果如图 8.1 所示。

视频讲解

图 8.1　map 运行效果

pages/map/map. wxml 的代码如下：

```
< map id = "tu" latitude = '23.099' longitude = '113.325'   scale = '15' class = "tu" controls =
'{{con}}' markers = '{{mar}}' bindcontroltap = 'con' bindmarkertap = 'mar'>
</map >
```

pages/map/map. js 的代码如下：

```
Page({
data: {
mar:[{
inconPath:"location.png",
id:0,
latitude:23.088994,
longitude:113.324520,
width:50,
heigth:50
    }
    ]
  },
con:[{
id:1,
    iconPath:'location.png',
positon:{left:0,
top:50,
width:50,
height:50},
clickable:true
  }],
  mar:function(e)
  {console.log("你点了标记点")},

  con: function (e) { console.log("你点了游标") }
})
```

map 只能简单地生成一幅地图，要对地图进行某些操作，如进行缩放和移动操作，开发者必须先在 JS 中获取 MapContext 对象，这时需要通过 wx. createMapContext('id')获取 MapContext 对象。读者可以理解为 wx. createMapContext('id')就是指向地图的一个指针。MapContext 对象常用操作如表 8.2 所示。

表 8.2　MapContext 对象常用操作函数

接　　口	功能和用途
MapContext. getCenterLocation()	获取当前地图中心的经度和纬度。返回的是 gcj02 坐标系，可以用于 wx. openLocation()
MapContext. moveToLocation()	将地图中心移动到当前定位点。需要配合 map 组件的 show-location 使用
MapContext. translateMarker(Object object)	平移 marker，带动画
MapContext. includePoints(Object object)	缩放视野展示所有经度和纬度
MapContext. getRegion()	获取当前地图的视野范围
MapContext. getScale()	获取当前地图的缩放级别

需要说明的是 MapContext.getRegion()接口获取图片的范围,即是经度和纬度的取值范围,取值范围是以地图的西南和东北两个顶点的经度和纬度来限定的。MapContext.translateMarker()和 MapContext.includePoints()两个接口中需要用到的经度和纬度不能超出 MapContext.getRegion()接口的经度和纬度取值范围。例 8-2 有助于理解这三个接口的含义。

视频讲解

例 8-2 MapContext 实例的常见操作,本例将对地图做获取地图中心经度和纬度、移动到当前位置、获取缩放比例等操作。程序的运行效果如图 8.2 所示。

图 8.2 Mapcontext 运行效果

pages/MapContext/MapContext.wxml 的代码如下：

```
<map
  id="ditu"
  style="width: 710rpx; height: 250px;"
  latitude="{{latitude}}"
  longitude="{{longitude}}"
  markers="{{markers}}"
  show-location>
</map>
<button bindtap="getCenterLocation" type="primary">获取地图中心的经度和纬度</button>
<button bindtap="moveToLocation"  type="primary">将地图中心移动到当前定位点</button>
<button bindtap="translateMarker" type="primary">平移 marker,带动画</button>
<button bindtap="includePoints" type="primary">缩放视野展示所有经度和纬度</button>
```

```
<button bindtap="scaleClick" type="primary">获取当前地图的缩放级别</button>
<button bindtap="getRegionClick" type="primary">获取当前地图的视野范围</button>
```

pages/MapContext/MapContext.js 的代码如下：

```
Page({
  data: {
    latitude: 22.557416086996245,
    longitude: 113.3832685578842,
    markers: [{
      id: 1,
      latitude: 22.557416086996245,
      longitude: 113.3832685578842,
      name: '中山北站'
    }],
  },
  onReady: function(e) {
    //创建 map 上下文 MapContext 对象
    this.zhizhen = wx.createMapContext('ditu')
  },
  //获取当前地图中心的经度和纬度
  getCenterLocation: function() {
    this.zhizhen.getCenterLocation({
      success: function(res) {
        console.log(res.longitude)
        console.log(res.latitude)
      }
    })
  },
  //将地图中心移动到当前定位点
  moveToLocation: function() {
    this.zhizhen.moveToLocation()
  },
  //平移 marker,带动画
  translateMarker: function() {
    this.zhizhen.translateMarker({
      markerId: 1,
      autoRotate: true,
      duration: 1000,
      destination: {
        latitude: 22.55229,
        longitude: 113.3845211,
      },
      animationEnd() {
        console.log('animation end')
      }
    })
  },
```

```
    //缩放视野展示所有经度和纬度
    includePoints: function() {
      this.zhizhen.includePoints({
        padding: [10],
        points: [{
          latitude: 22.54229,
          longitude: 113.3745211,
        }, {
            latitude: 22.55229,
            longitude: 113.3845211,
        }]
      })
    },
    //获取当前地图的缩放级别
    scaleClick: function() {
      this.zhizhen.getScale({
        success: function(res) {
          console.log(res.scale)
        }
      })
    },
    //获取当前地图的视野范围
    getRegionClick: function() {
      this.zhizhen.getRegion({
        success: function(res) {
          console.log(res.southwest)
          console.log(res.northeast)
        }
      })
    }
  })
```

【代码讲解】 this. zhizhen = wx. createMapContext('ditu')获取 MapContext 实例，此时在 JS 文件中，this. zhizhen 就代表了那幅地图，然后通过类似 this. zhizhen. getCenterLocation()和 this. zhizhen. moveToLocation()接口就可以对地图进行各种相关操作了。

通过控制台打印 this. zhizhen. getRegion()的经度和纬度取值，本例中的取值范围为(22. 547507698481496，113. 37030812392423)～(22. 56732376379575，113. 39622899184415)，this. zhizhen. includePoints()和 this. zhizhen. translateMarker()两个接口的经度和纬度取值范围就应该在这个区间里面。

8.1.2 位置

小程序提供了 3 个接口对位置进行操作：wx. getLocation(Object object)接口获取当前的地理位置；wx. openLocation(Object object)接口实现使用微信内置地图查看位置功能；wx. chooseLocation(Object object)接口选择一个位置来打开地图。

1. wx.getLocation(Object object)接口

常用的 wx.getLocation(Object object)接口的属性如表 8.3 所示。

表 8.3　wx.getLocation(Object object)属性表

属　性	类　型	必填	说　明
type	string	否	wgs84 返回 gps 坐标，gcj02 返回可用于 wx.openLocation()的坐标
altitude	string	否	传入 true 会返回高度信息，由于获取高度需要较高精确度，所以会减慢接口返回速度
success	function	否	接口调用成功的回调函数
fail	function	否	接口调用失败的回调函数
complete	function	否	接口调用结束的回调函数(调用成功、失败都会执行)

示例代码：

```
wx.getLocation({
  type: 'wgs84',
  success(res) {
    const latitude = res.latitude
    const longitude = res.longitude
    const speed = res.speed
    const accuracy = res.accuracy
  }
})
```

从示例代码可以发现，wx.getLocation()接口只是获取当前位置的经度和纬度。

2. wx.openLocation(Object object)接口

常用的 wx.openLocation(Object object)接口的属性如表 8.4 所示。

表 8.4　wx.openLocation(Object object)属性表

属　性	类　型	必填	说　明
latitude	number	是	纬度，范围为－90°～90°，负数表示南纬，使用 gcj02 国测局坐标系
longitude	number	是	经度，范围为－180°～180°，负数表示西经，使用 gcj02 国测局坐标系
scale	number	否	缩放比例，范围为 5～18
name	string	否	位置名
address	string	否	地址的详细说明
success	function	否	接口调用成功的回调函数
fail	function	否	接口调用失败的回调函数
complete	function	否	接口调用结束的回调函数(调用成功、失败都会执行)

示例代码：

```
wx.getLocation({
  type: 'gcj02',        //返回可以用于 wx.openLocation()的经度和纬度
  success(res) {
    const latitude = res.latitude
    const longitude = res.longitude
    wx.openLocation({
      latitude,
```

```
      longitude,
      scale: 18
    })
  }
})
```

从示例代码可以发现，wx.openLocation()接口是打开一个位置，它需要经度值和纬度值作为打开位置的参数。wx.getLocation()接口是获取用户当前所在位置的经度和纬度，可以作为wx.openLocation()接口的参数，当然也可以指定一个经度和纬度来打开位置地图。

例 8-3 wx.openLocation()接口的应用案例，程序运行效果如图8.3所示。

视频讲解

图 8.3 openLocation 运行效果

pages/openLocation/openLocation.wxml 的代码如下：

```
<view class = "container">
  <view  bindtap = "bindViewTap" class = "userinfo">
  <image class = "userinfo - avatar" src = "{{userInfo.avatarUrl}}"
   background - size = "cover"></image>
    <text class = "userinfo - nickname">{{userInfo.nickName}}</text>
  </view>
  <view class = "usermotto">
    <text class = "user - motto">{{motto}}</text>
  </view>
</view>
```

【代码讲解】 本例中 WXML 文件不起作用，因为 JS 代码写在了 onload 函数中，会自动加载 wx. openLacation()接口。

pages/openLocation/openLocation. wxml 的代码如下：

```
//获取应用实例
var app = getApp()
Page({
  data: {
    motto: 'Hello World',
    userInfo: {}
  },
  //事件处理函数
  bindViewTap: function() {
    wx.navigateTo({
      url: '../logs/logs'
    })
  },
  onLoad: function() {
    console.log('onLoad')
    var that = this
    //调用应用实例的方法获取全局数据
    app.getUserInfo(function(userInfo) {
      //更新数据
      that.setData({
        userInfo: userInfo
      })
    })
    wx.getLocation({
      type: 'gcj02',  //返回可以用于 wx.openLocation 的经度和纬度
      success(res) {
        const latitude = res.latitude
        const longitude = res.longitude
        wx.openLocation({
          latitude,
          longitude,
          scale: 18
        })
      }
    })
  }
})
```

【代码讲解】 本例先用 wx. getLoaction()接口获取经度和纬度，再用 wx. openLoaction()接口使用 wx. getLocation()接口获取的经度和纬度打开位置地图。

3. wx. chooseLocation(Object object)接口

常用的 wx. chooseLocation(Object object)的属性如表 8. 5 所示。

表 8.5　wx. chooseLocation(Object object)属性表

属　性	类　型	必填	说　明
success	function	否	接口调用成功的回调函数
fail	function	否	接口调用失败的回调函数
complete	function	否	接口调用结束的回调函数(调用成功、失败都会执行)

wx. chooseLocation(Object object)接口是选择一个地理位置，请注意区分 wx. chooseLocation(Object object)接口与其他接口的不同点。

例 8-4　wx. chooseLocation(Object object)接口的应用案例，程序运行效果如图 8.4 所示。

视频讲解

图 8.4　chooseLocation 运行效果

pages/chooseLocation/chooseLocation. js 的代码如下：

```
//获取应用实例
var app = getApp()
Page({
  data: {
    motto: 'Hello World',
    userInfo: {}
  },
  //事件处理函数
  bindViewTap: function() {
    wx.navigateTo({
      url: '../openLocation/openLocation'
    })
```

```
  },
  onLoad: function() {
    console.log('onLoad')
    var that = this
    //调用应用实例的方法获取全局数据
    app.getUserInfo(function(userInfo){
      //更新数据
      that.setData({
        userInfo:userInfo
      })
    })
    wx.getLocation({
      type: 'wgs84',  //默认为 wgs84 返回 gps 坐标,gcj02 返回可用于 wx.openLocation()的坐标
      success: function(res){
          var latitude = res.latitude
          var longitude = res.longitude
          wx.chooseLocation({
            latitude: latitude,
            longitude: longitude,
            success: function(res){
                console.log(res)
            },
            fail: function() {
              //fail
            },
            complete: function() {
              //complete
            }
          })
      },
      fail: function() {
        //fail
      },
      complete: function() {
        //complete
      }
    })
  }
})
```

【代码讲解】 本例先用 wx.getLocation()接口获取经度和纬度,再用 wx.chooseLocation()接口使用 wx.getLocation()接口获取的经度和纬度选择打开位置地图,运行效果如图 8.4 所示,注意在其右上角有一个"确定"按钮,该按钮即为 wx.chooseLocation()接口"选择"二字的含义所在。

8.2 图片

小程序提供了 6 个图片 API 接口,分别是: wx.chooseImage(Object object)接口,从本地相册选择图片或使用相机拍照; wx.chooseMessageFile(Object object)接口,从客户端会

话选择文件；wx. compressImage(Object object)接口，压缩图片接口，可选压缩质量；wx. getImageInfo(Object object)接口，获取图片信息；wx. previewImage(Object object)接口，在新页面中全屏预览图片；wx. saveImageToPhotosAlbum(Object object)接口，保存图片到手机本地相册。本节只介绍几个主要的接口。

1. wx. chooseImage(Object object)接口

wx. chooseImage(Object object)接口从本地相册选择图片或使用相机拍照，属性如表 8.6 所示。

表 8.6　wx. chooseImage()接口属性表

接　口	类　型	必填	说　明
count	number	否	最多可以选择的图片张数
sizeType	Array.<string>	否	所选图片的尺寸
sourceType	Array.<string>	否	选择图片的来源
success	function	否	接口调用成功的回调函数
fail	function	否	接口调用失败的回调函数
complete	function	否	接口调用结束的回调函数(调用成功、失败都会执行)

示例代码：

```
wx.chooseImage({
  count: 1,
  sizeType: ['original', 'compressed'],
  sourceType: ['album', 'camera'],
  success(res) {
    //tempFilePaths 可以作为 img 标签的 src 属性显示图片
    const tempFilePaths = res.tempFilePaths
  }
})
```

2. wx. getImageInfo(Object object)接口

wx. getImageInfo(Object object)接口获取图片信息，包括图片的类型、高度和宽度，属性如表 8.7 所示。

表 8.7　wx. getImageInfo()接口属性表

接　口	类　型	必填	说　明
src	string	是	图片的路径，可以是相对路径、临时文件路径、存储文件路径、网络图片路径
success	function	否	接口调用成功的回调函数
fail	function	否	接口调用失败的回调函数
complete	function	否	接口调用结束的回调函数(调用成功、失败都会执行)

【例】8-5　本例先调用 wx. chooseImage()接口，然后调用 wx. getImageInfo()接口，第一个接口获取的图片供第二个接口使用，输出选中的图片的信息，程序运行效果如图 8.5 所示。

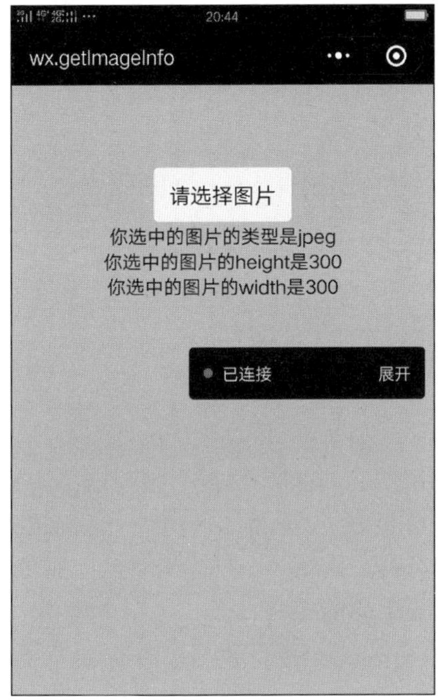

图 8.5　wx. getImageInfo()运行效果

视频讲解

pages/getImageInfo/getImageInfo. wxml 的代码如下：

```
<view class = "container">
  <view class = "userinfo">
    <button bindtap = "choose"> 请选择图片 </button>
  </view>
 <view > 你选中的图片的类型是{{type}}</view >
  <view > 你选中的图片的 height 是{{height}}</view >
  <view > 你选中的图片的 width 是{{width}}</view >
</view >
```

pages/getImageInfo/getImageInfo. js 的代码如下：

```
const app = getApp()
Page({
  data: {
    type:'',
    height:null,
    width:null,
  },
  onLoad: function() {
  },
  choose: function(e) {
    var that = this;
    wx.chooseImage({
```

```
          success(res) {
            wx.getImageInfo({
              src: res.tempFilePaths[0],
              success(res) {
                that.data.type = res.type,
                that.data.height = res.height,
                that.data.width = res.width
                console.log(res)
                that.setData({
                 type:res.type,
                 height:res.height,
                width:res.width
               })
              }
            })
          }
        })
      }
    })
```

【代码讲解】 本例先调用 wx.chooseImage()接口选择本地手机的一幅图片,然后调用 wx.getImageInfo()接口获取图片的信息,信息包括 type、width 和 height 等。图 8.5 显示了 wx.getImageInfo()获取的信息。

3. wx.previewImage(Object object)接口

wx.previewImage(Object object)接口对图片进行预览操作,小程序可以预览多张图片,属性如表 8.8 所示。

表 8.8 wx.previewImage()接口属性表

接　　口	类　　型	必填	说　　明
urls	Array.<string>	是	需要预览的图片链接列表。从 2.2.3 版本开始支持云文件 ID
current	string	否	当前显示图片的链接
success	function	否	接口调用成功的回调函数
fail	function	否	接口调用失败的回调函数
complete	function	否	接口调用结束的回调函数(调用成功、失败都会执行)

示例代码:

```
wx.previewImage({
  current: '',          //当前显示图片的 http 链接
  urls: []              //需要预览的图片 http 链接列表
})
```

例 **8-6** 本例实现点击一组网络图片进行预览的功能,程序运行效果如图 8.6 所示。

视频讲解

pages/previewImage/previewImage.wxml 的代码如下:

```
<view class = "container">
<scroll - view scroll - y = "true" class = "show - area" style = "height:{{scrollHeight}}px;">
    <view class = "photo - list" wx:for = "{{photoList}}" >
   <image id = "{{item}}"   bindtap = "previewPhoto"
```

```
            style = "width:{{imageWidth}}rpx;height:{{imageHeight}}rpx;"
            mode = "aspectFill" src = "{{item}}"/>
        </view>
    </scroll - view>
    </view>
```

图 8.6　预览效果

pages/previewImage/previewImage.js 的代码如下：

```
var app = getApp();
Page({
    data:{
        scrollHeight:569,
        imageWidth:250,
        imageHeight:250,
      photoList: ['https://timgsa.baidu.com/timg?image&quality = 80&size = b9999
_10000&sec = 1563898546498&di = b7326b07d8a4734fea1f5cbdd795ed99&imgtype = 0&src = http %
3A % 2F % 2Fimg.romzhijia.net % 2Farticlepic % 2F2019 % 2F3 % 2F12 % 2F15 % 2F50 % 2Fefeb6d52 -
d5ef - 4195 - a201 - c4b6309b6bbe.jpg', 'https://timgsa.baidu.com/timg?image&quality = 80&size
= b9999_10000&sec = 1563898547199&di = 7ece197fc1eec8b051a4b97bae022ac1&imgtype = 0&src =
http % 3A % 2F % 2Fwww.fabuzhe.com.cn % 2Fresources % 2Fupload % 2Fjsp % 2Fupload % 2Fimage %
2F20190418 % 2F1555554319079014095.jpg', 'https://timgsa.baidu.com/timg?image&quality =
80&size = b9999_10000&sec = 1563898547199&di = 8a7f838a1f73468b4a5c2f110b45b482&imgtype =
0&src = http % 3A % 2F % 2Fpic1.cxtuku.com % 2F00 % 2F09 % 2F36 % 2Fb562e69ba9a3.jpg']
    },
    previewPhoto:function(e){
        console.log(e);
```

```
        var curTarget = e.currentTarget.id
        console.log(curTarget);
        var shuzu = [e.currentTarget.id]
          wx.previewImage({
            urls:this.data.photoList,
            success: function(res){
              console.log(res);
            }
          })
      }
})
```

【代码讲解】 photoList 数组里面的图片地址是 previewImage.wxml 中需要显示的图片库，也是 previewImage.js 中 wx.previewImage()接口需要预览的图片库。注意 wx.previewImage()接口一般预览的是一组图片而很少是一张图片，故需要一个类似于 photoList 的数组。

4. wx.saveImageToPhotosAlbum(Object object)接口

wx.saveImageToPhotosAlbum(Object object)接口，保存图片到相册，具体属性如表 8.9 所示。

表 8.9 wx.saveImageToPhotosAlbum()接口属性表

接 口	类 型	必填	说 明
filePath	string	是	图片文件路径，可以是临时文件路径或永久文件路径，不支持网络图片路径
success	function	否	接口调用成功的回调函数
fail	function	否	接口调用失败的回调函数
complete	function	否	接口调用结束的回调函数(调用成功、失败都会执行)

示例代码：

```
wx.saveImageToPhotosAlbum({
  success(res) { }
})
```

视频讲解

例 8-7 本例实现点击一幅网络图片保存到本地手机，程序运行效果如图 8.7 所示。

pages/saveImage/saveImage.wxml 的代码如下：

```
<view class = "container">
<scroll - view scroll - y = "true" class = "show - area" style = "height:{{scrollHeight}}px;">
<view class = "photo - list" wx:for = "{{photoList}}" wx:key = " * this" wx:for - item =
"imageSrc">
<image id = "{{imageSrc.file}}"  bindtap = "saveImage" style = "width:{{imageWidth}}rpx;
height:{{imageHeight}}rpx;" mode = "aspectFill" src = "{{imageSrc.file}}"/>
 </view>
</scroll - view>
</view>
```

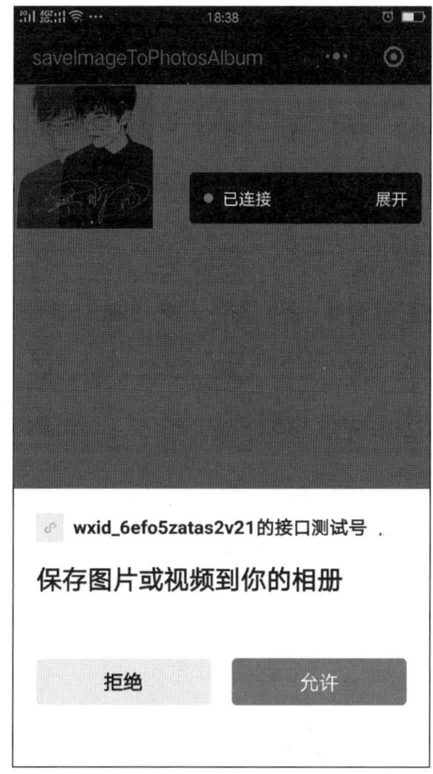

图 8.7　保存图片到相册效果

pages/saveImage/saveImage.js 的代码如下：

```
var app = getApp();
Page({
  data: {
    scrollHeight: 569,
    imageWidth: 250,
    imageHeight: 250,
    curTarget: "",
    photoList: [{
      file: "http://hbimg.b0.upaiyun.com/53ecac6cd550461d2596aee47
          f80a7d42ce990ab784ce - f7i61N_fw658"
    }]
  },
  saveImage: function(e) {
    wx.downloadFile({
      url: 'http://hbimg.b0.upaiyun.com/53ecac6cd550461d2596aee47
          f80a7d42ce990ab784ce - f7i61N_fw658',
      success: function(res) {
        console.log(res);
        //图片保存到本地
        wx.saveImageToPhotosAlbum({
```

```
            filePath: res.tempFilePath,
            success: function(data) {
              console.log(data);
            },
          })
        }
      })
    }
  })
```

【代码讲解】 wx.saveImageToPhotosAlbum()接口不能直接对网络图片进行保存操作,所以本例中先使用了wx.downloadFile()接口下载网络图片到本地形成临时文件e.tempFilePath,然后再把e.tempFilePath作为wx.saveImageToPhotosAlbum()接口的下载参数。

8.3 视频

小程序提供了wx.createVideoContext(string id, Object this)、wx.chooseVideo(Object object)、wx.saveVideoToPhotosAlbum(Object object)等接口对手机视频进行操作。

1. wx.createVideoContext(string id, Object this)接口

wx.createVideoContext(string id, Object this)接口负责创建video上下文VideoContext对象。其语法如下:

```
this.videoContext = wx.createVideoContext('myVideo')
```

获取VideoContext对象之后,就可以对视频做相关操作了。VideoContext对象常用函数如表8.10所示。

表8.10 VideoContext对象常用函数

接　口	功能和用途
VideoContext.play()	播放视频
VideoContext.pause()	暂停视频
VideoContext.stop()	停止视频
VideoContext.seek(number position)	跳转到指定位置
VideoContext.sendDanmu(Object data)	发送弹幕
VideoContext.playbackRate(number rate)	设置倍速播放
VideoContext.requestFullScreen(Object object)	进入全屏
VideoContext.exitFullScreen()	退出全屏
VideoContext.showStatusBar()	显示状态栏,仅在iOS全屏下有效
VideoContext.hideStatusBar()	隐藏状态栏,仅在iOS全屏下有效

例 **8-8**　本例使用 wx. createVideoContext()创建 video 上下文 VideoContext 对象，然后再对视频进行发送弹幕、播放、暂停、定位和回滚操作，程序运行效果如图 8.8 所示。

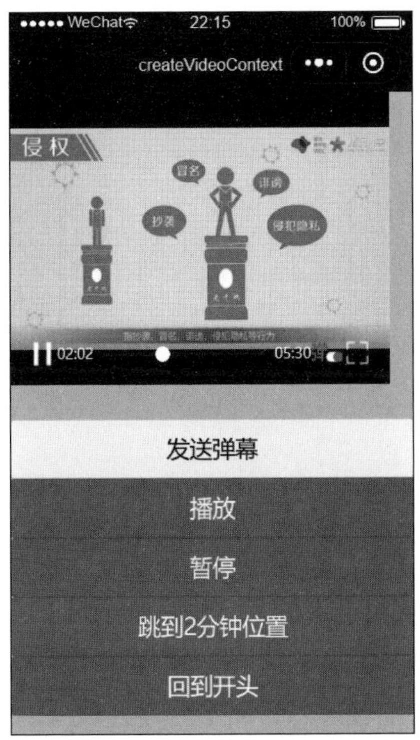

图 8.8　VideoContext 对象运行效果

视频讲解

pages/createVideoContext/createVideoContext. wxml 的代码如下：

```
< view class = "section tc">
  < video  id = "myVideo"  src = "http://wxsnsdy. tc. qq. com/105/20210/snsdy
videodownload?filekey = 30280201010421301f0201690402534804102ca905ce620b1241b7
26bc41dcff44e00204012882540400&bizid = 1023&hy = SH&fileparam = 302c020101042530230
204136ffd93020457e3c4ff02024ef202031e8d7f02030f42400204045a320a0201000400"
danmu - list = "{{danmuList}}" enable - danmu  danmu - btn  controls ></video >
  < view class = "btn - area">
    < input bindblur = "bindInputBlur" />
    < button bindtap = "bindSendDanmu">发送弹幕</button >
  </view >
</view >
< button type = "primary" bindtap = "audioPlay">播放</button >
< button type = "primary" bindtap = "audioPause">暂停</button >
< button type = "primary" bindtap = "audio14">跳到 2 分钟位置</button >
< button type = "primary" bindtap = "audioStart">回到开头</button >
```

pages/createVideoContext/createVideoContext. js 的代码如下：

```
function getRandomColor() {
  const rgb = []
  for (let i = 0; i < 3; ++i) {
```

```
    let color = Math.floor(Math.random() * 256).toString(16)
    color = color.length == 1 ? '0' + color : color
    rgb.push(color)
  }
  return '#' + rgb.join('')
}
Page({
  onReady(res) {
    this.videoContext = wx.createVideoContext('myVideo')
  },
  inputValue: '',
  data: {
    src: '',
    danmuList: [
      {
        text: '第 1s 出现的弹幕',
        color: '#ff0000',
        time: 1
      },
      {
        text: '第 3s 出现的弹幕',
        color: '#ff00ff',
        time: 3
      }]
  },
  bindInputBlur(e) {
    this.inputValue = e.detail.value
  },
  bindSendDanmu() {
    this.videoContext.sendDanmu({
      text: this.inputValue,
      color: getRandomColor()
    })
  },
  audioPlay: function() {
    this.videoContext.play()
  },
  audioPause: function() {
    this.videoContext.pause()
  },
  audio14: function() {
    this.videoContext.seek(120)
  },
  audioStart: function() {
    this.videoContext.seek(0)
  }
})
```

【代码讲解】 本例先获取 VideoContext 对象，然后通过 this.videoContext.sendDanmu()、this.videoContext.play()、this.videoContext.pause()、this.videoContext.seek(120)和 this

.videoContext.seek(0)等接口实现了发送弹幕、播放、暂停、定位和回滚操作。

2. wx.chooseVideo()接口

wx.chooseVideo()接口是在本地手机选择或者拍摄视频，选中的视频可以用来显示，也可以用来上传到网络。wx.chooseVideo()接口参数如表8.11所示。

表8.11 wx.chooseVideo()接口参数表

属 性	类 型	必填	说 明
sourceType	Array.<string>	否	视频选择的来源
compressed	boolean	否	是否压缩所选择的视频文件
maxDuration	number	否	视频最长拍摄时间，单位：s
camera	string	否	默认拉起的是前置或者后置摄像头。部分Android手机下由于系统ROM不支持，无法生效
success	function	否	接口调用成功的回调函数
fail	function	否	接口调用失败的回调函数
complete	function	否	接口调用结束的回调函数(调用成功、失败都会执行)

例8-9 本例使用wx.chooseVideo()接口选中手机上的某一视频，然后对选中的视频进行播放操作，程序运行效果如图8.9～图8.11所示。图8.10是小程序正在压缩选中的视频；图8.11是成功加载选中的视频的效果。

视频讲解

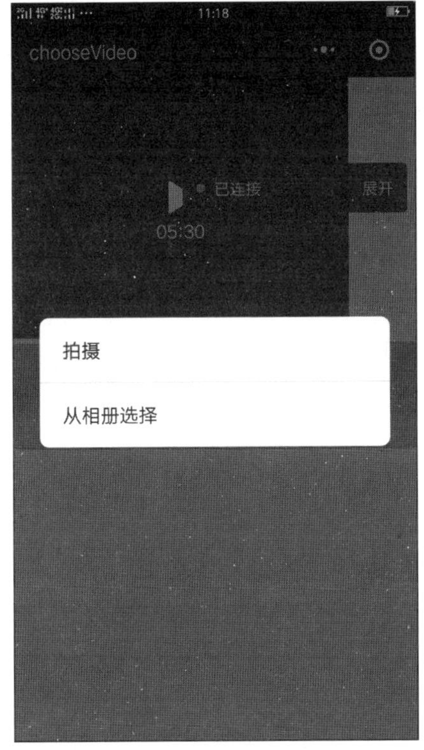

图8.9 视频上传和播放1

图8.10 视频上传和播放2

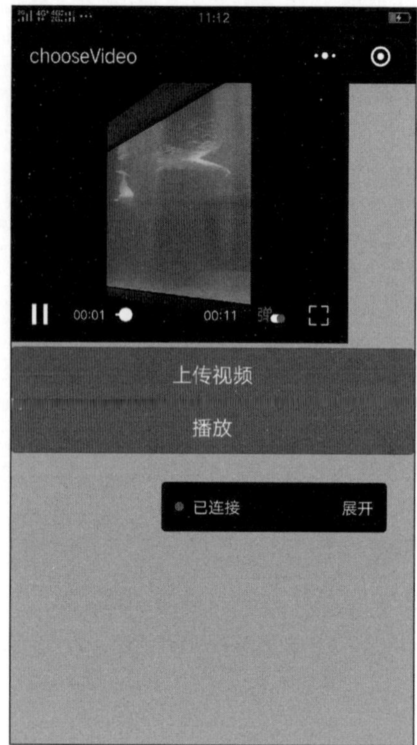

图 8.11　视频上传和播放 3

pages/chooseVideo/chooseVideo. wxml 的代码如下：

```
<view class = "section tc">
<video  id = "myVideo"  src = "{{src}}" danmu - list = "{{danmuList}}" enable - danmu
danmu - btn  controls ></video>
</view>
<button type = "primary" bindtap = "uploadvideo">上传视频</button>
<button type = "primary" bindtap = "audioPlay">播放</button>
```

pages/chooseVideo/chooseVideo. wxml 的代码如下：

```
Page({
  onReady(res) {
  },
  inputValue: '',
  data: {
    src: 'http://wxsnsdy.tc.qq.com/105/20210/snsdyvideodownload?filekey =
30280201010421301f0201690402534804102ca905ce620b1241b726bc41dcff44e002040128
82540400&bizid = 1023&hy = SH&fileparam = 302c020101042530230204136ffd93020457e3c4
ff02024ef202031e8d7f02030f42400204045a320a0201000400'
  },
  uploadvideo: function() {
      var that = this;
    wx.chooseVideo({
```

```
        sourceType: ['album', 'camera'],
        maxDuration: 60,
        camera: 'back',
        success(res) {
          that.setData({
            src: res.tempFilePath
          })
          console.log(that.data.src)
        }
      })
    },
    audioPlay: function() {
      this.videoContext = wx.createVideoContext('myVideo')
      this.videoContext.play()
    }
  })
```

【代码讲解】 sourceType:['album','camera']说明可以选择手机上的视频,也可以即时拍摄视频。在选择了新视频之后采用 wx.createVideoContext()来获取 VideoContext 对象,使用 this.videoContext.play()来播放选择的视频。

3. wx.saveVideoToPhotosAlbum(Object object)接口

wx.saveVideoToPhotosAlbum(Object object)保存视频到系统相册,支持 MP4 视频格式,详细属性如表 8.12 所示。

表 8.12 wx.saveVideoToPhotosAlbum()接口属性表

属 性	类 型	必填	说 明
filePath	string	是	视频文件路径,可以是临时文件路径,也可以是永久文件路径
success	function	否	接口调用成功的回调函数
fail	function	否	接口调用失败的回调函数
complete	function	否	接口调用结束的回调函数(调用成功、失败都会执行)

例 8-10 本例使用 wx.saveVideoToPhotosAlbum()接口保存一个视频到手机视频库中,程序运行效果如图 8.12 所示。

视频讲解

pages/savaVideo/savaVideo.wxml 的代码如下:

```
<view class="section tc">
 <video id="myVideo" src="{{src}}" danmu-list="{{danmuList}}" enable-danmu danmu-btn controls></video>
</view>
<button type="primary" bindtap="save">保存视频</button>
```

pages/savaVideo/savaVideo.js 的代码如下:

```
Page({
  inputValue: '',
  data: {
```

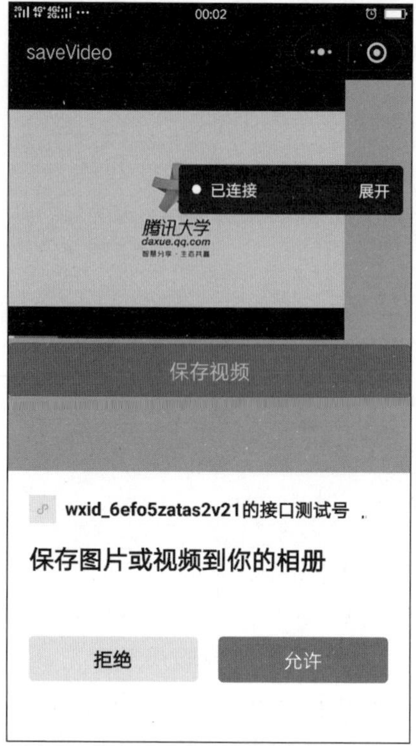

图 8.12　保存视频到手机视频库

```
      src: 'http://wxsnsdy.tc.qq.com/105/20210/snsdyvideodownload?filekey=
30280201010421301f0201690402534804102ca905ce620b1241b726bc41dcff44e002040128
82540400&bizid=1023&hy=SH&fileparam=302c020101042530230204136ffd93020457e3c4
ff02024ef202031e8d7f02030f42400204045a320a0201000400'
    },
    save: function() {
       var that = this;
      wx.downloadFile({
        url: 'http://wxsnsdy.tc.qq.com/105/20210/snsdyvideodownload?filekey=
30280201010421301f0201690402534804102ca905ce620b1241b726bc41dcff44e002040128
82540400&bizid=1023&hy=SH&fileparam=302c020101042530230204136ffd93020457e3c4
ff02024ef202031e8d7f02030f42400204045a320a0201000400',   //仅为示例,并非真实的资源
        success: function(res) {
          if (res.statusCode === 200) {
            wx.saveVideoToPhotosAlbum({
              filePath: res.tempFilePath,
              success(res) {
                wx.showToast({
                  title: '保存视频成功!',
                })
              },
              fail(res) {
```

```
                wx.showToast({
                  title: '保存图片失败!',
                })
              }
            })
          }
        }
      })
    }
  })
```

8.4 录音、音频播放控制以及背景音乐

与 QQ 相比，微信拥有更加强大的语音功能。小程序继承了微信强大的语音处理功能，提供了录音、音频播忙放制和背景音乐等功能，它们的功能不相同，但有相似性，本节将把这三类接口安排在一起进行介绍。

8.4.1 录音

小程序提供了 wx.startRecord(Object object)(开始录音)、wx.stopRecord()(停止录音)和 RecorderManager(录音管理器)等接口对录音功能进行控制。因为 RecorderManager 录音管理器包含前两个接口的功能，所以本节只对 RecorderManager 录音管理器进行讲解。它的常用函数如表 8.13 所示。

表 8.13 RecorderManager 常用函数表

接　　口	功能和用途
RecorderManager.resume()	继续录音
RecorderManager.stop()	停止录音
RecorderManager.onStart(function callback)	监听录音开始事件
RecorderManager.onResume(function callback)	监听录音继续事件
RecorderManager.onPause(function callback)	监听录音暂停事件
RecorderManager.onStop(function callback)	监听录音结束事件
RecorderManager.onFrameRecorded(function callback)	监听已录制完指定帧大小的文件事件。如果设置了 frameSize，则会回调此事件
RecorderManager.onError(function callback)	监听录音错误事件

注意：在使用录音接口时，需先授权开放录音功能。

视频讲解

【例】8-11　本例使用 RecorderManager 录音管理器实现录音、暂停、继续录音、停止录音和播放录音功能。图 8.13 是授权开放录音功能界面；图 8.14 是开始录音的界面；图 8.15 是控制台打印的信息。

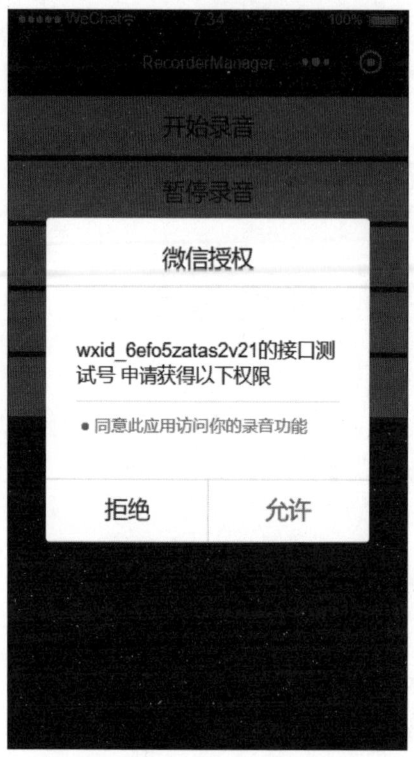

图 8.13　录音授权

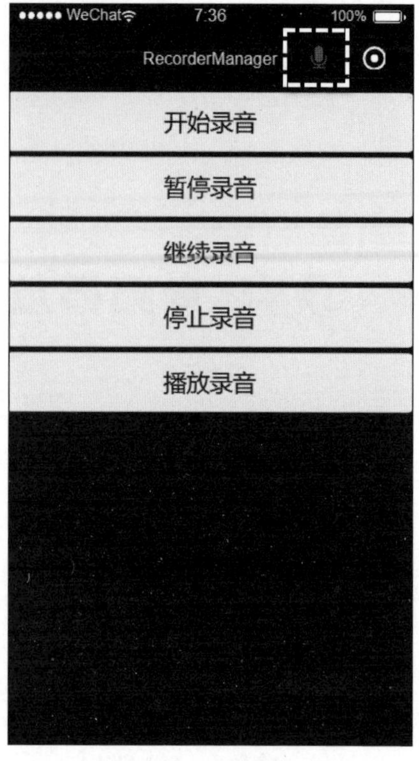

图 8.14　开始录音

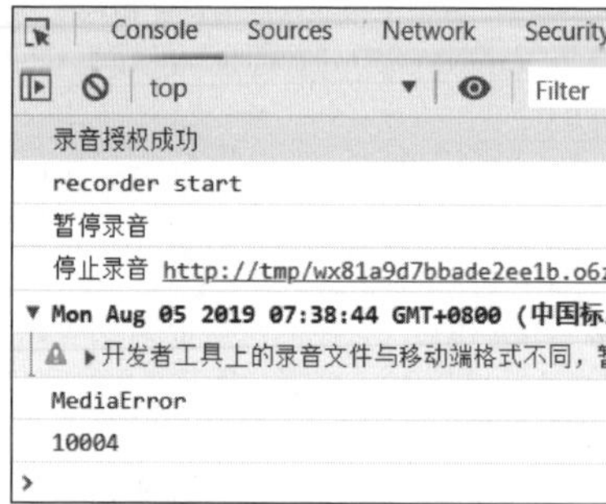

图 8.15　控制台信息

pages/recorderManager/recorderManager.wxml 的代码如下：

```
<button bindtap="start" class='btn'>开始录音</button>
<button bindtap="pause" class='btn'>暂停录音</button>
<button bindtap="resume" class='btn'>继续录音</button>
<button bindtap="stop" class='btn'>停止录音</button>
<button bindtap="play" class='btn'>播放录音</button>
```

pages/recorderManager/recorderManager.js 的代码如下：

```
const recorderManager = wx.getRecorderManager()
const innerAudioContext = wx.createInnerAudioContext()
Page({
  data: {
  },
  onLoad: function() {
  },
   start: function() {                           //开始录音的时候
    var that = this
    const options = {
      duration: 10000,                           //指定录音的时长,单位为 ms
      sampleRate: 16000,                         //采样率
      numberOfChannels: 1,                       //录音通道数
      encodeBitRate: 96000,                      //编码码率
      format: 'mp3',                             //音频格式,有效值为 aac/mp3
      frameSize: 50,                             //指定帧大小,单位为 KB
    }
       wx.authorize({                            //授权接口
      scope: 'scope.record',
      success() {
        console.log("录音授权成功");
           that.setData({
          status: 2,                             //第一次成功授权后 状态切换为 2
        })
        recorderManager.start(options);
        recorderManager.onStart(() => {
          console.log('recorder start')
        });
        recorderManager.onError((res) => {      //错误回调
          console.log(res);
        })
      },
      fail() {
        console.log("第一次录音授权失败");
        wx.showModal({
          title: '提示',
          content: '您未授权录音,功能将无法使用',
          showCancel: true,
          confirmText: "授权",
          confirmColor: "#52a2d8",
          success: function(res) {
            if (res.confirm) {                   //确认则打开设置页面(重点)
               wx.openSetting({
                success: (res) => {
                  console.log(res.authSetting);
                  if (!res.authSetting['scope.record']) {
                    //未设置录音授权
                    console.log("未设置录音授权");
                    wx.showModal({
```

```
                    title: '提示',
                    content: '您未授权录音,功能将无法使用',
                    showCancel: false,
                    success: function(res) {
                    },
                  })
                } else {
                  console.log("设置录音授权成功");      //第二次才成功授权
                  that.setData({
                    status: 2,
                  })
                  recorderManager.start(options);
                  recorderManager.onStart(() => {
                    console.log('recorder start')
                  });
                  //错误回调
                  recorderManager.onError((res) => {
                    console.log(res);
                  })
                }
              },
              fail: function() {
                console.log("授权设置录音失败");
              }
            })
          } else if (res.cancel) {
            console.log("cancel");
          }
        },
        fail: function() {
          console.log("openfail");
        }
      })
    }
  })
},
pause: function() {                                               //暂停录音
  recorderManager.pause();
  recorderManager.onPause((res) => {
    console.log('暂停录音')
  })
},
resume: function() {                                              //继续录音
  recorderManager.resume();
  recorderManager.onStart(() => {
    console.log('重新开始录音')
  });
  //错误回调
  recorderManager.onError((res) => {
    console.log(res);
  })
```

```
    },
    stop: function() {                                          //停止录音
      recorderManager.stop();
      recorderManager.onStop((res) => {
        this.tempFilePath = res.tempFilePath;
        console.log('停止录音', res.tempFilePath)
        const { tempFilePath } = res
      })
    },
    play: function() {                                          //播放声音
      innerAudioContext.autoplay = true
      innerAudioContext.src = this.tempFilePath,
        innerAudioContext.onPlay(() => {
          console.log('开始播放')
        })
      innerAudioContext.onError((res) => {
        console.log(res.errMsg)
        console.log(res.errCode)
      })
    },
  })
```

【代码讲解】 本例通过 recorderManager. wxml 中的 5 个按钮来调用 RecorderManager 录音管理器的录音、暂停、继续录音、停止录音和播放录音功能。在录制好音频之后也可以上传到服务器,本例只是把录制好的音频存放在手机临时目录,然后用来播放。

8.4.2 音频播放控制

小程序提供了 wx.stopVoice()、wx.setInnerAudioOption()、wx.playVoice()、wx.pauseVoice()、wx. getAvailableAudioSources()、wx. createInnerAudioContex()和 wx. createAudioContext()等接口对音频进行操作。其中,wx. createInnerAudioContex()接口和 wx. createAudioContext()接口包含前面几个接口的功能。wx. createAudioContext()接口是以组件< audio >为基础的操作,本节讲解 wx. createAudioContext()接口,而 wx. createInnerAudioContext()接口将在 8.6 节中以实训项目的形式介绍。

AudioContext 实例对象可通过 wx. createAudioContext()接口获取,它通过 id 跟一个< audio >组件绑定,操作对应的< audio >组件。AudioContext 对象常用的函数如表 8.14 所示。

表 8.14 AudioContext 对象常用函数

接　　口	功能和用途
AudioContext. setSrc(string src)	设置音频地址
AudioContext. play()	播放音频
AudioContext. pause()	暂停音频
AudioContext. seek(number position)	跳转到指定位置

例 **8-12** 本例首先通过 wx.createAudioContext()接口获取 AudioContext 实例，然后调用播放和暂停功能，最后用 slider 组件来定位播放位置，程序运行效果如图 8.16 所示。

视频讲解

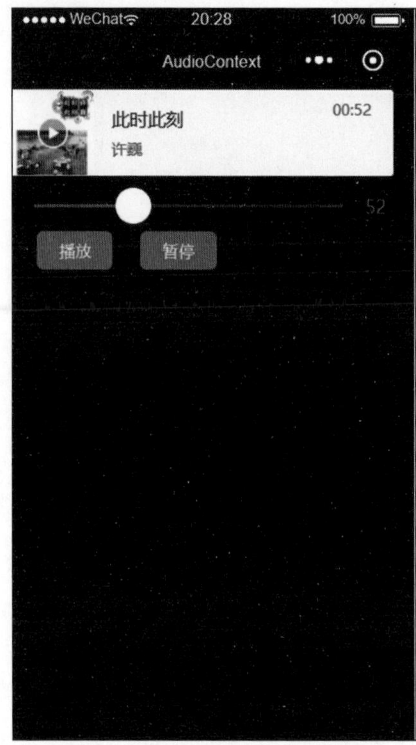

图 8.16　AudioContext 播放器实例

pages/AudioContext/AudioContext.wxml 的代码如下：

```
<audio poster = "{{poster}}" name = "{{name}}" author = "{{author}}" src = "{{src}}" id =
"myAudio" controls loop></audio>
<slider bindchange = 'change' min = "0" max = "160" value = "{{time}}" color = "blue" selected -
color = "red" show - value = "true"></slider>
<button class = 'b1' type = "primary" size = "mini" bindtap = "audioPlay">播放</button>
<button class = 'b1' type = "primary" size = "mini"  bindtap = "audioPause">暂停</button>
```

pages/AudioContext/AudioContext.wxml 的代码如下：

```
Page({
  onReady: function(e) {
    //使用 wx.createAudioContext 获取 audio 上下文 context
    this.audioCtx = wx.createAudioContext('myAudio') },
  data: {
    time:0,
    poster: 'http://y.gtimg.cn/music/photo_new/T002R300x300M000003rsKF44
GyaSk.jpg?max_age = 2592000',
    name: '此时此刻', author: '许巍',
src: 'http://aqqmusic.tc.qq.com/amobile.music.tc.qq.com/C400001VfvsJ21x
```

```
Fqb.m4a?guid=7755942796&vkey=DCE16865FD7290841AD5C3E90461ADA3FA0ECA135E00058
61E0A1D406642332094B84AC22FCDAF9A70C8A2EC729EECB010AE0E585F8CFB70&uin=0&from
tag=38',
  },
  audioPlay: function() {
    this.audioCtx.play()
  },
  audioPause: function() {
    this.audioCtx.pause()
  },
  audio14: function() {
    this.audioCtx.seek(0)
  },
  change: function(e) {
    console.log(e)
    this.audioCtx.seek(e.detail.value)
  }
})
```

【代码讲解】 本例调用了 AudioContext 对象的 AudioContext.play()、AudioContext.pause()和 AudioContext.seek() 3 个接口，其中，AudioContext.seek()是通过 AudioContext.wxml 中的 slider 组件的拉动来定位播放的时间位置。滑块条(slider)拉动的值被传递到了 e.detail.value 中，this.audioCtx.seek(e.detail.value)就定位到滑块条拉动的值对应的播放时间处进行播放。

8.4.3 背景音乐

小程序提供了 wx.stopBackgroundAudio()、wx.seekBackgroundAudio()、wx.playBackgroundAudio()、wx.pauseBackgroundAudio()、wx.onBackgroundAudioStop()、wx.onBackgroundAudioPlay()、wx.onBackgroundAudioPause()、wx.getBackgroundAudioPlayerState()、wx.getBackgroundAudioManager()和 BackgroundAudioManager 实例等接口对背景音乐进行操作，从接口的数量可以看出小程序对背景音乐功能的重视程度。

BackgroundAudioManager 实例对象是通过 wx.getBackgroundAudioManager()接口获取的，它是小程序官方发布的接口，功能强大，包含其他背景音乐处理接口的所有功能。本节将介绍 BackgroundAudioManager 实例对象，它的常见属性和功能函数分别如表 8.15 和表 8.16 所示。

表 8.15　BackgroundAudioManager 对象常见属性

接　　口	功能和用途
string src	音频的数据源(2.2.3 版本开始支持云文件 ID)，默认为空字符串，当设置了新的 src 时，会自动开始播放，目前支持的格式有 m4a、aac、mp3、wav
number startTime	音频开始播放的位置，单位：s

续表

接　　口	功能和用途
string title	音频标题，用于原生音频播放器音频标题(必填)。原生音频播放器中的分享功能，分享出去的卡片标题也将使用该值
string epname	专辑名，原生音频播放器中的分享功能，分享出去的卡片简介也将使用该值
string singer	歌手名，原生音频播放器中的分享功能，分享出去的卡片简介也将使用该值
string coverImgUrl	封面图 URL，用于做原生音频播放器背景图。原生音频播放器中的分享功能，分享出去的卡片配图及背景也将使用该图
string webUrl	页面链接，原生音频播放器中的分享功能，分享出去的卡片简介也将使用该值
string protocol	音频协议，默认值为'http'，设置'hls'可以支持播放 HLS 协议的直播音频
number duration	当前音频的长度(单位：s)，只有在有合法 src 时返回(只读)
number currentTime	当前音频的播放位置(单位：s)，只有在有合法 src 时返回(只读)
boolean paused	当前是否暂停或停止(只读)
number buffered	音频已缓冲的时间，仅保证当前播放时间点到此时间点内容已缓冲(只读)

表 8.16　BackgroundAudioManager 对象常见功能函数

接　　口	功能和用途
BackgroundAudioManager.play()	播放音乐
BackgroundAudioManager.pause()	暂停音乐
BackgroundAudioManager.stop()	停止音乐
BackgroundAudioManager.onCanplay(function callback)	监听背景音频进入可播放状态事件。但不保证后面可以流畅播放
BackgroundAudioManager.seek(number currentTime)	跳转到指定位置
BackgroundAudioManager.onWaiting(function callback)	监听音频加载中事件。当音频因为数据不足，需要停下来加载时会触发
BackgroundAudioManager.onError(function callback)	监听背景音频播放错误事件
BackgroundAudioManager.onPlay(function callback)	监听背景音频播放事件
BackgroundAudioManager.onPause(function callback)	监听背景音频暂停事件
BackgroundAudioManager.onSeeking(function callback)	监听背景音频开始跳转操作事件
BackgroundAudioManager.onEnded(function callback)	监听背景音频自然播放结束事件
BackgroundAudioManager.onSeeked(function callback)	监听背景音频完成跳转操作事件
BackgroundAudioManager.onStop(function callback)	监听背景音频停止事件
BackgroundAudioManager.onTimeUpdate(function callback)	监听背景音频播放进度更新事件，只有小程序在前台时会回调
BackgroundAudioManager.onNext(function callback)	监听用户在系统音乐播放面板点击下一曲事件(仅 iOS)
BackgroundAudioManager.onPrev(function callback)	监听用户在系统音乐播放面板点击上一曲事件(仅 iOS)

[例]8-13 本例在古诗《将进酒》页面中添加朗读古诗的背景音乐,实现开始朗读、暂停朗读和继续朗读 3 个功能。程序是通过 backgroundAudioManager 实例调用播放接口和暂停接口来实现上述功能,程序运行效果如图 8.17 所示。

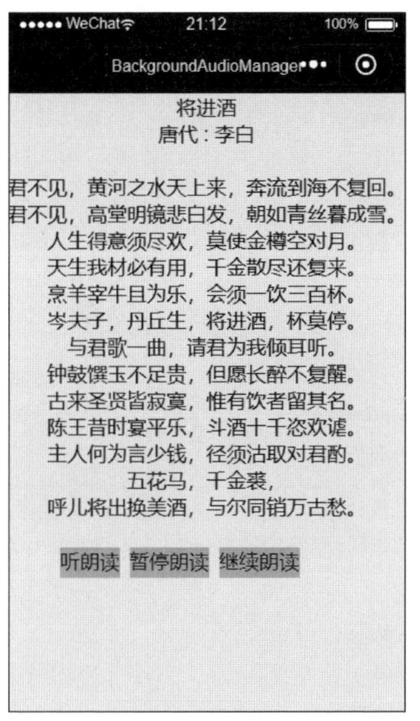

视频讲解

图 8.17 backgroundAudioManager 播放器实例

pages/backgroundAudioManager/backgroundAudioManager.wxml 的代码如下:

```
<view class = 'zong'>
<view>将进酒</view>
<view class = "zz">唐代 : 李白</view>
<view class = "content">
<view>君不见,黄河之水天上来,奔流到海不复回。</view>
<view>君不见,高堂明镜悲白发,朝如青丝暮成雪。</view>
<view>人生得意须尽欢,莫使金樽空对月。</view>
<view>天生我材必有用,千金散尽还复来。</view>
<view>烹羊宰牛且为乐,会须一饮三百杯。</view>
<view>岑夫子,丹丘生,将进酒,杯莫停。</view>
<view>与君歌一曲,请君为我倾耳听。</view>
<view>钟鼓馔玉不足贵,但愿长醉不复醒。</view>
<view>古来圣贤皆寂寞,惟有饮者留其名。</view>
<view>陈王昔时宴平乐,斗酒十千恣欢谑。</view>
<view>主人何为言少钱,径须沽取对君酌。</view>
<view>五花马,千金裘,</view>
<view>呼儿将出换美酒,与尔同销万古愁。</view>
</view>
</view>
<view class = "bg">
```

```
<view class = "bg1" bindtap = 'play'>听朗读</view><view class = "bg2" bindtap = 'pause'>暂停朗读</view><view class = "bg3" bindtap = 'continue'>继续朗读</view>
</view>
```

pages/backgroundAudioManager/backgroundAudioManager.js 的代码如下：

```
const backgroundAudioManager = wx.getBackgroundAudioManager()
backgroundAudioManager.title = '将进酒'
backgroundAudioManager.epname = '将进酒'
backgroundAudioManager.singer = '李白'
Page({
  /*** 页面的初始数据 */
  data: {
  },
  onShow: function() {
  },
  play:function(){
    //设置了 src 之后会自动播放
    backgroundAudioManager.src =
"http://isure.stream.qqmusic.qq.com/C4000013xY8G2lIong.m4a?guid = 7755942796&v
key = 7606A71BA1B1FE25153B1739773F3508E140C9241416E3F491BF664BC4293AF618B0876E
7B85B160A61B79E54187A0D5B5F288A386615D57&uin = 0&fromtag = 66"
  },
  pause:function(){
    backgroundAudioManager.pause();
  },
  continue:function(){
    backgroundAudioManager.play();
  }
})
```

【代码讲解】 本例先通过 wx.getBackgroundAudioManager()获取 backgroundAudioManager 实例，然后调用了 backgroundAudioManager.src、backgroundAudioManager.pause()和 backgroundAudioManager.play()三个接口。需要注意 backgroundAudioManager.src 设置了背景音乐的地址，因为背景音乐是会自动播放的，所以没有用 backgroundAudioManager.play()就实现了第一次播放功能。

8.5 设备操作

因为小程序应用经常会和手机硬件打交道，所以小程序提供了众多设备操作接口，本节将介绍获取系统信息、网络环境、电量等接口。

8.5.1 获取系统信息

小程序开发的过程中可能需要获取用户的手机信息，例如可能需要获取用户手机的高度和宽度来设置一个页面的布局，或者需要获取所有用户使用的手机型号来判断用户的人群画像。小程序提供了 wx. getSystemInfoSync()和 wx. getSystemInfo()两个接口来获取系统信息,前者是同步接口,后者是异步接口,二者没有本质区别。异步的含义是接口执行上需要比较长的时间才返回结果,本节只介绍异步接口 wx. getSystemInfo(),其属性如表 8. 17 所示。

表 8. 17　wx. getSystemInfo()接口属性表

接　口	类　型	功能和用途
brand	string	设备品牌
model	string	设备型号
pixelRatio	number	设备像素比
screenWidth	number	屏幕宽度,单位 px
screenHeight	number	屏幕高度,单位 px
windowWidth	number	可使用窗口宽度,单位 px
windowHeight	number	可使用窗口高度,单位 px
statusBarHeight	number	状态栏的高度,单位 px
language	string	微信设置的语言
version	string	微信版本号
system	string	操作系统及版本
platform	string	客户端平台

例 8-14　本例使用 wx. getSystemInfo()获取系统的手机型号、像素比、手机窗口宽度、手机窗口高度、微信语言和微信版本号等信息,程序运行效果如图 8. 18 所示。

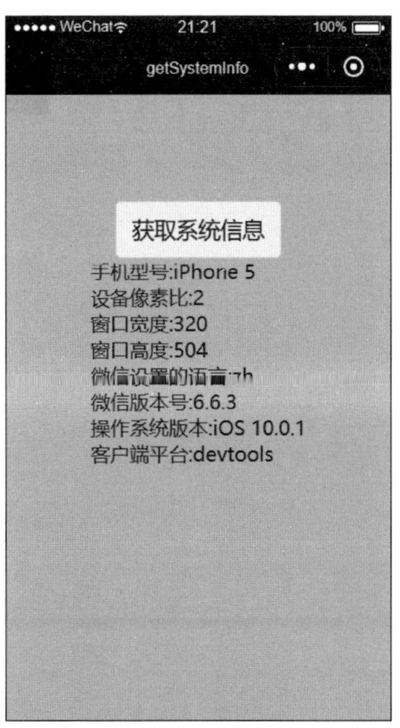

图 8. 18　getSystemInfo 获取系统信息

视频讲解

pages/getSystemInfo/getSystemInfo. wxml 代码如下：

```
<view class = "container">
  <button bindtap = 'getInfo'>获取系统信息</button>
  <view wx:if = "{{model != ''}}">
    <view>手机型号:{{model}}</view>
    <view>设备像素比:{{pixelRatio}}</view>
    <view>窗口宽度:{{windowWidth}}</view>
    <view>窗口高度:{{windowHeight}}</view>
    <view>微信设置的语言:{{language}}</view>
    <view>微信版本号:{{version}}</view>
    <view>操作系统版本:{{system}}</view>
    <view>客户端平台:{{platform}}</view>
  </view>
</view>
```

pages/getSystemInfo/getSystemInfo. js 代码如下：

```
const app = getApp()
Page({
  data: {
    model: '',
    pixelRatio: '',
    windowWidth: '',
    windowHeight: '',
    language: '',
    version: '',
    system: '',
    platform: ''
  },
  onLoad: function() {
  },
  getInfo: function() {
    var that = this;
    wx.getSystemInfo({
      success: function(res) {
        that.setData({
          model: res.model,
          pixelRatio: res.pixelRatio,
          windowWidth: res.windowWidth,
          windowHeight: res.windowHeight,
          language: res.language,
          version: res.version,
          system: res.system,
          platform: res.platform
        })
      },
      fail: function(res) {
      },
      complete: function(res) {
      }
```

```
        })
    }
})
```

【代码讲解】 小程序提供了 wx.getSystemInfoSync()和 wx.getSystemInfo()两个接口来获取系统信息,两个接口功能类似,选择其一即可。本例使用了 wx.getSystemInfo(),它获取的信息全部被封装在 res 对象中。

8.5.2 网络环境

在一些需要消耗流量的应用场景中,小程序需要先获取用户的网络环境,根据不同的网络环境做出不同的处理。wx.getNetworkType()接口的属性如表 8.18 所示。

表 8.18 wx.getNetworkType()接口属性表

属　性	类　型	说　明
success	function	接口调用成功的回调函数
fail	function	接口调用失败的回调函数
complete	function	接口调用结束的回调函数(调用成功、失败都会执行)

例 8-15 本例在例 8-13 的基础上加入网络环境判断,当处于 WiFi 条件下就播放背景音乐,在非 WiFi 条件下弹出一个窗口让用户决定是否继续,程序运行效果如图 8.19 所示。

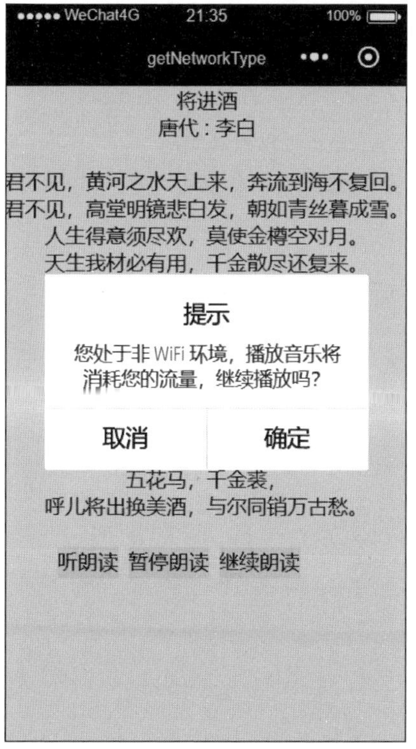

图 8.19 getNetworkType 获取网络条件

视频讲解

pages/getNetworkType/getNetworkType.js 代码如下：

```
const backgroundAudioManager = wx.getBackgroundAudioManager()
backgroundAudioManager.title = '将进酒'
backgroundAudioManager.epname = '将进酒'
backgroundAudioManager.singer = '李白'
Page({
  /*** 页面的初始数据 */
  data: {src:'http://isure.stream.qqmusic.qq.com/C4000013xY8G2lIong.m4a
?guid=7755942796&vkey=7606A71BA1B1FE25153B1739773F3508E140C9241416E3F491BF66
4BC4293AF618B0876E7B85B160A61B79E54187A0D5B5F288A386615D57&uin=0&fromtag=66'},
  onShow: function() {
  },
  play:function(){
    var that = this
    wx.getNetworkType({
      success(res) {
        const networkType = res.networkType
        console.log(networkType)
        if(networkType == "WiFi")                      //判断是不是 WiFi 环境
        {
          backgroundAudioManager.src = that.data.src
        }
        else{
          wx.showModal({
            title: '提示',
            content: '您处于非 WiFi 环境,播放音乐将消耗您的流量,继续播放吗?',
            success(res) {
              if(res.confirm) {
                console.log('用户点击确定')             //设置了 src 之后会自动播放
                backgroundAudioManager.src = that.data.src
              } else if (res.cancel) {
                console.log('用户点击取消')
              }
            }
          })
        }
      }
    })
  },
  pause:function(){
    backgroundAudioManager.pause();
  },
  continue:function(){
    backgroundAudioManager.play();
  }
})
```

【代码讲解】 本例和例 8-13 使用的是同一个 wxml 文件，js 文件中使用 wx.getNetworkType 获取用户网络环境，当处于非 WiFi 环境时程序调用 wx.showModal()弹窗让用户决定是否继续播放背景音乐。

8.5.3 电量

应用程序有时候需要像检测网络环境一样检测手机的电量情况,例如在播放视频的时候,检测到电量不足则让屏幕进入省电模式或者直接提示用户充电。在小游戏开发中 wx. getSystemInfo()接口的回调函数 success 的参数 res 中,res. battery 是代表电量的,但是小程序中的 wx. getSystemInfo()接口的回调函数 success 没有 res. battery 属性,小程序提供的接口是 wx. getBatteryInfo()接口和 wx. getBatteryInfoSync()接口。wx. getBatteryInfo()接口除了有电量值 level,还有 isCharging 这个属性,所以 wx. getBatteryInfo()接口相对来说更专业。wx. getBatteryInfoSync()接口是 wx. getBatteryInfo()接口的同步版本,同步和异步接口在电量检测方面差别不大,另外,wx. getBatteryInfoSync()接口在 iOS 上是不可用的。本节介绍 wx. getBatteryInfo()接口,属性如表 8.19 所示。

表 8.19 wx. getBatteryInfo()接口属性表

属　性	类　型	说　明
level	string	设备电量,范围为 1～100
isCharging	boolean	是否正在充电中

例 8-16 本例通过 wx. getBatteryInfo()接口获取手机电量和充电状态,并打印电量水平和充电状态,程序运行效果如图 8.20 所示。

图 8.20 手机电量和充电状态

视频讲解

pages/getBatteryInfo/getBatteryInfo. wxml 代码如下:

```
<view>你的电量是{{level}}%</view>
<view>你的充电状态是{{isCharging}}</view>
```

pages/getBatteryInfo/getBatteryInfo. js 代码如下:

```
Page({
  data: {
    level:100,
    isCharging:true
  },
  onReady(res) {
    var that = this
    wx.getBatteryInfo({
      success(res) {
        console.log('电量: ', res.level)
```

```
          console.log('是否正在充电：', res.isCharging)
          that.setData({
            level:res.level,
            isCharging: res.isCharging
          })
        }
      })
    },
  onShow(e){
  }
})
```

【代码讲解】 wx.getBatteryInfo()获取的电量 level 是不带百分比的 1～100 的整数，如果需要带上百分比显示电量，就需要程序另外补上。

8.6 实训项目——音乐播放器案例

视频讲解

本实训项目将实现一个简单的音乐播放器，具体实现以下功能：

(1) 播放和暂停功能之间的切换。

(2) 上一首和下一首。

(3) 播放列表的弹出和消失。

(4) 是否循环播放，通过按钮控制循环与否。

(5) 进度条随着播放器时间一秒一秒地自动移动。

播放器效果如图 8.21 所示。

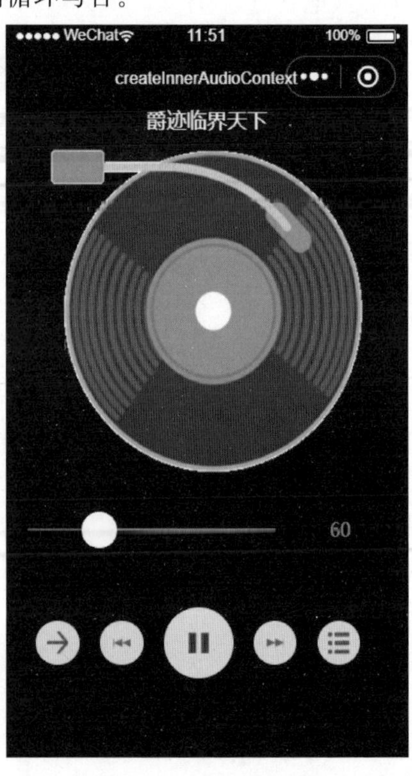

图 8.21 音乐播放器

在编写程序之前，开发者需要知道如何从网络上获取音乐的播放地址，以 QQ 音乐为例，用户在 QQ 音乐网站上访问的音乐地址如 https://y.qq.com/n/yqq/song/000oQ6Cm1OsjPd.html，直接把这个地址设置为小程序音乐的 src 属性是不能正常播放的。小程序需要形如"http://dl.stream.qqmusic.qq.com/C400003KgsVz3yPsO6.m4a?guid=7755942796&vkey=CE204E710F1DB40A1BDEEE08BD2578A9D1C97A57FFAD0BFE941273E85C08F1C46CD2CC1EAD101BD835FC0AF100D35B313220 6EDE31D9DD2C&uin=0&fromtag=38"的地址。

下面以歌曲《爵迹临界天下》为例介绍如何在 QQ 音乐网站上获取小程序可以正常识别的音乐 URL 地址。

首先访问歌曲《爵迹临界天下》地址 https://y.qq.com/n/yqq/song/000oQ6Cm1OsjPd.html，然后

按下键盘上的 F12 键，再按下 F5 键，播放页面如图 8.22 中所示，按照图 8.22 的"1""2""3"的步骤即可复制到小程序可以识别的音乐 URL 地址。

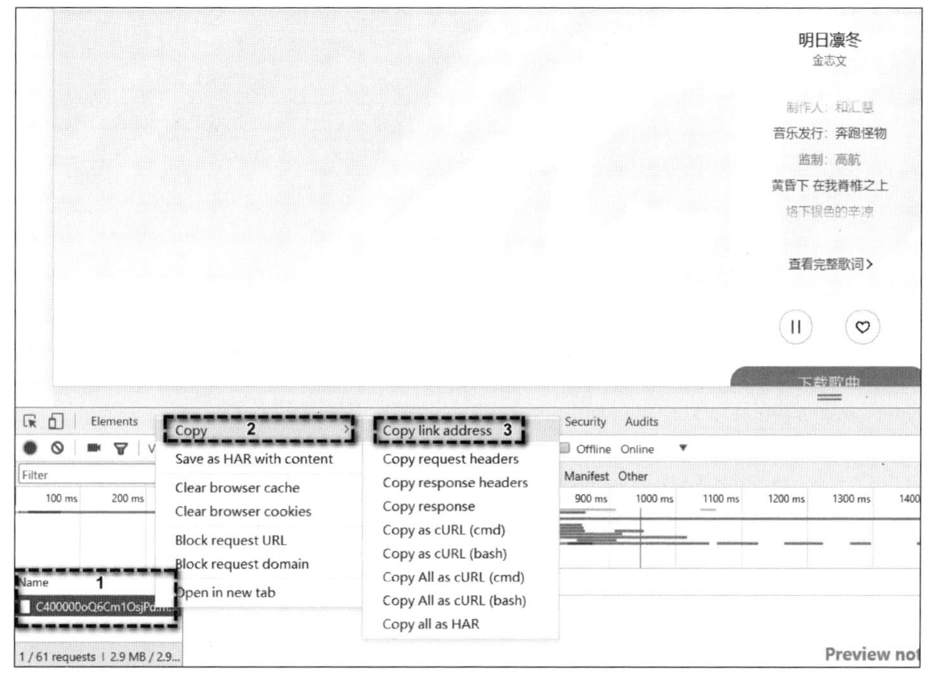

图 8.22　获取小程序可识别的音乐 URL 地址

接下来介绍播放器的具体实现，本实训项目使用的是 InnerAudioContext 对象，而不是例 8-12 中的 AudioContext 对象，前者的功能更加丰富。AudioContext 对象是通过 wx. createAudioContext ( ) 接口来创建的，而 InnerAudioContext 对象是通过 wx. createInnerAudioContext()接口来创建的。InnerAudioContext 对象的属性如表 8.20 所示，常用函数如表 8.21 所示。

表 8.20　InnerAudioContext 对象常用属性

属　　性	含义和取值
string src	音频资源的地址，用于直接播放，2.2.3 版本开始支持云文件 ID
number startTime	开始播放的位置(单位：s)，默认为 0
boolean autoplay	是否自动开始播放，默认为 false
boolean loop	是否循环播放，默认为 false
boolean obeyMuteSwitch	是否遵循系统静音开关，默认为 true。当此参数为 false 时，即使用户打开了静音开关，也能继续发出声音。从 2.3.0 版本开始此参数不生效，使用 wx. setInnerAudioOption 接口统一设置
number volume	音量，范围为 0～1，默认为 1
number duration	当前音频的长度(单位：s)。只有在当前有合法的 src 时返回(只读)
number currentTime	当前音频的播放位置(单位：s)。只有在当前有合法的 src 时返回，时间保留小数点后 6 位(只读)
boolean paused	是否暂停(只读)
number buffered	音频缓冲的时间点，仅保证当前播放时间点到此时间点内容已缓冲(只读)

表 8.21 InnerAudioContext 对象常见函数

函　　数	功能和用途
InnerAudioContext.play()	播放
InnerAudioContext.pause()	暂停。暂停后的音频再播放会从暂停处开始播放
InnerAudioContext.stop()	停止。停止后的音频再播放会从头开始播放
InnerAudioContext.seek(number position)	跳转到指定位置
InnerAudioContext.destroy()	销毁当前实例
InnerAudioContext.onCanplay(function callback)	监听音频进入可以播放状态的事件，但不保证后面可以流畅播放
InnerAudioContext.offCanplay(function callback)	取消监听音频进入可以播放状态的事件
InnerAudioContext.onPlay(function callback)	监听音频播放事件
InnerAudioContext.offPlay(function callback)	取消监听音频播放事件
InnerAudioContext.onPause(function callback)	监听音频暂停事件
InnerAudioContext.offPause(function callback)	取消监听音频暂停事件
InnerAudioContext.onStop(function callback)	监听音频停止事件
InnerAudioContext.offStop(function callback)	取消监听音频停止事件
InnerAudioContext.onEnded(function callback)	监听音频自然播放至结束的事件
InnerAudioContext.offEnded(function callback)	取消监听音频自然播放至结束的事件
InnerAudioContext.onTimeUpdate(function callback)	监听音频播放进度更新事件
InnerAudioContext.offTimeUpdate(function callback)	取消监听音频播放进度更新事件
InnerAudioContext.onError(function callback)	监听音频播放错误事件
InnerAudioContext.offError(function callback)	取消监听音频播放错误事件
InnerAudioContext.onWaiting(function callback)	监听音频加载中事件。当音频因为数据不足，需要停下来加载时会触发
InnerAudioContext.offWaiting(function callback)	取消监听音频加载中事件
InnerAudioContext.onSeeking(function callback)	监听音频进行跳转操作的事件
InnerAudioContext.offSeeking(function callback)	取消监听音频进行跳转操作的事件
InnerAudioContext.onSeeked(function callback)	监听音频完成跳转操作的事件
InnerAudioContext.offSeeked(function callback)	取消监听音频完成跳转操作的事件

从 InnerAudioContext 对象的属性和函数表可以看出，InnerAudioContext 对象拥有很多 AudioContext 对象没有的功能。下面是项目的程序部分。

pages/index/createInnerAudioContext.wxml 的代码如下：

```
<view  class = "top">
    <text class = "playmusicname">{{playmusicname}}</text>
    <view class = "music - player" style = "{{animation_pause}}">
      <image class = "topimg" src = "./img/play.gif"></image>
    </view>
  </view>
 <slider min = "0" max = "{{duration}}" value = "{{currentTime}}" color = "red"
 selected - color = "blue" show - value = "true"></slider>
  <view class = "mid">
    <view class = "btns pre_next" bindtap = "modetap">
    <image class = "modeimg" src = "{{mode.mode_img[mode.index]}}"></image>
```

```
        </view>
        <view class = "btns pre_next pre" bindtap = "Preaudio">
        <image class = "img_pre_next" src = "./img/pre.png"></image>
        </view>
        <view class = "btns play_pause" bindtap = "Playaudio">
          <image class = "img_play_pause" src = "{{model.mode_img[model.index]}}"></image>
        </view>
        <view class = "btns pre_next next" bindtap = "Nextaudio">
          <image class = "img_pre_next" src = "./img/next.png"></image>
        </view>
        <view class = "btns pre_next" catchtap  = "showlist">
        <image class = "modeimg" src = "./img/list.png"></image>
        </view>
      </view>
    <view bindtap = "closeList" style = "display:{{songList_show}}" class = "music_list">
    <text class = "list_li" style = "text-align:center;">播放列表
     ({{songList.length}})</text>
          <block wx:for = "{{songList}}">
          <text class = "list_li"
      style = "background-color:silver;">{{item.no}}.{{item.name}}</text>
        </block>
  </view>
```

【代码讲解】 本例的 createInnerAudioContext.wxml 从上到下主要分为四部分：第一部分是歌曲名字的显示；第二部分是一张滚动的动画图片；第三部分是进度条；第四部分是播放器的 5 个按钮。文件中< slider >组件随着歌曲播放的进度来修改 value 值。

pages/index/createInnerAudioContext.js 的代码如下：

```
const innerAudioContext  = wx.createInnerAudioContext();
innerAudioContext.loop = false
var app  = getApp()
Page({
  data: {
    id: "",
    songName: "",
    play_pause_src: './img/play.png',
    animation_pause: 'animation-play-state:paused',
    timer: null,
    touches: {},
    mode: {
      mode_img: ['./img/shunxu.png', './img/singlemusic.png'],
      index: 0
    },
    mode1: {
      mode_img: ['./img/play.png','./img/pause.png'],
      index: 0
    },
    songList: [{ no: 1, name: "慢半拍", url: "http://aqqmusic.tc.qq.com
/amobile.music.tc.qq.com/C400002qpjAV21Yx81.m4a?guid = 7755942796&vkey = 761C71F
AA22FB595EA9717ED744BAE5919879B74DF2579EB29F93BEBA1A02E5086663A948F19EA2EB11
```

```
B25A9595137A777D355D87747C54D&uin = 0&fromtag = 38" },
{no: 2, name: "木偶人", url: "http://aqqmusic.tc.qq.com/amobile.music.tc.qq.com
/C400003iHc0e2UIgMC.m4a?guid = 7755942796&vkey = 4AAA588B92E9BF4FC3B3AB069854334
9152EC7871C7139ADD967B94BB076B4018C907D389E1337884B65C54B0D30D3D07D798C913A8
7B9CE&uin = 0&fromtag = 38"},
{ no: 3, name: "我的朋友", url:"http://aqqmusic.tc.qq.com/amobile.music.tc.qq.com
/C400001nGinl3Tsk4H.m4a?guid = 7755942796&vkey = BC68A40B32AC814AAF389875DBD4537
82BDF1CEA44FA2F382E93F9A35CFAC2915C042CE485967353FD90958ED886E361C24587333F2
E8224&uin = 0&fromtag = 38"}],
    playmusicname:"木偶人",
    songList_show: "none",
    duration:208,
    currentTime:0
  },
  onLoad: function() {
  },
  onReady: function() {
    console.log(this.data.songList[1].url)
    innerAudioContext.src = this.data.songList[1].url
    console.log("已经有了 innerAudioContext" + innerAudioContext.duration)
  },
  modetap:function(){                                        //播放模式的设置
    if (innerAudioContext.loop = true)
    { innerAudioContext.loop = false}
    else{
      innerAudioContext.loop = true}
    var mode = this.data.mode
    mode.index++
    if(mode.index >= 2){
      mode.index = 0
    }
    this.setData({
      mode:mode
    })
  },
  Preaudio: function() {                                     //前一首歌
    innerAudioContext.src = this.data.songList[0].url
    innerAudioContext.play();
    this.setData({
      playmusicname: "慢半拍"
    })
  },
  Playaudio: function() {                                    //播放当前歌曲
    var that = this
    var mode1 = this.data.mode1
    mode1.index++
    if (mode1.index >= 2) {
      mode1.index = 0
    }
    this.setData({
      mode1: mode1
```

```
      })
      if (innerAudioContext.paused) {
        innerAudioContext.play();
        innerAudioContext.onPlay(() => {
          console.log('audioContext.paused', innerAudioContext.paused)
        })
        innerAudioContext.onTimeUpdate(() =>{
          that.setData({
            duration:innerAudioContext.duration,
            currentTime: innerAudioContext.currentTime
          })
        })
      } else {
        innerAudioContext.pause();
        innerAudioContext.onPause(() => {
          console.log('audioContext.paused', innerAudioContext.paused)
        })
      }
    },
    Nextaudio: function() {                                   //后一首歌
      innerAudioContext.src = this.data.songList[2].url
      innerAudioContext.play();
      this.setData({
        playmusicname:"我的朋友"
      })
    },
    showlist: function(e) {                                   //显示播放列表
      this.setData({
        songList_show: "block"
      })
    },
    closeList: function() {                                   //关闭播放列表
      this.setData({
        songList_show: "none"
      })
    }
  })
```

【代码讲解】 songList 数组存放了音乐库，本例中只设置了 3 首歌曲，其中歌曲的 URL 地址的获取很关键，可以根据本实训项目附带的视频来获取小程序可以识别的 URL 地址。文件中，modeChoose 函数控制歌曲播放方式循环与否；audioPlay 函数控制歌曲播放和暂停；audioPre 和 audioNext 函数显示上一首和下一首的切换；showList 和 closeList 函数控制播放列表的显示与不显示。

pages/index/createInnerAudioContext.wxss 的代码如下：

```
/** index.wxss **/
.top {
  display: flex;
  flex-direction: column;
  align-items: center;
```

```
}
.playmusicname{
 height:30rpx;
 color:#ffffff;
}
.music-player {
 margin: 40rpx;
 border-radius: 50%;
 animation: trun_around 20s linear 1s infinite;
 position: relative;
}
.topimg{
 width: 600rpx; height: 600rpx;
}
@keyframes trun_around{
 0% {transform: rotate(0deg);transition:transform 0s;}
 100% {transform: rotate(360deg);}
}
.mid {
 width:600rpx;
 margin-top: 50px;
 padding:0;
 display: flex;
 justify-content: space-between;
 align-items: center;
 margin-left:60rpx;
}
.btns {
 border-radius: 50%;
 text-align: center;
 background-color: rgb(206,206,231);
 display:flex;
 justify-content:center;
 align-items: center;
}
.play_pause {
 width:130rpx;
 height:130rpx;
}
.pre_next {
 width:80rpx; height:80rpx;
}
.img_pro_next{
 width:40%;
 height:40%;
}
.img_play_pause{
 width:30%; height:30%;
}
.modeimg{
 width:50%; height:50%;
```

```
}
.progress_bar{
  width:190px; height:2px;
  padding:0;
  background-color: rgb(106,136,131);
  display:flex;
  justify-content: flex-start;
  align-items: stretch;
}
.passed_time {
  width:10%;
  background-color: red;
}
.point{
  width:31rpx; height:31rpx;
  margin:30rpx 0;
  border-radius: 50%;
  background-color: white;
  position:relative;
  top: -40rpx;
}
.time {
  font-size:10px;
  width:70rpx; color:white;
}
.music_list{
  width:100%; height: 400rpx;
  position:fixed;
  bottom:0; right:0;
  color:black;
  background-color: white;
  opacity:0.8;
  overflow: scroll;
  display: flex;
}
.list_li{
  display:block;
  width:100%;
  padding:20rpx;
  border-bottom:1px dotted silver;
}
page{background-color: #000000}
```

第9章 交互接口和开放接口

交互接口主要实现用户和小程序之间的交互功能，例如 wx. showToast()接口可以弹出提示消息给用户，wx. showModal()和 wx. showActionSheet()接口可以让用户做出选择操作。开放接口则实现了让小程序调用第三方服务的功能，例如微信支付接口让小程序可以调用微信支付功能，模板消息接口让小程序可以调用公众平台来发送消息。可以说交互接口和开放接口是商业开发必然会使用到的功能性接口，其中微信登录接口、微信支付接口和模板消息接口必须配合后端程序才能实现功能，它们属于难度较大的高级接口。

本章主要目标

- 了解微信小程序常见的交互接口和开放接口；
- 熟练掌握 wx. showToast()、wx. showLoading()、wx. showModal()和 wx. showActionSheet() 4 个交互接口的应用；
- 熟练掌握微信登录接口、微信支付接口、模板消息接口、获取用户信息接口、小程序间跳转接口、获取用户收货地址接口和 SOTER 指纹认证接口等开放接口的应用。

9.1 交互反馈

用户在使用小程序的时候，通常需要和程序进行交互操作，小程序提供了 wx. showToast()、wx. showModal()、wx. showLoading()、wx. showActionSheet()、wx. hideToast()和 wx. hideLoading() 6 个交互接口。

9.1.1 消息提示框 wx. showToast()和加载提示框 wx. showLoading()

wx. showToast(Object object)和 wx. showLoading(Object object)分别提示普通信息和加载信息。这两个接口的相同之处是用户没有选择的权限，只能被动接受。

wx. showToast()接口一般显示文字信息，给用户提醒。wx. showLoading()接口一般显示状态信息，例如"正在下载"的状态、"正在登录"的状态或者"正在处理"的状态。它们的属性如表 9.1 和表 9.2 所示。

表 9.1　wx. showToast()接口属性表

属　　性	类　　型	必填	说　　明
title	string	是	提示的内容
icon	string	否	图标
image	string	否	自定义图标的本地路径，image 的优先级高于 icon
duration	number	否	提示的延迟时间
mask	boolean	否	是否显示透明蒙层，防止触摸穿透
success	function	否	接口调用成功的回调函数
fail	function	否	接口调用失败的回调函数
complete	function	否	接口调用结束的回调函数（调用成功、失败都会执行）

表 9.2　wx. showLoading()接口属性表

属　　性	类　　型	必填	说　　明
title	string	是	提示的内容
mask	boolean	否	是否显示透明蒙层，防止触摸穿透
success	function	否	接口调用成功的回调函数
fail	function	否	接口调用失败的回调函数
complete	function	否	接口调用结束的回调函数（调用成功、失败都会执行）

例 9-1　本例通过一个登录的案例来说明 wx. showToast()和 wx. showLoading()接口的使用，程序运行效果如图 9.1 和图 9.2 所示。

视频讲解

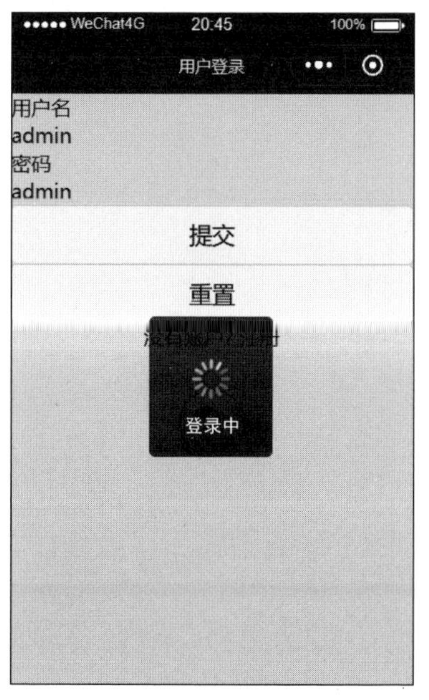

图 9.1　登录中

图 9.2　登录成功

pages/login/login.wxml 的代码如下：

```
<form report-submit bindsubmit='formsubmit' bindreset='formreset'>
    <view>用户名<input placeholder='请输入用户名' name="yhm"></input></view>
     <view>密码<input placeholder='请输入密码' name="mm"></input></view>
     <view><button form-type='submit'>提交</button>
     <button form-type='reset'>重置</button></view>
  </form>
    <view class="item bottom-info">
       <view> 没有账户? <text> 注册 </text> </view>
       <view class="wechat-icon" bindtap='dian'>
            <image src="/image/wechat.png"  />
       </view>
    </view>
```

pages/login/login.js 的代码如下：

```
var app = getApp();
Page({
  data:{
  },
  onLoad:function(options){
  },
  formsubmit:function(e){
    console.log(e)
    if (e.detail.value.yhm == "admin"&e.detail.value.mm=="admin")
    {
      wx.setStorageSync("name", e.detail.value.yhm);
      wx.showLoading({
        title: '登录中',
        duration: 10000
      })
      setTimeout(function() {
        wx.redirectTo({
          url: '../user/user',
        })
      }, 10000)
    }
      setTimeout(function() {
        wx.showToast({
          title: '登录成功',
          icon: 'success',
          duration: 10000
        })
      }, 10000)
    }
})
```

【代码讲解】 为了简化程序，本例设置当判断账号和密码同时等于“admin”的时候则登录成功，先调用 wx.showLoading()接口显示“登录中”，再调用 wx.showToast()接口显示“登录成功”。但是两个接口不能被同时激发，因为如果被同时激发，wx.showToast()接口会覆盖wx.showLoading()接口，所以wx.showToast()接口被写在了setTimeout()延时函数里面，从时间上把两个接口错开。

9.1.2 模态对话框wx.showModal()和操作菜单wx.showActionSheet()

当用户做出选择的时候，9.1.1节中的wx.showToast(Object object)和wx.showLoading(Object object)接口就不够用了，小程序提供了wx.showModal(Object object)显示模态对话框和wx.showActionSheet(Object object)显示操作菜单两个接口，它们的属性如表9.3和表9.4所示。

表9.3 wx.showModal()接口属性表

属性	类型	必填	说明
title	string	是	提示的标题
content	string	是	提示的内容
showCancel	boolean	否	是否显示取消按钮
cancelText	string	否	取消按钮的文字，最多4个字符
cancelColor	string	否	取消按钮的文字颜色，必须是十六进制格式的颜色字符串
confirmText	string	否	确认按钮的文字，最多4个字符
confirmColor	string	否	确认按钮的文字颜色，必须是十六进制格式的颜色字符串
success	function	否	接口调用成功的回调函数
fail	function	否	接口调用失败的回调函数
complete	function	否	接口调用结束的回调函数(调用成功、失败都会执行)

表9.4 wx.showActionSheet()接口属性表

属性	类型	必填	说明
itemList	Array.<string>	是	按钮的文字数组，数组长度最大为6
itemColor	string	否	按钮的文字颜色
success	function	否	接口调用成功的回调函数
fail	function	否	接口调用失败的回调函数
complete	function	否	接口调用结束的回调函数(调用成功、失败都会执行)

wx.showModal(Object object)显示模态对话框接口是弹出包括“确定”和“取消”两个按钮的对话框，在第8章的例8-15中已经使用过了该接口，这里不再赘述。

wx.showModal()接口只能接受用户的两个选择操作，当交互选择有两项以上的时候则需要用wx.showActionSheet(Object object)显示操作菜单接口，它可以弹出多项选择供用户操作，比wx.showModal()接口更加丰富。

例 9-2　本例对图片做“保存图片”“预览图片”“复制图片地址”等操作，这时 wx.showModal()已经不能胜任了，需要使用 wx.showActionSheet()接口，程序运行效果如图 9.3 所示。

视频讲解

图 9.3　wx.showActionSheet()运行效果

pages/showActionSheet/showActionSheet.wxml 的代码如下：

```
<view id="{{src[0]}}" bindtap="share">
<image  src="{{src[0]}}"></image>
</view>
```

pages/showActionSheet/showActionSheet.js 的代码如下：

```
const app = getApp()
Page({
  data: {
    tapIndex: 0,
    src:
["https://timgsa.baidu.com/timg?image&quality=80&size=b9999_10000&sec=155946
9676437&di=2ec38f7708431273d174cd783bc1bc58&imgtype=0&src=http%3A%2F%2Fwx1.
sinaimg.cn%2Flarge%2F007aKGvQgy1fv0x6gx8rzj30140bv44c.jpg"]
  },
share: function(e) {
  var that=this
  console.log(e.currentTarget.id)
    wx.showActionSheet({
```

```
        itemList: ['保存图片', '预览图片', '复制图片地址'],
        success(res) {
          console.log(res.tapIndex)
          console.log(res)
          if (res.tapIndex == 0) {
            wx.downloadFile({
              url: that.data.src[0],
              success(res) {
                wx.saveImageToPhotosAlbum({
                  filePath: res.tempFilePath
                })
                wx.showToast({
                  title: '保存成功',
                  icon: 'success',
                  duration: 1500
                });
              }
            })
          }
          if (res.tapIndex == 1) {
            wx.previewImage({
              current: that.data.src[0],
              urls: that.data.src
            })
          }
        },
        fail(res) {
          console.log(res.errMsg)
        }
        });
    }
  })
```

【代码讲解】 在使用 wx. showActionSheet()接口的过程中需要注意：用户做出的选择被赋值为"res. tapIndex"。当用户选中了第一个选项，"res. tapIndex"等于 0；当用户选中了第二个选项，"res. tapIndex"等于 1；以此类推。

本例中使用的 wx. saveImageToPhotosAlbum()接口不能直接保存网络图片，需要通过 wx. downloadFile()接口获取一个临时图片地址。

另外，程序调用 wx. previewImage()接口来预览图片时，因为该接口的 urls 参数需要是一个数组，所以 data 数据中的 src 写成了数组。

9.2 微信登录接口 wx. login()

小程序是运行在微信基础之上的应用程序，当一个小程序需要登录功能的时候，开发者首先想到的应该是使用微信自带的登录接口来登录小程序应用系统，而不是让用户注册并登录。wx. login(Object object)正是小程序提供的开放登录接口。

9.2.1 微信登录前端

wx.login(Object object)调用接口获取登录凭证(code),通过凭证换取用户登录态信息,包括用户的唯一标识(openid)及登录的会话密钥(session_key)等。使用微信登录接口登录小程序的步骤如下:

(1) 通过 wx.login()来获取 code。

(2) 如果成功获取,那么返回 code。

(3) 调用 wx.request()向服务端发起一个请求,即向登录 api 接口发送 code。

(4) 换取 openid 和 session_key。

其中,第 4 步用 code 换取 openid 和 session_key,需要后端程序向第三方登录凭证校验接口 https://api.weixin.qq.com/sns/jscode2session?appid=APPID&secret=SECRET&js_code=JSCODE&grant_type=authorization_code 发送请求来完成,接口地址中的 AppID 和 AppSecret 是开发者管理后台获取的,具体请参考图 1.13,JSCODE 是由 wx.login()获取的 code。

第 4 步正常返回的 JSON 数据如下:

```
{
 "openid": "USERID",
 "session_key": "kJtdi6RF + Dv67QkbLlPGjw == ",
}
```

第 4 步错误返回的 JSON 数据如下:

```
{
"errcode": 40029,
"errmsg": "invalid code"
}
```

因为每一个用户微信是不一样的,即 JSCODE 不一样,每个应用的 AppID 和 secret 也不一样,导致返回的 openid 和 session_key 也具有唯一性,所以小程序使用 openid 和 session_key 来标识用户是否成功登录了小程序。

例 9-3 本例是使用 wx.login()开放接口实现登录的前端部分,其功能为向后端发起 request 请求。当成功返回 openid 的时候,程序在 Console 控制台打印结果,如图 9.4 所示。

视频讲解

```
Console Sources Network Security AppData Audits Sensor Storage Trace Wxml
top ▼ Filter Default levels ▼
code:033VI7gV1ilUhZ0bV4gV1cv8gV1VI7gf
▼{data: {…}, header: {…}, statusCode: 200, cookies: Array(0), errMsg: "request:ok"}
 ▶cookies: []
 ▶data: {session_key: "5GZan3QHT6eDpuucsEr5ZA==", openid: "on4SK5Zphzx9hyaWFEj42RKBitZo"}
  errMsg: "request:ok"
 ▶header: {Server: "Apache-Coyote/1.1", catch-control: "no-catch", Content-Type: "applicati
  statusCode: 200
 ▶__proto__: Object
```

图 9.4 登录成功效果

pages/wxlogin/wxlogin.wxml 的代码如下：

```
<form report-submit bindsubmit='formsubmit' bindreset='formreset'>
    <view>用户名<input placeholder='请输入用户名' name="yhm"></input></view>
     <view>密码<input placeholder='请输入密码' name="mm"></input></view>
     <view><button form-type='submit'>提交</button><button form-type='reset'>重置
</button></view>
</form>
  <view class="item bottom-info">
        <view> 没有账户? <text> 注册 </text> </view>
        <view class="wechat-icon"  bindtap='login'>
             <image src="/image/wechat.png"  />
        </view>
   </view>
```

【代码讲解】 本例是对 wxlogin.wxml 中的图片绑定了点击事件，该事件调用微信登录功能。

pages/wxlogin/wxlogin.js 的代码如下：

```
const app = getApp()
Page({
 data: {
 },
 //登录获取 code
 login: function() {
 wx.login({
 success: function(res) {
 console.log('code:' + res.code)
       //发送请求
 wx.request({
 url: 'http://localhost:8080/minilogin/loginservlet',  //改成自己的服务器地址
      data: {
         code: res.code,                              //上面的 wx.login()成功获取的 code
         operFlag: 'getOpenid',
          },
      header: {
             'content-type': 'application/json'         //默认值
          },
     success: function(res) {
             console.log(res)
          }
        })
      }
    })
  }
})
```

【代码讲解】 本例先使用 wx.login()接口获取用户的 code，然后使用 wx.request()向后端程序发起请求，返回的结果在 res 对象中。后端程序是采用 Servlet 形式来编写的。图 9.4 显示程序已经成功获取了 session_key 和 openid，它们可以用作成功登录前端的凭证。

9.2.2 微信登录后端

视频讲解

例 9-4 本例是使用 wx. login()开放接口实现登录的后端部分，其功能是接收前端传递过来的 code 参数，向第三方登录验证接口发送请求，并向前端返回 session_key 和 openid 两个结果。

src/servlet/loginservlet. java 的代码如下：

```
package servlet;
import java.io.BufferedReader;
import java.io.IOException;
import java.io.InputStreamReader;
import java.io.PrintWriter;
import java.net.URL;
import java.net.URLConnection;
import java.util.List;
import javax.servlet.ServletException;
import javax.servlet.annotation.WebServlet;
import javax.servlet.http.HttpServlet;
import javax.servlet.http.HttpServletRequest;
import javax.servlet.http.HttpServletResponse;
/**
 * Servlet implementation class loginservlet
 */
@WebServlet("/loginservlet")
public class loginservlet extends HttpServlet {
    private static final long serialVersionUID = 1L;
    private String appid = "你的 appid";
    private String secretKey = "你的 secretKey";
    @Override
    protected void doGet(HttpServletRequest req, HttpServletResponse resp) throws ServletException,
IOException {
        System.out.println("doGet --------------------- ");
        doPost(req, resp);
    }
    @Override
    protected void doPost(HttpServletRequest request, HttpServletResponse response)
            throws ServletException, IOException {
        System.out.println("doPost --------------------- ");
        //获取操作类型,根据类型执行不同操作
        String operFlag = request.getParameter("operFlag");
        System.out.println("operFlag" + operFlag);
        String results = "";
        if ("getOpenid".equals(operFlag)) {
        String code = request.getParameter("code");    //获得微信小程序传过来的 code
            System.out.println(code);
            String url = "https://api.weixin.qq.com/sns/jscode2session?
              appid = " + appid + "&secret = " + secretKey + "&js_code = " + code + "&grant_type =
              authorization_code";
          //接口地址 https://api.weixin.qq.com/sns/jscode2session?appid =
          APPID&secret = SECRET&js_code = JSCODE&grant_type = authorization_code;
```

```
            System.out.println("url" + url);
            results = sendGetReq(url);                  //发送 http 请求
            System.out.println("results" + results);
        }
        response.setContentType("application/json;charset = UTF - 8");
        response.setHeader("catch - control", "no - catch");
        PrintWriter out = response.getWriter();
        out.write(results);
        out.flush();
        out.close();
    }
    private String sendGetReq(String url) {
        String result = "";
        BufferedReader in = null;
        try {
            String urlNameString = url;
            URL realUrl = new URL(urlNameString);
            //打开和 URL 之间的链接
            URLConnection connection = realUrl.openConnection();
            //设置通用的请求属性
            connection.setRequestProperty("accept", "*/*");
            connection.setRequestProperty("connection", "Keep - Alive");
            connection.setRequestProperty("user - agent", "Mozilla/4.0
            (compatible; MSIE 6.0; Windows NT 5.1;SV1)");
            //建立实际的连接
            connection.connect();
            //获取所有响应头字段
            java.util.Map<String, List<String>> map = connection.getHeaderFields();
            //遍历所有响应头字段
            for (String key : map.keySet()) {
                System.out.println(key + "--->" + map.get(key));
            }
            //定义 BufferedReader 输入流来读取 URL 的响应
            in = new BufferedReader(new InputStreamReader(connection.getInputStream()));
            String line;
            while ((line = in.readLine()) != null) {
                result += line;
            }
        } catch (Exception e) {
            System.out.println("发送 GET 请求出现异常!" + e);
            e.printStackTrace();
        }                                               //使用 finally 块来关闭输入流
        finally {
            try {
                if (in != null) {
                    in.close();
                }
            } catch (Exception e2) {
                e2.printStackTrace();
            }
        }
        return result;
    }
}
```

【代码讲解】 本例和例 6-3 有所不同，例 6-3 是向 MySQL 数据库发起数据访问请求，本例是向微信的第三方登录验证接口发起请求。但结果都返回了 JSON 格式的数据。从图 9.4 可以看到小程序已经成功获取了用户登录应用返回的 session_key 和 openid，它们可以作为判断是否登录的标识，用户就不需要输入账号、密码登录应用程序了。

9.3 微信支付接口 wx.requestPayment()

微信支付是腾讯公司的支付业务产品，微信支付商户平台支持线下场所、公众号、小程序、PC 网站、App、企业微信等经营场景快速接入微信支付。微信支付全面打通 O2O 生活消费领域，提供专业的互联网+行业解决方案，微信支付支持微信红包和微信理财通，是移动支付的首选。微信小程序提供了 wx.requestPayment(Object object)来发起微信支付。

使用微信支付功能之前，应先开通微信支付功能，目前微信支付功能只对商户开放申请，申请地址为 https://pay.weixin.qq.com/。申请之前需准备好以下材料。

(1) 营业执照：彩色扫描件或数码照片。

(2) 组织机构代码证：彩色扫描件或数码照片，若已三证合一，则无须提供。

(3) 对公银行账户：包含开户行省市信息，开户账号。

(4) 法人身份证：彩色扫描件或数码照片。

商户在微信公众平台或开放平台提交微信支付开通申请，操作界面如图 9.5 所示。

图 9.5 申请开通微信支付功能操作界面

微信支付工作人员审核资料无误后，为商户开通相应的微信支付权限。微信支付申请审核通过后，商户申请资料时填写的邮箱会收到由微信支付小助手发送的邮件，此邮件包含商户的支付账户信息，如表 9.5 所示。表中的 4 个参数是在微信支付后端开发时需要用的参数，其中涉及隐私信息，需妥善保管。

表 9.5　邮件收到的参数表

邮件中的参数	参数名	说　　明
AppID	appid	appid 是微信小程序后台 App 的唯一标识，在小程序后台申请小程序账号后，微信会自动分配对应的 appid，用于标识该应用，可在小程序→设置→开发设置中查看
微信支付商户号	mch_id	商户申请微信支付后，由微信支付分配的商户收款账号
API 密钥	key	交易过程生成签名的密钥，仅保留在商户系统和微信支付后台，不会在网络中传播。商户妥善保管该密钥，切勿在网络中传输，不能在其他客户端中存储，保证密钥不会被泄露。商户可根据邮件提示登录微信商户平台进行密钥设置，也可按以下路径设置：微信商户平台(pay.weixin.qq.com)→账户设置→安全→密钥设置
AppSecret	secret	AppSecret 是 AppID 对应的接口密码，用于获取接口调用凭证时使用

9.3.1　微信支付前端

微信支付属于小程序众多接口中比较复杂的一个，它需要后端的支持，其前端接口 wx.requestPayment()属性如表 9.6 所示。

表 9.6　wx.requestPayment()接口属性表

属　　性	类　　型	必填	说　　明
timeStamp	string	是	时间戳，从 1970 年 1 月 1 日 00:00:00 至今的秒数，即当前的时间
nonceStr	string	是	随机字符串，长度为 32 个字符以下
package	string	是	统一下单接口返回的 prepay_id 参数值，提交格式如：prepay_id= ***
signType	string	否	签名算法
paySign	string	是	签名，具体签名方案参见小程序支付接口文档
success	function	否	接口调用成功的回调函数
fail	function	否	接口调用失败的回调函数
complete	function	否	接口调用结束的回调函数(调用成功、失败都会执行)

示例代码如下：

```
wx.requestPayment({
  timeStamp: '',
  nonceStr: '',
  package: '',
  signType: 'MD5',
  paySign: '',
  success(res) { },
```

```
  fail(res) { }
})
```

视频讲解

例 9-5 本例以小程序前端实现支付购买一件商品为例，展示微信支付的过程和代码实现，为减少交易费用，本例把金额固定为1分钱，程序运行效果如图9.6～图9.8所示。

图 9.6 发起支付

图 9.7 输入支付密码或调用指纹

图 9.8 支付成功

pages/pay/pay.wxml 的代码如下：

```
<view class="top">
<image src="../img/p30.png"></image>
</view>
<view class='product'>
<view class="title">HUAWEI P30 麒麟 980 超感光徕卡三摄 屏内指纹 双景录像 8GB+64GB 全网通版(天空之境)</view>
<view class="id">商品编码  2601010102103</view>
<view class="price">价    格 ¥3988.00</view>
</view>
<view class="buttonarea" >
    <button class="btn" type="primary" bindtap="orderpay">去结算</button>
    <button class="btn" type="warn" bindtap="clear">清空</button>
</view>
```

pages/pay/pay.js 的代码如下：

```
//获取应用实例
var app = getApp()
Page({
  //
  orderpay: function(e) {
```

```
        var that = this;
        wx.login({
          success: function(res) {
                wx.request({
                  url: 'https://skill.qiujikeji.com/mini/getOpenid',
                  method: 'POST',
                  data: { 'js_code': res.code },
                     success: function(res) {
                       console.log(res.data)
                       that.xiadan(res.data);
                     }
                   })
                }
        });
      },
      //下单
      xiadan: function(openid) {
        var that = this;
        wx.request({
          url: 'https://skill.qiujikeji.com/mini/pay/teacher',
          method: 'POST',
          data: {'openid':openid},
          success: function(res) {
            var res = res.data
            wx.requestPayment({
              timeStamp: res.timeStamp,
              nonceStr: res.nonceStr,
              package: res.package,
              signType: 'MD5',
              paySign: res.paySign,
            })
          }
        })
      },
    })
```

【代码讲解】 微信支付的前端实际上代码相对简单，文件 pay.js 的微信支付前端部分包含 3 个步骤：步骤一是向后端发起请求，获取 openid，此步骤和 9.2 节内容相同；步骤二是利用返回的 openid 向后端发起下单请求，下单成功返回如图 9.9 所示的 5 个参数，而这 5 个参数是 wx.requestPayment()接口需要的参数；步骤三利用返回的 5 个参数同微信支付接口 wx.requestPayment()发起支付请求。有了步骤一和步骤二，后端就知道步骤三这笔支付是谁发起的了。

9.3.2　微信支付后端

微信支付是小程序接口中最为复杂的接口之一，其复杂度体现在后端的开发难度上。接口前端和后端的工作流程包含以下 7 个步骤：

(1) 通过 wx.login()接口获取 code(一个 code 只能用一次)。

(2) 通过 code 获取 openid，详见 9.2 节。

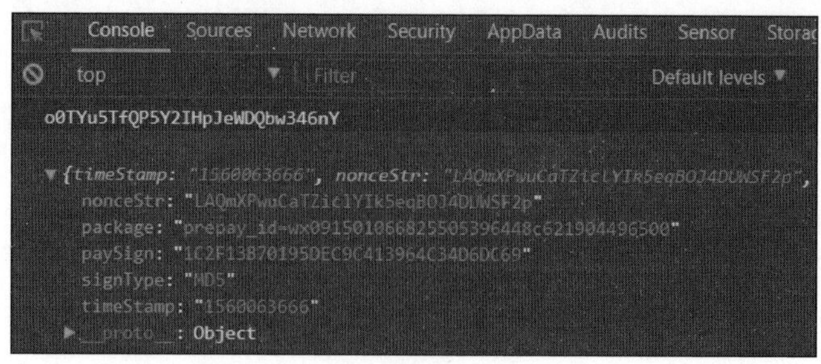

图 9.9 微信支付过程

(3) 小程序发送 openid 到后端程序，后端程序利用 openid 和表 9.7 中的 appid、mch_id、随机字符串、签名、商品描述、商品订单号、金额、下单 IP、通知地址和交易类型等参数一起组装成订单数据。

(4) 后端程序向微信统一下单接口 https://api.mch.weixin.qq.com/pay/unifiedorder 提交订单，并发送步骤(3)中组装的订单数据，后端程序接收返回结果后，加工得到预付款编号 prepay_id。

(5) 后端程序利用 AppID、timeStamp、nonceStr、prepayid 和 signType 5 个参数加工得到 wx.requestPayment()接口需要的 JSON 格式的 timeStamp、nonceStr、package、signType 和 sign 参数，并返回给小程序。

(6) 小程序前端调用 wx.requestPayment()接口发起支付，并完成支付。

(7) 服务器收到回调。

微信统一订单接口属性如表 9.7 所示。

表 9.7 微信统一订单接口属性表

字段名	变 量 名	必填	类型(长度)	示 例	描 述
小程序 ID	appid	是	String(32)	wxd678efh567hg6787	微信分配的小程序 ID
商户号	mch_id	是	String(32)	1230000109	微信支付分配的商户号
随机字符串	nonce_str	是	String(32)	5K8264ILTKCH16CQ2502SI8ZNMTM67VS	随机字符串，长度要求在 32 位以内。推荐随机数生成算法
签名	sign	是	String(32)	C380BEC2BFD727A4B6845133519F3AD6	通过签名算法计算得出的签名值，详见签名生成算法
商品描述	body	是	String(128)	腾讯充值中心-QQ 会员充值	商品简单描述，该字段请按照规范传递，具体请见参数规定
商户订单号	out_trade_no	是	String(32)	2.01508E+13	商户系统内部订单号，要求 32 个字符内，只能是数字、大小写字母和符号"_-| *"，且在同一个商户号下唯一。详见商户订单号

续表

字段名	变 量 名	必填	类型（长度）	示 例	描 述
标价金额	total_fee	是	Int	88	订单总金额，单位为分，详见支付金额
终端 IP	spbill_create_ip	是	String (64)	123.12.12.123	支持 IPv4 和 IPv6 两种格式的 IP 地址。调用微信支付 API 的机器 IP
通知地址	notify_url	是	String (256)	http://www.weixin.qq.com/wxpay/pay.php	异步接收微信支付结果通知的回调地址，通知 URL 必须为外网可访问的 URL，不能携带参数
交易类型	trade_type	是	String (16)	JSAPI	小程序取值如下：JSAPI，详细说明见参数规定
用户标识	openid	否	String (128)	oUpF8uMuAJO_M2pxb1Q9zNjWeS6o	trade_type=JSAPI，此参数必传，是用户在商户 appid 下的唯一标识，openid 的获取可参考【获取 openid】

表 9.7 中只列出了 11 个属性，统一下单接口参数共有 23 个属性，其他参数请参看网址 https://pay.weixin.qq.com/wiki/doc/api/wxa/wxa_api.php?chapter=9_1。

例 9-6 本例为例 9-4 的后端代码部分，项目结构图如图 9.10 所示。

视频讲解

图 9.10 微信支付后端项目结构图

src/servlet/order.java 的代码如下：

```
protected void doPost ( HttpServletRequest request, HttpServletResponse response) throws
ServletException, IOException {
        JSONArray results = null;
         JSONObject obj = new JSONObject();
         obj.put("return_code", "fail");
```

```
        results.add(obj);
            String openid = request.getParameter("openid");
            System.out.println("openid = " + openid);
            String reqStr = tool.getReqStr(openid);          //组装预下单的请求数据
            String url = "https://api.mch.weixin.qq.com/pay
                        /unifiedorder" + "?order" + reqStr;
            System.out.println("reqStr = " + reqStr);
            String results1 = tool.sendGetReq(url);          //发送 post 数据到微信预下单
            System.out.println("prepay from weixin: \n " + results1);
            Map<String,String> return_data = null;
            try {
                return_data = WXPayUtil.xmlToMap(results1);//微信的一个工具类
            } catch (Exception e) {
                //TODO Auto - generated catch block
                e.printStackTrace();
                System.out.println(e.getMessage());
            }
            String return_code = return_data.get("return_code");
            System.out.println("return_code = " + return_code);
            if("SUCCESS".equals(return_code)){
                String prepay_id = return_data.get("prepay_id");
                results = tool.conPayParam(prepay_id);       //组装返回数据
            }
            response.setContentType("application/json;charset = UTF - 8");
            response.setHeader("catch - control", "no - catch");
            PrintWriter out = response.getWriter();
            out.write(results);
            out.flush();
            out.close();
    }
```

【代码讲解】 order.java 是一个 Servlet 程序，其功能主要是接收 openid，通过 tool.getReqStr()把 openid 转化成统一下单接口需要的如表 9.7 所示的参数 reqStr，然后向统一下单接口下单返回 results1，再通过 WXPayUtil.xmlToMap()函数用 results1 获取 prepay_id，最后使用 tool.conPayParam(prepay_id)返回给小程序 wx.requestPayment()需要的 5 个参数 results。图 9.10 中 com.github.wxpay.sdk 包下的 8 个工具类是微信支付官方提供的 SDK 案例代码，可在 https://pay.weixin.qq.com/wiki/doc/api/jsapi.php?chapter=11_1 下载得到。

src/servlet/tool.java 的代码如下：

```
public class tool {
    public static String getReqStr(String openid){
        Map<String,String> data = new HashMap<String,String>();
        String out_trade_no = setTradeNo();                    //
        data.put("appid", "你的 appid");
        data.put("mch_id","你的 mer_id");
        data.put("nonce_str", WXPayUtil.generateNonceStr());
        data.put("sign_type", "MD5");
        data.put("body", "spy test");
```

```
        data.put("out_trade_no", out_trade_no);
        data.put("device_info", "");
        data.put("fee_type", "CNY");
        data.put("total_fee", "1");                        //1 分钱
        data.put("spbill_create_ip", "123.12.12.123");
        data.put("notify_url", "http://xxx/wxpay/notify");
        data.put("trade_type", "JSAPI");
        data.put("product_id", "12");
        data.put("openid", openid);
        try {
            String sign = WXPayUtil.generateSignature(data, "你的 merKey", SignType.MD5);
            data.put("sign", sign);
        } catch (Exception e) {
            e.printStackTrace();
            System.out.println("sign error");
        }
        String reqBody = null;
        try {
            reqBody = WXPayUtil.mapToXml(data);
        } catch (Exception e) {
            //TODO Auto - generated catch block
            e.printStackTrace();
        }
        return reqBody;
    }

//保证唯一
    public static String setTradeNo(){
     Random rdm = new Random(6);
        String orderid = "20211909105011" + rdm.nextInt();
        System.out.println("orderid = " + orderid);
        return orderid;
    }

    //组装返回客户端的请求数据
    public static  JSONArray conPayParam(String prepayid){
        String results = "";
        Map<String,String> map = new HashMap<String,String>();
        map.put("appId", "你的 appid");
        String timeStamp = Long.toString(WXPayUtil.getCurrentTimestamp());
        map.put("timeStamp", timeStamp);
        map.put("nonceStr", WXPayUtil.generateNonceStr());
        map.put("package", "prepay_id = " + prepayid);
        map.put("signType", "MD5");                        //5 个参数
        String sign = null;
        try {
            sign = WXPayUtil.generateSignature(map,"你的 merKey", SignType.MD5);
                                                           //第 6 个参数
            map.put("sign", sign);
        } catch (Exception e) {
            //TODO Auto - generated catch block
```

```
            e.printStackTrace();
            System.out.println(e.getMessage());
        }
        JSONArray jsonarray = new JSONArray();
        JSONObject jsonobj = new JSONObject();
        jsonobj.put("timeStamp", timeStamp);
        jsonobj.put("nonceStr", WXPayUtil.generateNonceStr());
        jsonobj.put("package", "prepay_id=" + prepayid);
        jsonobj.put("signType", "MD5");
        jsonobj.put("paySign", sign);
        jsonarray.add(jsonobj);
        return jsonarray;
    }
public static String sendGetReq(String url) {
    String result = "";
    BufferedReader in = null;
    try {
        String urlNameString = url;
        url realUrl = new url(urlNameString);
        //打开和 URL 之间的链接
        URLConnection connection = realUrl.openConnection();
        //设置通用的请求属性
        connection.setRequestProperty("accept", "*/*");
        connection.setRequestProperty("connection", "Keep-Alive");
        connection.setRequestProperty("user-agent", "Mozilla/4.0
        (compatible; MSIE 6.0; Windows NT 5.1;SV1)");
        //建立实际的连接
        connection.connect();
        //获取所有响应头字段
        java.util.Map<String, List<String>> map = connection.getHeaderFields();
        //遍历所有的响应头字段
        for (String key : map.keySet()) {
            System.out.println(key + "--->" + map.get(key));
        }
        //定义 BufferedReader 输入流来读取 URL 的响应
        in = new BufferedReader(new InputStreamReader(connection.getInputStream()));
        String line;
        while ((line = in.readLine()) != null) {
            result += line;
        }
    } catch (Exception e) {
        System.out.println("发送 GET 请求出现异常!" + e);
        e.printStackTrace();
    }  //使用 finally 块来关闭输入流
    finally {
        try {
            if (in != null) {
                in.close();
            }
        } catch (Exception e2) {
            e2.printStackTrace();
        }
    }
```

```
        return result;
    }
}
```

【代码讲解】 tool. java 是 order. java 的工具类，其中，getReqStr(String openid)函数把 openid 转化成统一下单接口需要的数据；conPayParam(String prepayid)函数将 prepayid 转化成 wx. requestPayment()接口需要的数据；sendGetReq(String url)函数是 Servlet 向外部 URL(这里是腾讯统一下单接口)发送请求并返回结果。

项目中的 com 包下的多个 Java 程序是微信支付官方提供的 demo 工具类，下载地址为 https://pay. weixin. qq. com/wiki/doc/api/jsapi. php?chapter=11_1。

9.4 获取用户信息接口 wx. getUserInfo()

小程序提供了 wx. getUserInfo()接口来获取用户的微信账号信息，该接口一般伴随 wx. login()接口出现，它可以获取的信息属性如表 9.8 所示。

表 9.8 wx. getUserInfo()属性表

字 段 名	含 义
UserInfo	用户信息
string nickName	用户昵称
string avatarUrl	用户头像图片的 URL，URL 最后一个数值代表正方形头像大小(有 0、46、64、96、132 数值可选，0 代表 640×640 的正方形头像，46 表示 46×46 的正方形头像，剩余数值以此类推，默认为 132)
number gender	用户性别
string country	用户所在国家
string province	用户所在省份
string city	用户所在城市
string language	显示 country、province、city 所用的语言

例 9-7 本例使用 wx. getUserInfo()接口获取用户的信息，在 Console 控制台打印的个人微信账号信息如图 9.11 所示。

视频讲解

图 9.11 Console 控制台打印的个人微信信息

pages/getUserInfo/getUserInfo.wxml 的代码如下：

```
<view class="container">
  <view class="userinfo">
      <image bindtap="bindViewTap" class="userinfo-avatar" src="{{userInfo.avatarUrl}}" mode="cover"></image>
      <text class="userinfo-nickname">{{userInfo.nickName}}</text>
      <text class="userinfo-nickname">{{userInfo.gender}}</text>
      <text class="userinfo-nickname">{{userInfo.city}}</text>
      <text class="userinfo-nickname">{{userInfo.province}}</text>
      <text class="userinfo-nickname">{{userInfo.country}}</text>
      <text class="userinfo-nickname">{{userInfo.language}}</text>
  </view>
</view>
```

pages/getUserInfo/getUserInfo.js 的代码如下：

```
const app = getApp()
Page({
  data: {
    motto: 'Hello World',
    userInfo: {}
  },
  onLoad: function() {
    var that = this;
    wx.getUserInfo({
      success: function(res) {
        console.log(res.userInfo)
        that.setData({
          userInfo: res.userInfo
        })
      }
    })
  }
})
```

【代码讲解】 wx.getUserInfo()获取的信息封装在 UserInfo 对象中，通过形如 userInfo.nickName 的属性可以引用到个人的微信账号信息。

9.5 模板消息 template

小程序不可以和微信一样发送可能打扰用户的消息，但是可以发送模板消息。模板消息是基于微信的通知渠道，小程序官方为开发者提供了可以高效触达用户的模板消息能力，以便实现服务的闭环并提供更佳的体验。只有以下情况才允许发送模板消息。

1. 支付

当用户在小程序内完成过支付行为，可允许小程序在 7 天内向用户推送有限条数的模板消息(1 次支付可下发 3 条，多次支付下发条数独立，互相不影响)。

2. 提交表单

当用户在小程序内发生过提交表单行为且该表单声明为要发模板消息的，开发者需要向用户提供服务时，可允许小程序在 7 天内向用户推送有限条数的模板消息(1 次提交表单可下发 1 条，多次提交下发条数独立，相互不影响)。

在开发模板消息功能之前，开发者需要先进入小程序管理后台创建消息模板。进入小程序管理后台之后，单击左侧菜单“模板消息”会出现如图 9.12 所示的模板消息栏目，其中的“模板 ID”是开发模板消息后端程序时需要使用到的。

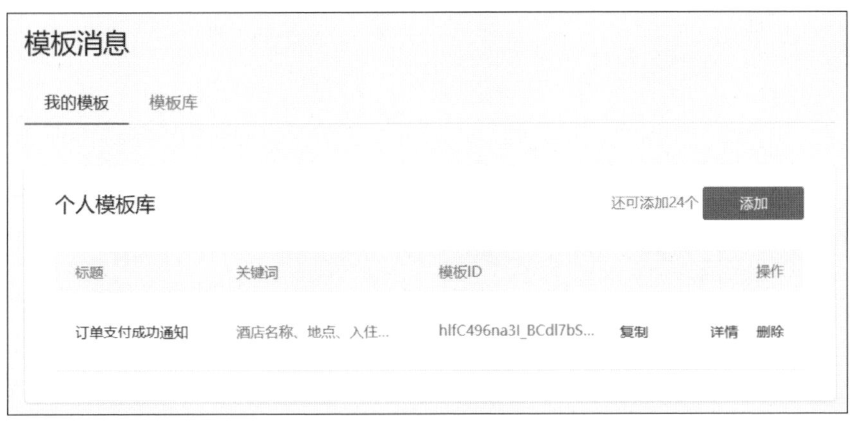

图 9.12 “模板消息”栏目

单击图 9.12 中的“添加”按钮，按照图 9.13 所示进行模板消息的创建，配置关键词的时候只能选中系统默认已经存在的关键词，不能新建关键词。被选中的关键词会出现在“已选中的关键词”栏目中，可以通过长按鼠标左键进行拖动的方式来调整已选中的关键词的上下顺序，这个顺序就决定了图 9.14 中关键词的顺序。在选中所有意向关键词之后，单击“提交”按钮，即完成模板消息在小程序管理账号中的创建操作。

图 9.13 创建模板消息

图 9.14　创建好的一个模板消息

9.5.1　模板消息前端

小程序模板消息接口地址为 https://api. weixin. qq. com/cgi-bin/message/wxopen/template/send? access_token=ACCESS_TOKEN,该模板消息接口的参数格式如下：

```
{
  "touser": "openid",
  "template_id": "template_id ",
  "page": "index",
  "form_id": "formId",
  "data": {
      "keyword1": {
          "value": "339208499"
      },
      "keyword2": {
          "value": "2015 年 01 月 05 日 12:30"
      },
      "keyword3": {
          "value": "粤海喜来登酒店"
      } ,
      "keyword4": {
          "value": "广州市天河区天河路 208 号"
      }
  },
  "emphasis_keyword": "keyword1.DATA"
}
```

图 9.14 是在小程序管理后台创建好的一条微信模板消息,以发送图 9.14 中的消息为例,前端需要提供图 9.14 中的 keyword1、keyword2、keyword3、keyword4、keyword5 和 keyword6 6 个关键词；从模板消息接口地址看,小程序前端需要提供 ACCESS_TOKEN；从模板消息接口参数来看,小程序前端需要提供 openid、template_id、page 和 formId 等参数,这 3 组参数共计 11 个,绝大部分应该由前端提供并发送给后端程序,其中 template_id 和 page 可以在后端程序中编写,也可以由前端小程序传递。

视频讲解

例 9-8 本例是发送图 9.14 所示的模板消息的前端小程序部分。图 9.15 是项目前端效果图，图 9.16 为模板消息成功发送之后微信消息提示的效果图，图 9.17 是 Console 控制台打印的结果。

图 9.15 模板消息案例前端图

图 9.16 模板消息成功发送后微信消息提示

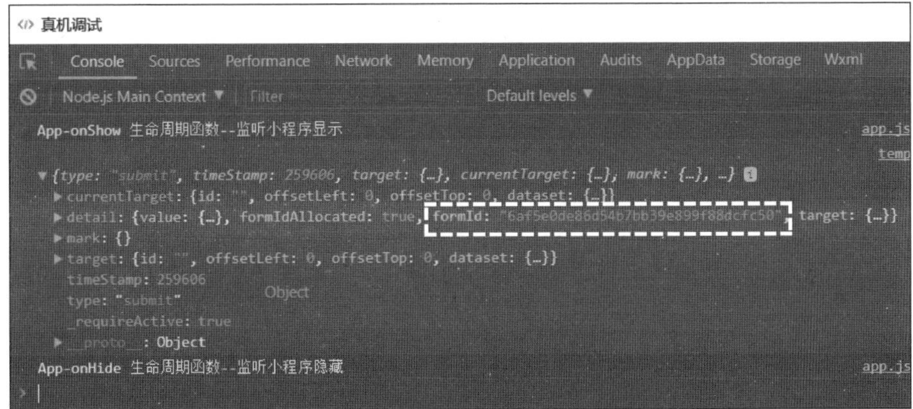

图 9.17 Console 控制台打印的结果

pages/template/template.wxml 的代码如下：

```
< form report - submit = 'true' bindsubmit = "formSubmit" bindreset = "formReset">
< view class = "add - page">
```

```
<view class = "mod - a">
    <! -- 酒店名称 -->
    <view class = "mod t - name">
        <text class = "key">酒店名称</text>
      <input class = "input" maxlength = "100" placeholder = "请输入酒店名称"
      bindKeyInput = "bindKeyInput" name = "hotelname"/>
        <image class = "arrow - r" src = "../../image/arrow - r - 0.png"></image>
    </view>
</view>
<view class = "mod - a mt20">
    <! -- 门店 -->
    <view class = "mod t - address" bindtap = "chooseLocation">
        <text class = "key">门店地址</text>
        <text class = "value">{{address}}</text>
        <image class = "arrow - r" src = "../../image/arrow - r - 0.png"></image>
    </view>
</view>
<view class = "mod - a mt20">
    <! -- 入住时间 -->
    <view class = "mod t - time">
        <view class = "start">
            <text class = "key">入住于</text>
          <picker mode = "date" value = "{{startDay}}" start =
        "{{startDay}}" bindchange = "startDateChange">
                <view class = "date">{{startDay}}< image class = "arrow - d"
        src = "../../image/arrow - d - 0.png"></image></view>
            </picker>
        </view>
        <view class = "pipe"></view>
        <view class = "end">
            <text class = "key">离店</text>
          <picker mode = "date" value = "{{endDay}}" start = "{{endDay}}"
            bindchange = "endDateChange">
              <view class = "date">{{endDay}}< image class = "arrow - d"
          src = "../../image/arrow - d - 0.png"></image></view>
            </picker>
        </view>
    </view>
</view>
 <view class = "mod - a">
    <! -- 姓名 -->
    <view class = "mod t - name">
        <text class = "key">您的姓名</text>
      <input class = "input" maxlength = "100" placeholder = "请输入您的姓名"
      bindKeyInput = "bindKeyInput" name = "name"/>
        <image class = "arrow - r" src = "../../image/arrow - r - 0.png"></image>
    </view>
</view>
<view class = "mod - a">
    <! -- 联系电话 -->
    <view class = "mod t - name">
```

```
            < text class = "key">联系电话</text >
          < input class = "input" maxlength = "100" placeholder = "请输入您的电话"
            bindKeyInput = "bindKeyInput" name = "tel"/>
            < image class = "arrow - r" src = "../../image/arrow - r - 0.png"></image >
        </view >
    </view >
    <! -- 创建按钮 -- >
    < view class = "create">
        < button class = "btn"   form - type = "submit">提交</button >
    </view >
</view ></form >
```

【代码讲解】 为了不打扰用户，微信小程序只能在提交表单或者完成支付之后才可以向用户发送模板消息，其他情况下系统是不允许它发送模板消息的。本例以提交表单为例调用模板消息接口。页面的< form >组件的 report-submit 属性为 true 时，表明需要发送模板消息，此时单击"提交"按钮提交表单可以获取 formid 属性，如图 9.17 所示，formid 属性是发送模板消息必需的参数。当小程序发生支付行为而调用模板消息时，应先获取 prepay_id 作为发送模板消息的参数，formId 和 prepay_id 是提交表单和完成支付两种情况下发送模板消息所需要的参数。

本例中"酒店名称""客户姓名""电话"由< input >组件实现，"门店地址"则采用 wx .chooseLocation()接口由客户选择地理位置，"入住时间"和"离店时间"则由< picker >组件实现。

需要注意的是 formid 有效期为 7 天，模板消息需要在表单提交之后的 7 天之内发送，超过这个期限则不可以再发送该表单对应的模板消息。

pages/template/template.js 的代码如下：

```
Page({
  data: {
    hotelname: '',
    address: '点击选择地点',
    startDay: '2019 - 11 - 00',
    endDay: '2019 - 11 - 01',
    name:null,
    tel:null,
    formid:null,
    openid: '',
    AccessToken:null,
  },
  //设置地点
  chooseLocation: function() {
    var that = this;
    wx.chooseLocation({
      success: function(res){
        console.log(res)
        that.setData({
          address: res.address,
        })
```

```
      },
      fail: function() {
        //fail
      },
      complete: function() {
        //complete
      }
    })
  },
  //设置开始日期
  startDateChange: function(e) {
    this.setData({
      startDay: e.detail.value
    })
  },
  //设置结束日期
  endDateChange: function(e) {
    this.setData({
      endDay: e.detail.value
    })
  },
  onLoad:function()
  {
      var that = this;
     wx.request({
     url: 'http://localhost:8080/wxtemplate/getAccessToken',
           //改成自己的服务器地址
      data: {
      },
      header: {
        'content - type': 'application/json'                    //默认值
      },
      success: function(res) {
        console.log("getAccessToken 的 ---- " + res.data.access_token)
        that.setData({
          AccessToken: res.data.access_token
        })
      }
    })
      wx.login({
        success: function(res) {
          console.log('code:' + res.code)
          //发送请求
          wx.request({
            url: 'http://localhost:8080/wxtemplate/loginservlet',
              //改成自己的服务器地址
            data: {
              code: res.code,//上面的 wx.login()成功获取的 code
              operFlag: 'getOpenid',
            },
            header: {
```

```
              'content-type': 'application/json'          //默认值
            },
            success: function(res) {
              console.log("getopenid----" + res.data.openid)
              that.setData({
                openid: res.data.openid
              })
            }
          })
        }
      })
    },
    //发送模板
    formSubmit: function(e) {
      console.log(e)
      this.setData({
        name:e.detail.value.name,
        hotelname: e.detail.value.hotelname,
        tel: e.detail.value.tel,
        formid: e.detail.formId
      })
      wx.request({
    url: 'http://localhost:8080/wxtemplate/sendmessage',          //改成自己的服务器地址
       //method: "POST",
        data: {
          ACCESS_TOKEN:this.data.AccessToken,
          openid:this.data.openid,
          formid:"e2e30c878c3a4cd189e4199416319021",
          keyword11: this.data.hotelname,
          keyword21: this.data.address,
          keyword31: this.data.startDay,
          keyword41: this.data.endDay,
          keyword51: this.data.name,
          keyword61: this.data.tel
        },
        header: {

          'content-type': 'application/json'          //默认值
        },
        success: function(res) {
          console.log("模板消息发送成功----" + res.data)
        }
      })
    }
  })
```

【代码讲解】 本例使用 wx.chooseLocation()接口获取地理位置的名称 res.address 作为“门店地址”，函数 startDateChange 和 endDateChange 获取< picker >组件的值赋值给入

住时间 startDay 和离店时间 endDay。

onLoad 函数完成了 openid 和 access_token 的获取。openid 的获取可以参看例 9-3。access_token 是后台接口调用凭据，小程序调用绝大多数后台接口时都需使用 access_token，开发者需要妥善保存。access_token 的小程序官方获取接口地址为 https://api.weixin.qq.com/cgi-bin/token?grant_type=client_credential&appid=APPID&secret=APPSECRET，其请求方式和 openid 的请求方式类似，其返回数据示例如下：

正常时返回：

```
{"access_token":"ACCESS_TOKEN","expires_in":7200}
```

错误时返回：

```
{"errcode":40013,"errmsg":"invalid appid"}
```

在调试本例程序的时候，需要掌握一定的技巧才能出现如图 9.16 所示的发送模板消息成功的结果，技巧在于要先使用真机调试得到如图 9.17 所示的 formid，然后再把得到的具体的 formid 作为发送模板消息的参数，并且要用模拟器调试。其中的道理在于模拟器调试是不会产生 formid 的，要获取 formid 只能进行真机调试，而真机调试又不能访问本例中 JS 文件中的后端接口，因为本例中的所有的后端接口都没有部署在网络上，而是在本地的 Tomcat 地址，所以真机调试或者模拟器调试在本例中都不能看到模板消息发送成功的结果，但如果程序部署到了网络上就不存在此类问题了。

9.5.2 模板消息后端

视频讲解

例 9-9 模板消息的发送是需要后端程序配合的，本例是例 9-8 的后端部分。图 9.18 为后端项目结构图，项目中包含 getAccessToken.java、loginservlet.java、sendmessage.java 和 tool.java 4 个文件，其中前 3 个文件为 Servlet 文件，tool.java 为普通工具类。后端项目在 Console 中的打印结果如图 9.19 所示。后端程序正确的情况下，模板消息可以如图 9.16 所示成功发送。

```
wxtamplate
  Deployment Descriptor: wxtamplate
  Deployed Resources
  JavaScript Resources
  src
    servlet
      getAccessToken.java
      loginservlet.java
      sendmessage.java
      tool.java
  JRE System Library [jdk17]
  Apache Tomcat v8.0 [Apache Tomcat v8
  JSTL 1.2.1 Library
  WebRoot
```

图 9.18 模板消息案例后端

```
Problems  Tasks  Console  Workspace Migration
Tomcat v8.0 Server at localhost [MyEclipse Server] D:\java
Content-Length--->[82]
Content-Type--->[text/plain]
Connection--->[keep-alive]
results{"session_key":"QN+Hg4Vj+D7NTF2BXYRKxw=
url=https://api.weixin.qq.com/cgi-bin/message/
keyword21====广东省广州市越秀区东风中路259号
keyword31====2019-11-00
keyword61====18888888888
keyword21++++广东省广州市越秀区东风中路259号
keyword31++++2019-11-00
```

图 9.19 Console 控制台中的打印结果

loginservlet.java 的代码和例 9-4 中的 loginservlet.java 代码一样，读者可以参考例 9-4 代码。

src/servlet/getAccessToken.java 的代码如下：

```
package servlet;
import java.io.BufferedReader;
import java.io.IOException;
import java.io.InputStreamReader;
import java.io.PrintWriter;
import java.net.URL;
import java.net.URLConnection;
import java.util.List;
import javax.servlet.ServletException;
import javax.servlet.annotation.WebServlet;
import javax.servlet.http.HttpServlet;
import javax.servlet.http.HttpServletRequest;
import javax.servlet.http.HttpServletResponse;
/*** Servlet implementation class loginservlet */
@WebServlet("/getAccessToken")
public class getAccessToken extends HttpServlet {
    private static final long serialVersionUID = 1L;
    private String appid = "wx800eb4451b223c35";
    private String secretKey = "c618f0e1d9038d956bc405c9a0a837d1";
    @Override
    protected void doGet (HttpServletRequest req, HttpServletResponse resp) throws
ServletException, IOException {
        System.out.println("doGet--------------------");
        doPost(req, resp);
    }
    @Override
    protected void doPost(HttpServletRequest request, HttpServletResponse response)
            throws ServletException, IOException {
        System.out.println("doPost--------------------");
        String results = "";
        String url = "https://api.weixin.qq.com/cgi-bin/token?grant_type=
                client_credential&appid=" + appid + "&secret=" + secretKey;
        System.out.println("url" + url);
        results = tool.sendGetReq(url);                      //发送 HTTP 请求
        System.out.println("AccessToken 是" + results);
        response.setContentType("application/json;charset=UTF-8");
        response.setHeader("catch-control", "no-catch");
        PrintWriter out = response.getWriter();
        out.write(results);
        out.flush();
        out.close();
    }
}
```

【代码讲解】 getAccessToken.java是一个Servlet文件。文件中tool.sendGetReq()函数和例9-4中的tool.sendGetReq()是同一函数。本文件功能是获取access_token，access_token的接口地址是https://api.weixin.qq.com/cgi-bin/token?grant_type=client_credential&appid=APPID&secret=APPSECRET，该接口需要appid和secretKey两个对象作为参数。

src/servlet/sendmessage.java的代码如下：

```
protected void doPost(HttpServletRequest request, HttpServletResponse response) throws
    ServletException, IOException {
          String ACCESS_TOKEN = request.getParameter("ACCESS_TOKEN");
          String tmpurl  =  "https://api.weixin.qq.com/cgi - bin/message/wxopen
           /template/send?access_token = " + ACCESS_TOKEN;
        System.out.println("url = " + tmpurl);
        JSONObject json  =  new JSONObject();
        String openid = request.getParameter("openid");
        String templat_id = "hlfC496na3I_BCdl7bSA4a2 - Z_mOg2qjFko - hGkMtf4";
        String pageurl = "m.baidu.com";
        String formId = request.getParameter("formId");
        String keyword11 = request.getParameter("keyword11");
        String keyword21 = request.getParameter("keyword21");
        System.out.println("keyword21 ==== " + keyword21);
        String keyword31 = request.getParameter("keyword31");
        System.out.println("keyword31 ==== " + keyword31);
        String keyword41 = request.getParameter("keyword41");
        String keyword51 = request.getParameter("keyword51");
        String keyword61 = request.getParameter("keyword61");
        System.out.println("keyword61 ==== " + keyword61);
        JSONObject data = tool.packJsonmsg(keyword11, keyword21, keyword31, keyword41,
        keyword51, keyword61);
        try {
            json.put("touser", openid);
            json.put("template_id", templat_id);
            json.put("page", pageurl);
            json.put("form_id", formId);
            json.put("data",data);
        } catch (JSONException e) {
            e.printStackTrace();
        }
        String result  =  tool.httpsRequest(tmpurl, "POST", json.toString());
       //System.err.println(result);
       //return "success";
        response.setContentType("application/json;charset = UTF - 8");
        response.setHeader("catch - control", "no - catch");
        PrintWriter out  =  response.getWriter();
        out.write(result);
        out.flush();
```

```
        out.close();
    }
```

【代码讲解】 sendmessage.java 的代码由两部分组成。前半部分是接收小程序前端传递过来的多个参数，并形成 9.5.1 节中的模板消息标准格式的参数。在形成模板消息标准格式参数的过程中，调用了 tool.java 中的 packJsonmsg 函数，packJsonmsg 负责把小程序前端传递过来的 keyword11、keyword21、keyword31、keyword41、keyword51 和 keyword61 封装成标准格式参数的 data 部分。后半部分是向模板消息接口发起请求，并形成 JSON 格式结果返回给前端小程序。

src/servlet/tool.java 的代码如下：

```
public static JSONObject packJsonmsg(String keyword11, String keyword21, String keyword31,
    String keyword41, String keyword51, String keyword61){
        JSONObject json = new JSONObject();
        try {
            JSONObject keyword1 = new JSONObject();
            keyword1.put("value", keyword11);
            keyword1.put("color", "#173177");
            json.put("keyword1", keyword1);
            JSONObject keyword2 = new JSONObject();
            keyword2.put("value", keyword21);
            keyword2.put("color", "#173177");
            json.put("keyword2", keyword2);
            System.out.println("keyword21++++" + keyword21);
            JSONObject keyword3 = new JSONObject();
            keyword3.put("value", keyword31);
            keyword3.put("color", "#173177");
            json.put("keyword3", keyword3);
            System.out.println("keyword31++++" + keyword31);
            JSONObject keyword4 = new JSONObject();
            keyword4.put("value", keyword41);
            keyword4.put("color", "#173177");
            json.put("keyword4", keyword4);
            JSONObject keyword5 = new JSONObject();
            keyword5.put("value", keyword51);
            keyword5.put("color", "#173177");
            json.put("keyword5", keyword5);
            JSONObject keyword6 = new JSONObject();
            keyword6.put("value", keyword61);
            keyword6.put("color", "#173177");
            json.put("keyword6", keyword6);
        } catch (JSONException e) {
            e.printStackTrace();
        }
        return json;
    }
```

【代码讲解】 tool.java 的核心函数为 packJsonmsg(),packJsonmsg()函数,它把小程序前端传递过来的 keyword11、keyword21、keyword31、keyword41、keyword51 和 keyword61 封装成模板消息标准格式参数的 data。

9.6 权限接口

部分接口需要经过用户授权同意才能调用。我们把这些接口按使用范围分成多个 scope,用户选择对 scope 进行授权,当授权给一个 scope 之后,其对应的所有接口都可以直接使用。此类接口调用时:

(1) 如果用户未接受或拒绝过此权限,会弹窗询问用户,用户单击同意后方可调用接口。

(2) 如果用户已授权,可以直接调用接口。

(3) 如果用户已拒绝授权,则不会出现弹窗,而会直接进入接口 fail 回调。

此类接口在权限中的对象 scope 的字段和接口的对应关系如表 9.9 所示。

表 9.9 scope 列表

scope	对应接口	描述
scope.userInfo	wx.getUserInfo	用户信息
scope.userLocation	wx.getLocation, wx.chooseLocation	地理位置
scope.userLocationBackground	wx.userLocationBackground	后台定位
scope.address	wx.chooseAddress	通信地址
scope.invoiceTitle	wx.chooseInvoiceTitle	发票抬头
scope.invoice	wx.chooseInvoice	获取发票
scope.werun	wx.getWeRunData	微信运动步数
scope.record	wx.startRecord	录音功能
scope.writePhotosAlbum	wx.saveImageToPhotosAlbum, wx.saveVideoToPhotosAlbum	保存到相册
scope.camera	camera 组件	摄像头

小程序提供了 3 组接口对接口权限进行相应的操作:wx.getSetting()获取用户当前的授权状态;wx.openSetting()打开设置界面以引导用户开启授权;wx.authorize()改变授权状态。

9.6.1 用户授权接口 wx.authorize()

该接口调用后会立刻弹窗询问用户是否同意授权小程序,使用某项功能或获取用户的某些数据,但不会实际调用对应接口。如果用户之前已经同意授权,则不会出现弹窗,直接返回成功。

示例代码:

```
wx.getSetting({
```

```
  success(res) {
    if (!res.authSetting['scope.record']) {
      wx.authorize({
        scope: 'scope.record',
        success() {
          //用户已经同意小程序使用录音功能,后续调用 wx.startRecord 接口不会弹窗询问
          wx.startRecord()
        }
      })
    }
  }
})
```

9.6.2　获取用户权限设置接口 wx.getSetting()

该接口获取用户的当前设置。返回值中只会出现小程序已经向用户请求过的权限。

示例代码：

```
wx.getSetting({
  success(res) {
    console.log(res.authSetting)
    //res.authSetting = {
    //   "scope.userInfo": true,
    //   "scope.userLocation": true
    //}
  }
})
```

9.6.3　打开用户权限设置界面接口 wx.openSetting()

该接口调用客户端小程序设置界面,返回用户设置的操作结果。设置界面只会出现小程序已经向用户请求过的权限。

注意：2.3.0 版本开始，用户发生点击行为后，才可以跳转打开设置页，管理授权信息详情。

示例代码：

```
wx.openSetting({
  success(res) {
    console.log(res.authSetting)
    //res.authSetting = {
    //   "scope.userInfo": true,
    //   "scope.userLocation": true
    //}
  }
})
```

视频讲解

【例】9-10 本例使用获取地理位置接口 wx.getLocation()和开始录音接口 wx.startRecord()进行相关操作，而这两个接口都需要设置操作权限。图 9.20 是使用 wx.authorize()接口对位置操作进行授权；图 9.21 是使用 wx.openSetting()接口对录音操作进行授权。

图 9.20 wx.authorize()接口授权

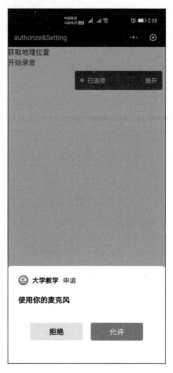

图 9.21 wx.openSetting()接口授权

pages/authorize&Setting/authorize&Setting.wxml 的代码如下：

```
<view class = "body" bindtap = "location1">获取地理位置</view>
<view  class = "body">{{context}}</view>
<view class = "body" bindtap = "location2">开始录音</view>
```

pages/authorize&Setting/authorize&Setting.js 的代码如下：

```
Page({
  onLoad: function() {
    context:''
  },
  location1:function(){
    var that = this
    wx.getSetting({
      success(res) {
        console.log(res)
        if (!res.authSetting['scope.userLocation']) {
          wx.authorize({
            scope: 'scope.userLocation',
            success() {
```

```
                wx.getLocation({
                  success: function(res) {
                   console.log(res)
                    that.setData({ context: "你所在的经度是" + res.latitude + "你所在的纬度
                    是" + res.longitude})
                  },
                })
              }
            })
          }
        }
      })
    },
    location2: function() {
              wx.getSetting({
            success(res) {
              console.log(res.authSetting)
              if (!res.authSetting['scope.record']) {
                wx.openSetting({
                  success(res) {
                    console.log(res)
                    wx.startRecord({
                      success(res) {
                        const tempFilePath = res.tempFilePath
                        console.log("录音结束")
                      }
                })
              }
        })
        }}
        })
    }
  })
```

【代码讲解】 location1()函数实现获取地理位置的功能，该函数先调用 wx.getSetting()接口获取权限状态，然后调用 wx.authorize()接口修改地理位置权限 scope.userLocation。location2()函数实现录音功能，该函数先调用 wx.getSetting()接口获取权限状态，然后调用 wx.openSetting()接口打开录音权限设置界面来修改录音权限。从本例可以看出设置权限的时候应该先调用 wx.getSetting()接口获取权限状态，在权限没有打开的情况下可以调用 wx.authorize()接口或者 wx.openSetting()接口来修改权限状态，wx.authorize()接口不出现修改权限的操作界面，而 wx.openSetting()接口会出现修改权限的操作界面。本案例请使用手机真机调试，否则会出现"openSetting:fail can only be invoked by user TAP gesture."报错。

9.7 微信运动接口 wx.getWeRunData()

该接口获取用户过去 30 天微信运动步数。该接口比较复杂，具体操作步骤如下：

(1) 先调用 wx.login()接口获取 code。

(2) 使用 wx.request()接口向地址 https://api.weixin.qq.com/sns/jscode2session?appid=appid&secret=SECRET&js_code=JSCODE&grant_type=authorization_code 发送网络请求,此时需要发送 appid、secret 和 code 3 个参数,返回 session_key。

(3) 调用 wx.getWeRunData()接口返回 encryptedData 和 iv,其中 encryptedData 是加密的步数信息,需要解密为明文。

(4) 在小程序中使用 CryptoJS 组件对 encryptedData 进行解密,此时需要用到 appid、session_key 和 iv,解密得到明文的步数信息是小程序的可读信息。

例 9-11 本例使用 wx.getWeRunData()接口获取用户的运动步数并显示,程序运行结果如图 9.22 所示,如图 9.23 所示的项目结构中的虚线框选中的文件是引入的 CryptoJS 组件文件,该组件可以在 https://www.npmjs.com/package/crypto-js 下载得到。

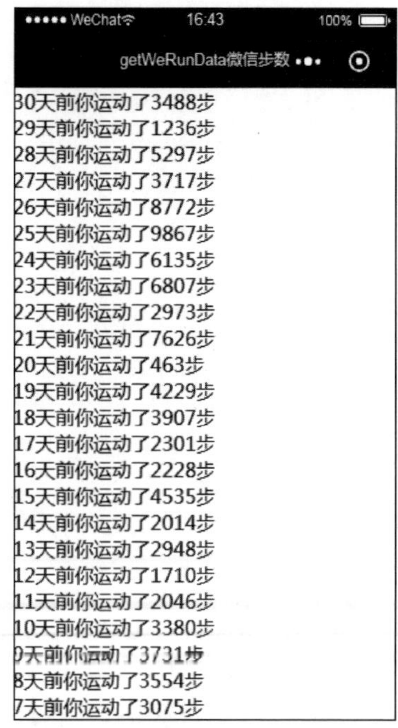

图 9.22 获取的微信运动步数

视频讲解

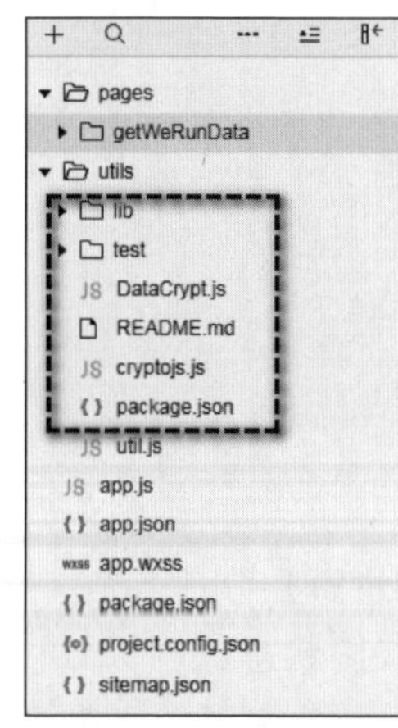

图 9.23 项目结构

../../utils/DataCrypt.js 的代码如下:

```
var Crypto = require('cryptojs.js').Crypto;
var app = getApp();
function RdWXBizDataCrypt(appid, sessionKey) {
  this.appid = appid
  this.sessionKey = sessionKey
}
RdWXBizDataCrypt.prototype.decryptData = function (encryptedData, iv) {
  var encryptedData = Crypto.util.base64ToBytes(encryptedData)
  var key = Crypto.util.base64ToBytes(this.sessionKey);
  var iv = Crypto.util.base64ToBytes(iv);
  var mode = new Crypto.mode.CBC(Crypto.pad.pkcs7);
  try {
```

```
    var bytes = Crypto.AES.decrypt(encryptedData, key, {
        asBpytes:true,
        iv: iv,
        mode: mode
    });
    var decryptResult = JSON.parse(bytes);
  } catch (err) {
    console.log(err)
  }
  if (decryptResult.watermark.appid !== this.appid) {
    console.log(err)
  }
  return decryptResult
}
module.exports = RdWXBizDataCrypt
```

pages/pages/getWeRunData/getWeRunData.wxml 的代码如下：

```
<view wx:for="{{stepInfoList}}">{{30 - index}}天前你运动了{{item.step}}步</view>
```

pages/getWeRunData/getWeRunData.js 的代码如下：

```
var DataCrypt = require('../../utils/DataCrypt.js');
Page({
  data: {
    stepInfoList:[]
  },
  onLoad: function(options) {
    var that = this;
     wx.login({
       success: function(res) {
         var appid = "wx800eb4451b223c35";
         var secret = "8a854735c1dea5dab2b0c09b6c4ab6b0";
         if (res.code) {
           wx.request({
             url: 'https://api.weixin.qq.com/sns/jscode2session?appid=' +
             appid + '&secret=' + secret + '&js_code=' + res.code +
             '&grant_type=authorization_code',
             header: { 'content-type': 'json' },
             success: function(res) {
                  var session_key = res.data.session_key;
                  console.log(session_key);
               that.getWeRunData(appid, session_key); }
                  })
                 }
             }
             })
   },
    //获取 encryptedData(没有解密的步数)和 iv(加密算法的初始向量)
  getWeRunData: function(appid,session_key) {
      var that = this
```

```
        wx.getSetting({ success: function (res) {
          console.log(res);
          if (!res.authSetting['scope.werun'])
          {wx.showModal
           ({
              title: '权限提示'
              content: '获取微信运动步数需要开启计步权限'
              success: function(res)
               {
                if (res.confirm) {
                  //跳转去设置
                  wx.openSetting({
                    success: function(res)
                    {}
                })
                }
               }
           })
         } else {
                wx.getWeRunData({
                  success: function(res) {
                    console.log(res);
                    var encryptedData = res.encryptedData;
                    var iv = res.iv;
                    var pc = new DataCrypt(appid, session_key);
                    console.log(pc);
                    var data = pc.decryptData(encryptedData, iv)
                   console.log("------" + data.stepInfoList[0].step)
                  that.setData({
                   stepInfoList: data.stepInfoList})
                  },
           fail: function(res) { console.log("获取数据失败") }
         })
       }
     }
     })
   }
})
```

【代码讲解】 本例中 CryptoJS 组件属于网络开源项目，下载之后需要按图 9.23 所示引入到小程序项目中，DataCrypt.js 文件的作用是把 CryptoJS 组件引入到小程序项目。wx.getWeRunData()接口获取的 encryptedData 经过 CryptoJS 组件解密之后得到的明文数据 data 中包含了微信运动数据，具体的数据在 data 的 stepInfoList 数组中，该数组共 31 条记录，对应当日和前 30 天的运动步数数据。其中，stepInfoList[0]是 30 天前的运动步数数据，stepInfoList[30]是当日的运动步数数据。

9.8 其他常见开放接口

本章前七节介绍了微信登录 wx. login()、微信支付 wx. requestPayment()、获取微信信息 wx. getUserInfo()和模板消息 template 等高级开放接口。除此之外,小程序还有多个开放接口,因为篇幅所限,本节将介绍小程序间跳转 wx. navigateToMiniProgram()、获取用户收货地址 wx. chooseAddress()和指纹认证 SOTER 3 个开放接口。

9.8.1 小程序间跳转接口 wx. navigateToMiniProgram()

小程序可以通过 wx. navigateTo()和 wx. redirectTo()接口实现页面之间的跳转,小程序还提供了 wx. navigateToMiniProgram()接口实现当前小程序跳转到另一个小程序的功能。表 9.10 是 wx. navigateToMiniProgram()的属性表。

表 9.10 wx. navigateToMiniProgram()属性表

属性	类型	必填	说明
appid	string	是	要打开的小程序 appid
path	string	否	打开页面路径,如果为空则打开首页。path 中“?”后面的部分会成为 query,在小程序的 App. onLaunch、App. onShow 和 Page. onLoad 的回调函数或小游戏的 wx. onShow 回调函数、wx. getLaunchOptionsSync 中可以获取 query 数据。对于小游戏,可以只传入 query 部分,来实现传参效果,例如传入"? foo=bar"
extraData	object	否	需要传递给目标小程序的数据,目标小程序可在 App. onLaunch、App. onShow 中获取这份数据;如果跳转的是小游戏,可以在 wx . onShow、wx. getLaunchOptionsSync 中获取这份数据
envVersion	string	否	要打开的小程序版本,仅在当前小程序为开发版或体验版时此参数有效。如果当前小程序是正式版,则打开的小程序必定是正式版
success	function	否	接口调用成功的回调函数
fail	function	否	接口调用失败的回调函数
complete	function	否	接口调用结束的回调函数(调用成功、失败都会执行)

appid 属性是需要打开的小程序的 appid,如果是打开自身的页面则 appid 留空。path 属性为要打开的小程序的具体页面,即打开操作可以具体到被打开的小程序的某一页面。extraData 是需要传递给目标小程序的参数。envVersion 的取值为 develop、trial 和 release,它们分别代表开发版、体验版和正式版,因为一个 appid 代表一个小程序,而小程序有开发版、体验版和正式版 3 个版本。

例 9-12 本例 wx.navigateToMiniProgram()接口从当前小程序跳转到另外一个小程序,程序运行效果如图 9.24 所示。

视频讲解

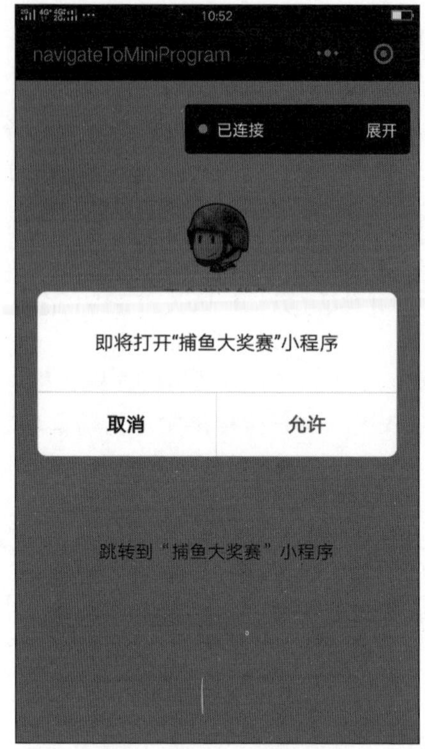

图 9.24 小程序间跳转

pages/navigateToMiniProgram/navigateToMiniProgram.wxml 的代码如下:

```
<view class="container">
  <view class="userinfo">
    <button wx:if="{{!hasUserInfo && canIUse}}" open-type="getUserInfo"
      bindgetuserinfo="getUserInfo"> 获取头像昵称 </button>
    <block wx:else>
      <image bindtap="bindViewTap" class="userinfo-avatar"
        src="{{userInfo.avatarUrl}}" mode="cover"></image>
      <text class="userinfo-nickname">{{userInfo.nickName}}</text>
    </block>
  </view>
  <view class="usermotto" bindtap='dian'>
    <text class="user-motto">跳转到"捕鱼大奖赛"小程序</text>
  </view>
</view>
```

pages/navigateToMiniProgram/navigateToMiniProgram.js 的代码如下:

```
Page({
dian:function(e){
    wx.navigateToMiniProgram({
      appid: 'wx3efb95b9c5579418',
```

```
      path: '',
      extraData: {
        foo: 'bar'
      },
      envVersion: 'release',
      success(res) {
        //打开成功
      }
    })
  }
})
```

【代码讲解】 在项目中要打开的外部小程序需要在全局 JSON 文件 app. json 中添加 navigateToMiniProgramAppIdList 属性的配置信息，把需要跳转过去的小程序的 appid 列在该属性内，如下面代码所示：

```
"navigateToMiniProgramAppIdList":[
   "wx3efb95b9c5579418",
   "wxc1039e003593f9b4",
   "wxc75cac912af33647"
   ]
```

9.8.2 获取用户收货地址接口 wx.chooseAddress()

wx. chooseAddress(Object object)获取用户收货地址，此接口调用的是微信的收货地址。微信收货地址不同于微信设置的省和市地址，它默认是不存在的。在某个小程序中，当用户第一次调用 wx. chooseAddress()接口时，会被要求填写微信收货地址；当用户第二次访问 wx. chooseAddress()接口时，则可以看到自己以前填写的收货地址。这里说的第二次可以是不同的小程序，例如用户在访问 A 小程序时填写了微信收货地址，则用户在访问带有 wx. chooseAddress()接口的 B 或者 C 小程序时都可以看到以前填写的微信收货地址。wx. chooseAddress()接口属性如表 9. 11 所示。

表 9. 11　wx. chooseAddress()属性表

属　　性	类型	说　　明
userName	string	收货人姓名
postalCode	string	邮编
provinceName	string	国标收货地址第一级地址
cityName	string	国标收货地址第二级地址
countryName	string	国标收货地址第三级地址
detailInfo	string	详细收货地址信息
nationalCode	string	收货地址国家码
telNumber	string	收货人手机号码
errMsg	string	错误信息

视频讲解

例 9-13　本例使用 wx.chooseAddress(Object object)获取微信收货地址,并把获取的收货地址信息填入图 9.25 的表单。因为是第一次访问 wx.chooseAddress()接口,所以点击“获取收货地址”按钮之后,出现了新建并保存收货地址页面,如图 9.26 所示。当用户点击图 9.26 右上角的“保存”按钮时,系统获取了微信收货地址并填入表单,如图 9.27 所示。

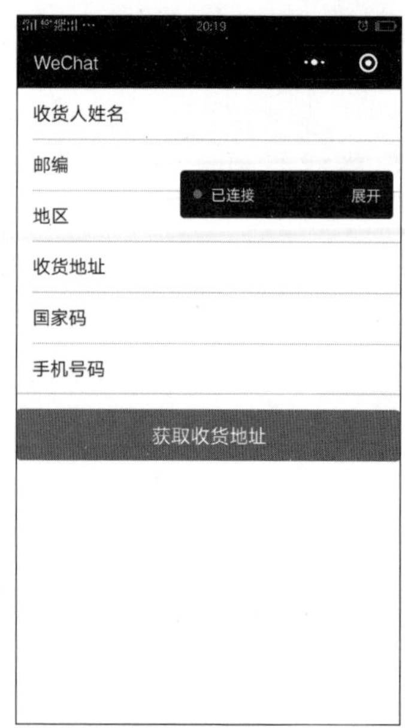

图 9.25　收货地址信息表单

图 9.26　新建并保存微信收货地址

图 9.27　微信收货地址被填入表单

pages/chooseAddress/chooseAddress.wxml 的代码如下:

```
<view class = "container">
  <form>
    <view class = "page - section">
      <view class = "weui - cells weui - cells_after - title">
        <view class = "weui - cell weui - cell_input">
          <view class = "weui - cell_hd">
            <view class = "weui - label">收货人姓名</view>
          </view>
          <view class = "weui - cell_bd">
            {{ addressInfo.userName }}
          </view>
        </view>
        <view class = "weui - cell weui - cell_input">
          <view class = "weui - cell_hd">
            <view class = "weui - label">邮编</view>
          </view>
          <view class = "weui - cell_bd">
            {{ addressInfo.postalCode }}
          </view>
        </view>
        <view class = "weui - cell weui - cell_input region">
```

```
            < view class = "weui - cell_hd">
              < view class = "weui - label">地区</view >
            </view >
            < view class = "weui - cell_bd">
              {{ addressInfo.provinceName }}
              {{ addressInfo.cityName }}
              {{ addressInfo.countyName }}
            </view >
          </view >
          < view class = "weui - cell weui - cell_input detail">
            < view class = "weui - cell_hd">
              < view class = "weui - label">收货地址</view >
            </view >
            < view class = "weui - cell_bd">
              {{ addressInfo.detailInfo }}
            </view >
          </view >
            < view class = "weui - cell weui - cell_input">
            < view class = "weui - cell_hd">
              < view class = "weui - label">国家码</view >
            </view >
            < view class = "weui - cell_bd">
              {{ addressInfo.nationalCode }}
            </view >
          </view >
          < view class = "weui - cell weui - cell_input">
            < view class = "weui - cell_hd">
              < view class = "weui - label">手机号码</view >
            </view >
            < view class = "weui - cell_bd">
              {{ addressInfo.telNumber }}
            </view >
          </view >
        </view >
      </view >
    </form >
    < view class = "btn - area"  hidden = "{{chooseAddress}}">
      < button type = "primary" bindtap = "chooseAddress">获取收货地址</button >
    </view >
     < view class = "btn - area"  hidden = "{{!chooseAddress}}">
      < button type = "primary" bindtap = "pay">提交</button >
    </view >
</view >
```

pages/chooseAddress/chooseAddress.js 的代码如下：

```
Page({
  data: {
    addressInfo: null,
```

```
      chooseAddress:false
    },
    chooseAddress:function(e){
      var that = this;
      wx.chooseAddress({
        success: function(res){
          that.setData({
            addressInfo: res,
            chooseAddress: true
          })
        },
        fail: function(err) {
          console.log(err)
        }
      })
    },
    pay:function(e){
      wx.navigateTo({
        url: '../pay/pay'
      })
    }
})
```

【代码讲解】 通过本例可以发现收货地址信息的 userName、postalCode、provinceName、cityName、countryName、detailInfo、nationalCode、telNumber 和 errMsg 属性都包含在 addressInfo 对象中。wx.chooseAddress()常在电商类小程序中被引用。

wx.chooseInvoiceTitle()接口和 wx.chooseAddress(Object object)接口类似,它用来获取用户的发票消息,包含抬头类型、抬头名称、抬头税号、单位地址、手机号码、银行名称和银行账号等属性。

9.8.3 SOTER 指纹认证

Tencent SOTER 是腾讯于 2015 年开始制定的生物认证平台与标准,通过与厂商合作,目前已经在一百余款、数亿部 Android 设备上得到支持,并且这个数字还在快速增长。Tencent SOTER 设计支持指纹认证、人脸认证和声纹认证,目前小程序只支持指纹认证。本节的 SOTER 认证特指指纹认证。

目前,Tencent SOTER 已经在微信指纹支付、微信公众号和小程序指纹授权接口等场景使用。接入 Tencent SOTER,小程序可以在不获取用户指纹图案的前提下,在 Android 设备上实现可信的指纹认证,获得与微信指纹支付一致的安全快捷认证体验。

小程序提供 wx.checkIsSupportSoterAuthentication()、wx.checkIsSoterEnrolledInDevice()和 wx.startSoterAuthentication()3 个接口来实现认证。wx.checkIsSupportSoterAuthentication()实现检测手机是否支持指纹认证的功能; wx.checkIsSoterEnrolledInDevice()实现检测手机安全设置系统是否已经录入过指纹信息; wx.startSoterAuthentication()判断当前录入的指纹是否与手机安全指纹一致。

wx.checkIsSupportSoterAuthentication()接口的 success 函数的回调结果保存在 res.supportMode 中。res.supportMode 的取值如表 9.12 所示。

表 9.12　res. supportMode 取值表

属　　性	说　　明
fingerPrint	指纹识别
facial	人脸识别(暂未支持)
speech	声纹识别(暂未支持)

wx. checkIsSoterEnrolledInDevice()检测手机是否保存了指纹、人脸和声纹信息,该接口的属性如表 9.13 所示。

表 9.13　wx. checkIsSoterEnrolledInDevice()属性表

属　　性	类　　型	必填	说　　明
checkAuthMode	string	是	认证方式
success	function	否	接口调用成功的回调函数
fail	function	否	接口调用失败的回调函数
complete	function	否	接口调用结束的回调函数(调用成功、失败都会执行)

因为前面只支持指纹认证,所以表 9.13 中的 checkAuthMode 的取值就只能是 fingerPrint。

wx. startSoterAuthentication()接口是真正执行 SOTER 认证的接口,也就是执行指纹认证,该接口属性如表 9.14 所示。

表 9.14　wx. startSoterAuthentication()属性表

属　　性	类　　型	必填	说　　明
requestAuthModes	Array. < string >	是	请求使用可接受的生物认证方式
challenge	string	是	挑战因子。挑战因子是调用者为此次生物鉴权准备的用于签名的字符串关键识别信息,将作为 resultJSON 的一部分,供调用者识别本次请求。例如,如果场景为请求用户对某订单进行授权确认,则可以将订单号填入此参数
authContent	string	否	验证描述,即识别过程中显示在界面上的对话框提示内容
success	function	否	接口调用成功的回调函数
fail	function	否	接口调用失败的回调函数
complete	function	否	接口调用结束的回调函数(调用成功、失败都会执行)

视频讲解

例 9-14　本例使用 SOTER 认证的 3 个接口验证小程序的指纹识别功能,程序运行效果如图 9.28 和图 9.29 所示。

pages/SoterAuthentication/SoterAuthentication. wxml 的代码如下:

```
<view class = "container">
  <view class = "page - body">
    <view class = "btn - area">
      <button type = "primary" bindtap = "startAuth">指纹认证</button>
    </view>
  </view>
</view>
```

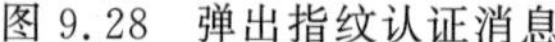

图 9.28　弹出指纹认证消息

图 9.29　认证成功

pages/SoterAuthentication/SoterAuthentication.js 的代码如下：

```
const AUTH_MODE = 'fingerPrint'
Page({
  startAuth() {
    wx.checkIsSupportSoterAuthentication({
      success: (res) => {
        console.log(res)
        checkIsEnrolled()
      },
      fail: (err) => {
        console.error(err)
        wx.showModal({
          title: '错误'
          content: '您的设备不支持指纹识别'
          showCancel: false
        })
      }
    })
    const checkIsEnrolled = () => {
      wx.checkIsSoterEnrolledInDevice({
        checkAuthMode: AUTH_MODE,
        success: (res) => {
          console.log(res)
```

```
            if (parseInt(res.isEnrolled) <= 0) {
              wx.showModal({
                title: '错误'
                content: '您暂未录入指纹信息,请录入后重试'
                showCancel: false
              })
              return
            }
            startSoterAuthentication();
          },
          fail: (err) => {
            console.error(err)
          }
        })
      }
      const startSoterAuthentication = () => {
        wx.startSoterAuthentication({
          requestAuthModes: [AUTH_MODE],
          challenge: 'test'
          authContent: '小程序示例'
          success: (res) => {
            console.log(res)
            wx.showToast({
              title: '认证成功',
              duration: 10000
            })
          },
          fail: (err) => {
            console.error(err)
            wx.showModal({
              title: '失败'
              content: '认证失败'
              showCancel: false
            })
          }
        })
      }
    }
  })
```

【代码讲解】 本例一次调用 wx. checkIsSupportSoterAuthentication()、wx. checkIsSoterEnrolledInDevice()和 wx. startSoterAuthentication()3 个接口,是规范的指纹认证流程。

9.9 实训项目——购物车与结算功能

视频讲解

本实训项目介绍一个简单的购物车与结算功能案例。项目共由 3 个页面构成:商品列表和购物车页面 goods. wxml、填写收货地址页面 chooseAddress

. wxml 及支付页面 pay. wxml。goods. wxml 页面通过“结算”按钮调用 chooseAddress 函数跳转到 chooseAddress. wxml。chooseAddress. wxml 页面通过“提交”按钮调用 pay 函数跳转到 pay. wxml 页面。pay. wxml 页面通过“支付”按钮调用例 9-5 的微信支付功能完成支付流程。项目执行效果如图 9.30～图 9.32 所示。

图 9.30　商品列表和购物车

图 9.31　填写收货地址

图 9.32　支付

pages/goods/goods. wxml 的代码如下：

```
<view class = "costumerinfo">
  <view class = "info">
    <view class = "info1">{{name}} {{tel}}</view>
    <view class = "info2">{{address}}</view>
  </view>
  <view class = "info3">
  </view>
</view>
<view class = "wxpay"> 支付方式：微信支付</view>
<view class = "cartlist">
  <view class = "table">
   <view class = "item" wx:for = "{{cart.list}}" wx:for - index = "id" wx:for - item = "num" wx:
key = "id">
    <view class = "tr">
      <view class = "td" class = "itemname">{{foodlist[id].name}}</view>
      <view class = "td" class = "itemnum">×{{cart.list[id]}}</view>
      <view class = "td" class = "itemprice">¥{{foodlist[id].price * cart.list[id]}}</view>
    </view>
    </view>
    <view></view>
    <view class = "trbom">
      <view class = "Discount1"> 满减活动</view>
```

```
      <view class="Discount2">- ¥10</view>
    </view>
  </view>
  <view class="total0">
    <view class="total1">订单 ¥{{cart.total}}</view>
    <view class="total2">优惠 ¥10</view>
    <view class="total3">待支付 ¥{{cart.total-10}}</view>
  </view>
</view>
<view class="pay" bindtap='orderpay'>
  <view class="pay1">待支付 ¥{{cart.total-10}} </view>
  <view class="pay2">| 优惠 ¥10</view>
  <button class="paybtn"> 支付</button>
</view>
```

【代码讲解】 good.wxml 主要使用页面的样式与布局知识点，整个 wxml 文件分为商品显示部分、隐藏的购物车部分和底部结算栏三部分。本例通过 hidden 属性的 false 和 true 的取值变化来控制购物车的隐藏和显示。商品显示部分使用了 wx:for 循环渲染来显示多个商品的"图片""价格"和"标题"字段。

pages/goods/goods.js 的代码如下：

```
var app = getApp();
Page({
  data: {
    filterId: 1,
   showhiddencart:false,
    foodlist:
    [
      {
        id: 1,
        name: '海南13.5度甜玉米新鲜现摘',
        pic: '/imgs/good/1.jpg',
        price:29.8
      },
      {
        id: 2,
        name: '枣阳黄桃新鲜当季水蜜桃',
        pic: '/imgs/good/2.jpg',
        price:59.8
      },
      {
        id: 3,
        name: '墨西哥进口牛油果宝宝辅食',
        pic: '/imgs/good/3.jpg',
        price:39.8
      },
      {
        id: 4,
        name: '正宗土鸡散养农家老母鸡',
        pic: '/imgs/good/4.jpg',
```

```
        price:58
      },
      {
        id: 5,
        name: '麻辣小龙虾 1.8 斤装十三香',
        pic: '/imgs/good/5.jpg',
        price:58
      },
      {
        id: 6,
        name: '洪泽湖大闸蟹鲜活现货',
        pic: '/imgs/good/6.jpg',
        price:148
      }
    ],
    cart: {
      count: 0,
      total: 0,
      list: {}
    },
  },
  onLoad: function() {
    console.log("onLoad");
    this.setData({
      id: 0
    });
  },
  //计算
  tapAddCart: function(e) {
    console.log(e);
    console.log(e.target.id);
    this.addCart(e.target.id);
  },
  tapReduceCart: function(e) {
    console.log(e);
    this.reduceCart(e.target.id);
  },
  addCart: function(id) {
    console.log(this.data.cart.list);
    var num = this.data.cart.list[id] || 0;
    console.log("num" + num);
    this.data.cart.list[id] = num + 1;
    this.countCart();
  },
  reduceCart: function(id) {
    var num = this.data.cart.list[id] || 0;
    if (num <= 1) {
      delete this.data.cart.list[id];
    } else {
      this.data.cart.list[id] = num - 1;
    }
```

```
    this.countCart();
  },
  countCart: function() {
    var count = 0,
      total = 0;
    for (var id in this.data.cart.list) {
      var goods = this.data.foodlist[id];
      count += this.data.cart.list[id];
      total += goods.price * this.data.cart.list[id];
    }
    this.data.cart.count = count;
    this.data.cart.total = total;
    this.setData({
      cart: this.data.cart
    });
  },
  //显示隐藏购物车
  showhiddencart: function() {
    this.setData({
      showhiddencart: !this.data.showhiddencart
    });
  },
  closehiddencart: function() {
    this.setData({
      showhiddencart: false
    });
  },
  chooseAddress: function() {
    try {
      wx.setStorageSync('cart',this.data.cart)
    } catch (e) { console.log("good.js 的 wx.setStorageSync 出错") }
    wx.navigateTo({
      url: '../chooseAddress/chooseAddress'
    })
  },
});
```

【代码讲解】 goods.js 通过 wx.setStorageSync('cart',this.data.cart)将购物车的内容存储到数据缓存区，以供 pay.wxml 文件使用。本例核心功能是通过 addCart 和 reduceCart 两个函数实现商品在购物车中的添加和删除功能。

pages/goods/goods.wxss 的代码如下：

```
page{font-family: "微软正黑体";}
/* 主体部分 */
.body{position:relative;top:10rpx;padding: 8rpx 0;margin-bottom:100rpx;}
.body .item{width: 50%;float:left;text-align: center;}
.body .item .goodsbox{background: #fff;width:90%;margin:0 auto;
       height:400rpx;position:relative;margin-top:30rpx;box-shadow:0 0 5px  #eee;}
.body .item image{width: 100%;height:250rpx;}
.body .item .price{color: #3c3c3c;font-size:
```

```
    24rpx;height:100rpx;width:100rpx;line-height:65rpx;background:#fff;
    border-radius:50%;text-align:center;position:absolute;top:200rpx;left:50%;
    margin-left:-50rpx;}
.body .item .goodsname{width:100%;color:#3c3c3c;font-size:
   30rpx;position:absolute;top:270rpx;left:0;z-index: 99;text-align: center}
.body .item .add{width:150rpx;background:#f39303;color:#fff;font-size:24 rpx;
   position:absolute;top:320rpx;left:30%;z-index: 99;text-align: center;
   border-radius:5px;font-size: 24rpx;height:50rpx;line-height: 50rpx;}
.gekai{clear:both;}
/*下方购物车*/
.hiddencart, .cart-detail .mask{position: fixed;left: 0;top: 0;width: 100%;height: 100%;
   z-index: 999999}
.hiddencart .cartheader{background: rgba(0,0,0,0.7);}
.hiddencart .list{position: absolute;left: 0;bottom: 100rpx;width: 100%;
  background: #f7f7f7;padding:0 0 70rpx;z-index:9999999}
.hiddencart .list .item{display: -webkit-flex;color: #333;font-size:
  36rpx;line-height: 50rpx;padding: 20rpx 40rpx;border-bottom:1px solid #d5d5d5}
.hiddencart .list .item .itemname{-webkit-flex: 1;font-size:30rpx;color:#606060}
.hiddencart .list .item .itemtotal{width: 120rpx;font-size:36rpx;color:#373737;
  font-weight: bold}
.hiddencart .list .item .itemreduce,
.hiddencart .list .item .itemadd{font-size: 50rpx;background: #4a4a4a;width: 50rpx;
  height: 50rpx;text-align: center;border-radius: 50%;color:#fff;line-height:
  40rpx;border:1px solid #4a4a4a}
.hiddencart .list .item .itemreduce{background: #ffffff;color:#4a4a4a;border:1px
  solid #4a4a4a}
.hiddencart .list .item .itemnum{width: 50rpx;text-align: center;margin: 0 5rpx;
  color:#4a4a4a}
.bottomcart{display: -webkit-flex;position: fixed;left: 0;bottom: 0;width: 100%;
  height: 100rpx;background: #3c3d41;z-index: 999999}
.bottomcart .data{-webkit-flex: 1;/*border-top: 1rpx solid #e7e7e7;*/}
.bottomcart .data .icon{position: absolute;left: 40rpx;top: -40rpx;width: 100rpx;
  height: 100rpx;background: #393939;border-radius: 50%;border:5px solid #3c3d41;
  box-shadow:0 0 5px #000}
.bottomcart .data .icon image{position: absolute;left: 15rpx;top: 15rpx;width: 70rpx;
  height: 70rpx;}
.bottomcart .data .icon .count{position: absolute;left: 70rpx;top: -20rpx;
  font-size: 30rpx;width: 50rpx;height: 50rpx;line-height: 50rpx;color: #fff;
  background: #f45044;border-radius: 50%;text-align: center}
.bottomcart .data .total{color: #f45044;font-size: 36rpx;line-height: 100rpx;
  padding-left: 160rpx;}
.bottomcart button{width: 200rpx;height: 100%;font-size: 30rpx;background:#f38815;
  color:#fff;line-height:100rpx;border-radius:0}
.bottomcart button[disabled][type="default"], wx-button[disabled]:not([type])
  {color:#fff;background-color:#333333;}
.goodstittle{background:#e5e5e5;padding:0 20rpx;height:100rpx;line-height:100rpx;}
.cartheaderleft{width:200rpx;height:50rpx;font-size: 30rpx;float:left;color:#4a4a4a;
  margin-top:25rpx;line-height:50rpx;border-left:3px solid #ffd600;
  text-indent:15rpx;}
.cartheaderright{font-size: 30rpx;float:right;color:#4a4a4a;text-align: right}
.cartitem{padding:0 20rpx;height:80rpx;line-height:80rpx;border-bottom:1px solid #d5d5d5}
```

```
.cartitem0{width:100rpx;height:50rpx;font - size:24rpx;float:left;color: #f38d1e;margin - top:15rpx;line - height:50rpx;border:1px solid #f38d1e;text - align: center;
  border - radius:3px;margin - right:5px;}
.cartitem1{float:left;font - size:24rpx;color: #4a4a4a;text - align: right}
.delimg{width:30rpx;height:40rpx;position:relative;top:8rpx;margin - right:3px;}
```

pages/chooseAddress/chooseAddress.wxml 的代码如下：

```
<view class = "container">
  <form>
    <view class = "page - section">
      <view class = "weui - cells weui - cells_after - title">
        <view class = "weui - cell weui - cell_input">
          <view class = "weui - cell_hd">
            <view class = "weui - label">收货人姓名</view>
          </view>
          <view class = "weui - cell_bd">
            {{ addressInfo.userName }}
          </view>
        </view>
        <view class = "weui - cell weui - cell_input">
          <view class = "weui - cell_hd">
            <view class = "weui - label">邮编</view>
          </view>
          <view class = "weui - cell_bd">
            {{ addressInfo.postalCode }}
          </view>
        </view>
        <view class = "weui - cell weui - cell_input region">
          <view class = "weui - cell_hd">
            <view class = "weui - label">地区</view>
          </view>
          <view class = "weui - cell_bd">
            {{ addressInfo.provinceName }}
            {{ addressInfo.cityName }}
            {{ addressInfo.countyName }}
          </view>
        </view>
        <view class = "weui - cell weui - cell_input detail">
          <view class = "weui - cell_hd">
            <view class = "weui - label">收货地址</view>
          </view>
          <view class = "weui - cell_bd">
            {{ addressInfo.detailInfo }}
          </view>
        </view>
          <view class = "weui - cell weui - cell_input">
          <view class = "weui - cell_hd">
            <view class = "weui - label">国家码</view>
          </view>
          <view class = "weui - cell_bd">
```

```
            {{ addressInfo.nationalCode }}
          </view>
        </view>
        <view class = "weui - cell weui - cell_input">
          <view class = "weui - cell_hd">
            <view class = "weui - label">手机号码</view>
          </view>
          <view class = "weui - cell_bd">
            {{ addressInfo.telNumber }}
          </view>
        </view>
      </view>
    </view>
  </form>
  <view class = "btn - area"   hidden = "{{chooseAddress}}">
    <button type = "primary" bindtap = "chooseAddress">获取收货地址</button>
  </view>
   <view class = "btn - area"   hidden = "{{!chooseAddress}}">
    <button type = "primary" bindtap = "pay">提交</button>
  </view>
</view>
```

【代码讲解】 chooseAddress.wxml 文件和例 9-13 类似，实现获取微信收货地址。因为微信收货地址基于微信的三方共享性，强烈推荐在电商系统中调用 wx.chooseAddress()接口来填写用户信息。

pages/chooseAddress/chooseAddress.js 的代码如下：

```
Page({
  data: {
    addressInfo: null,
    chooseAddress:false
  },
  chooseAddress:function(e){
    var that = this;
    wx.chooseAddress({
      success: function(res){
        that.setData({
          addressInfo: res,
          chooseAddress: true
        })
        try {
          wx.setStorageSync('name', that.data.addressInfo.userName)
          wx.setStorageSync('tel', that.data.addressInfo.telNumber)
          wx.setStorageSync('address1', that.data.addressInfo.provinceName)
          wx.setStorageSync('address2', that.data.addressInfo.cityName)
          wx.setStorageSync('address3', that.data.addressInfo.detailInfo)
        } catch (e) { console.log("wx.setStorageSync 出错") }
      },
      fail: function(err) {
        console.log(err)
```

```
        }
      })
    },
    pay:function(e){
      wx.navigateTo({
        url: '../pay/pay'
      })
    }
})
```

【代码讲解】 chooseAddress. js 5 次调用 wx. setStorageSync()接口将收货地址的姓名、电话、省、市和详细地址信息写入数据缓存区，以供 pay. wxml 文件使用。

pages/chooseAddress/chooseAddress. wxss 的代码如下：

```
@import "../lib/weui.wxss";
form {
  margin - top: 30rpx;
}
.weui - cell_bd {
  display: flex;
  justify - content: flex - start;
  padding: 20rpx 0;
  min - height: 60rpx;
}
.btn - area {
  margin - top: 30rpx;
}
.page - section{
  width: 100 %;
  margin - left: 0rpx;
}
```

pages/pay/pay. wxml 的代码如下：

```
< view class = "costumerinfo">
  < view class = "info">
    < view class = "info1">{{name}} {{tel}}</view >
    < view class = "info2">{{address1}}{{address2}}{{address3}}</view >
  </view >
  < view class = "info3">
  </view >
</view >
< view class = "wxpay"> 支付方式: 微信支付</view >
< view class = "cartlist">
  < view class = "table">
< view class = "item" wx:for = "{{cart.list}}" wx:for - index = "id"
 wx:for - item = "num" wx:key = "id">
    < view class = "tr">
      < view class = "td" class = "itemname">{{foodlist[id].name}}</view >
      < view class = "td" class = "itemnum">× {{cart.list[id]}}</view >
      < view class = "td" class = "itemprice">¥ {{foodlist[id].price * cart.list[id]}}</view >
```

```
        </view>
        </view>
        <view></view>
        <view class="trbom">
          <view class="Discount1"> 满减活动</view>
          <view class="Discount2">- ¥10</view>
        </view>
      </view>
      <view class="total0">
        <view class="total1">订单¥{{cart.total}}</view>
        <view class="total2">优惠¥10</view>
        <view class="total3">待支付 ¥{{cart.total - 10}}</view>
      </view>
    </view>
    <view class="pay" bindtap='orderpay'>
      <view class="pay1">待支付¥{{cart.total - 10}} </view>
      <view class="pay2">| 优惠 ¥10</view>
      <button class="paybtn"> 支付</button>
    </view>
```

pages/pay/pay.js 的代码如下：

```
Page({
  data:{
    name: '张三',
    tel: "18888888888",
    address1: '广东省',
    address2:'中山市',
    address3:'中山北站 1 号',
    imageUrl: "/imgs/index/bottom.jpg",
    foodlist:
      [
        {
          id: 1,
          name: '海南 13.5 度甜玉米新鲜现摘',
          pic: '/imgs/good/1.jpg',
          price: 29.8
        },
        {
          id: 2,
          name: '枣阳黄桃新鲜当季水蜜桃',
          pic: '/imgs/good/2.jpg',
          price: 59.8
        },
        {
          id: 3,
          name: '墨西哥进口牛油果宝宝辅食',
          pic: '/imgs/good/3.jpg',
          price: 39.8
        },
        {
```

```
          id: 4,
          name: '正宗土鸡散养农家老母鸡',
          pic: '/imgs/good/4.jpg',
          price: 58
        },
        {
          id: 5,
          name: '麻辣小龙虾 1.8 斤装十三香',
          pic: '/imgs/good/5.jpg',
          price: 58
        },
        {
          id: 6,
          name: '洪泽湖大闸蟹鲜活现货',
          pic: '/imgs/good/6.jpg',
          price: 148
        }
      ],
    cart: {
      count: 0,
      total: 0,
      list: {}
    },
  },
  onLoad:function(options){
    this.setData({
      name:wx.getStorageSync('name'),
      tel: wx.getStorageSync('tel'),
      address1: wx.getStorageSync('address1'),
      address2: wx.getStorageSync('address2'),
      address3: wx.getStorageSync('address3'),
      cart: wx.getStorageSync('cart')
    })
  },
 orderpay: function(e) {
    var that = this;
    wx.login({
      success: function(res) {
        wx.request({
          url: 'https://skill.qiujikeji.com/mini/getOpenid',
          method: 'POST',
          data: { 'js_code': res.code },
          success: function(res) {
            console.log(res.data)
            that.xiadan(res.data);
          }
        })
      }
    });
  },
  //下单
```

```
  xiadan: function(openid) {
    var that = this;
    wx.request({
      url: 'https://skill.qiujikeji.com/mini/pay/teacher',
      method: 'POST',
      data: { 'openid': openid },
      success: function(res) {
        var res = res.data
        wx.requestPayment({
          timeStamp: res.timeStamp,
          nonceStr: res.nonceStr,
          package: res.package,
          signType: 'MD5',
          paySign: res.paySign,
        })
      }
    })
  }
})
```

【代码讲解】 pay.js 6 次调用 wx.getStorageSync()接口将 chooseAddress.js 和 goods.js 写入数据缓存的信息全部读出，并通过数据绑定的方式在 pay.wxml 中显示。

pages/pay/pay.wxss 的代码如下：

```
page {
  font-family: "微软正黑体";
  background: #f5f5f5;
  padding-bottom: 150rpx;
}
.costumerinfo{font-size:35rpx;color:#606060;height:100rpx;background-siz e:auto 1px}
.info{width:80%;margin-left:20rpx;padding-left:50rpx;height:150rpx;
  float:left;background-size:auto 50rpx}
.info1{font-size:35rpx;color:#4a4a4a;margin-top:18rpx}
.info2{font-size:34rpx;color:#a4a4a4;margin-top:5rpx}
.info3{width:10%;height:150rpx;float:right;background-size:auto 30rpx}
.wxpay{background:#fff;padding:30rpx 20rpx;margin-top:30rpx;font-size:35rpx;
  color:#4a4a4a;}
.cartlist{background: #fff;margin-top:30rpx;padding:30rpx 0 0}
.cartlist .table{width: 100%;font-size:40rpx;color:#606060;text-align: center;
  border-bottom:1px solid #e8e8e8}
.tr{display: flex;padding:15rpx 20rpx;}
.th,.td{border:0;width:100%;}
.itemname{width:70%;text-align: left}
.itemnum{width:15%;}
.itemprice{width:15%;text-align: right}
.trbom{display: flex;padding:15rpx 20rpx;}
.tdbomtit{text-align: left;width:85%;font-size:28rpx;}
.tdbomprice{text-align: right;width:15%;color:#f38815}
.Summary0{font-size:35rpx;color:#4a4a4a;text-align: right;padding:20rpx;}
.Summary1{display:inline;font-size:35rpx;color:#606060;margin-right:10px}
.Summary2{display:inline;font-size:35rpx;color:#606060;margin-right:10px}
```

```
.Summary3{display:inline;font-size:35rpx;color:#f38815;}
.pay{
position:fixed;left:0;bottom:0;width:100%;height:100rpx;background:#3c3d41;
  z-index:999999;
}
.pay1{font-size:40rpx;color:#fff;display: inline-block;float:left;line-height:100rpx;
  margin-left:10px}
.pay2{font-size:34rpx;color:#eee;display: inline-block;float:left;line-height:100rpx;
  margin-left:5px}
.paybtn{
width:200rpx;
height:100%;
font-size:40rpx;
background:#f38815;
color:#fff;
line-height:100rpx;
border-radius:0;
float:right;
display: inline-block;
}
.Discount1{font-size:40rpx;color:rgb(30, 226, 128);display: inline-block;float:left;
  line-height:100rpx;margin-left:10px}
.Discount2{font-size:34rpx;color:rgb(46, 202, 59);display: inline-block;float:left;
  line-height:100rpx;margin-left:5px}
```

第10章

小程序云开发

第6～9章介绍了一些需要后端程序配合的高级接口，需要开发者部署在自己的服务器上，如第6章的数据库操作需要在服务器上搭建MySQL环境为小程序前端开发提供后端JSON接口数据。为了减少小程序对后端服务器的依赖，腾讯公司于2018年9月上线了小程序云开发服务，它加快了小程序开发的效率，降低了成本，为小程序前后端开发提供了一种全新的解决方案。

本章主要目标

- 了解小程序云开发产生的意义；
- 了解云函数、云存储和云数据库的概念和应用场景；
- 熟练掌握云开发控制台对云函数、云存储和云数据库的操作，掌握云数据库操作权限问题；
- 熟练掌握云函数、云存储和云数据库的小程序端API和服务端API，掌握云数据库增删改查操作。

10.1 云开发

微信小程序云开发是2018年9月腾讯上线的集云函数、云数据库、云存储和云调用等功能于一身的开放服务。云开发为开发者提供完整的原生云端支持和微信服务支持，弱化后端和运维概念，无须搭建服务器，使用平台提供的API进行核心业务开发，即可实现快速上线和迭代，同时这一能力同开发者使用的云服务相互兼容，并不互斥。

云开发提供了几大基础能力支持，如表10.1所示。

表 10.1 云开发基础能力

能　　力	作　　用	说　　明
云函数	无须自建服务器	在云端运行的代码，微信私有协议天然鉴权，开发者只需编写自身业务逻辑代码
云数据库	无须自建数据库	一个既可在小程序前端操作，也能在云函数中读写的 JSON 数据库
云存储	无须自建存储和 CDN	在小程序前端直接上传/下载云端文件，在云开发控制台可视化管理
云调用	原生微信服务集成	基于云函数免鉴权使用小程序开放接口的能力，包括服务端调用、获取开放数据等能力

可以简单地理解为：云开发是腾讯为小程序开发者在腾讯云上开辟了一片空间，本来需要用后端程序语言编写并部署在服务器上的后端功能函数中，现在可以在本地开发之后一键部署到云端（目前只支持 JavaScript，还不支持 Java 等）；本来需要在后端服务器创建的数据库，现在可以在云端创建；本来需要保存在后端服务器的程序素材文件，现在可以通过云存储免费存放在云端，并在其需要使用时，开发者只需使用云调用即可实现和调用服务器端资源一样调用云端资源。

注意：请确保 7.1.2 节中已经成功安装了 Node.js，否则本章中部分云函数操作将失败。

10.1.1 开通云开发功能

使用云开发之前需要先开通云开发功能，开通云开发功能可以按照以下步骤来进行：

（1）新建一个项目：在创建新项目的时候必须填写真实的 AppID，使用虚拟的测试 AppId 将不能开通云开发。

（2）开通云开发功能：在开发者工具菜单栏，如图 10.1 所示，单击"云开发"按钮即可打开图 10.2 所示界面，该界面显示了云开发控制台拥有的云函数、数据库、文件存储和数据分析的基本功能。

图 10.1 单击"云开发"按钮

（3）创建环境：单击图 10.2 中的"开通"按钮，将进入图 10.3 所示的创建云环境界面，填写"环境名称"和"环境 ID"两项内容。默认配额下可以创建两个环境，环境间相互隔离，每个环境都包含独立的数据库实例、存储空间、云函数配置等资源。每个环境都有唯一的环境 ID 标识，初始创建的环境自动成为默认环境。

（4）完成创建环境：单击图 10.3 中的"确定"按钮，将成功开通云开发功能并进入云开发控制台，如图 10.4 所示。

图 10.3 和图 10.4 都显示了云开发基础资源配额，例如存储空间、CDN 流量、云函数调用次数和云函数同时连接数等，这些资源配置都是腾讯云免费开放给小程序开发者的。除

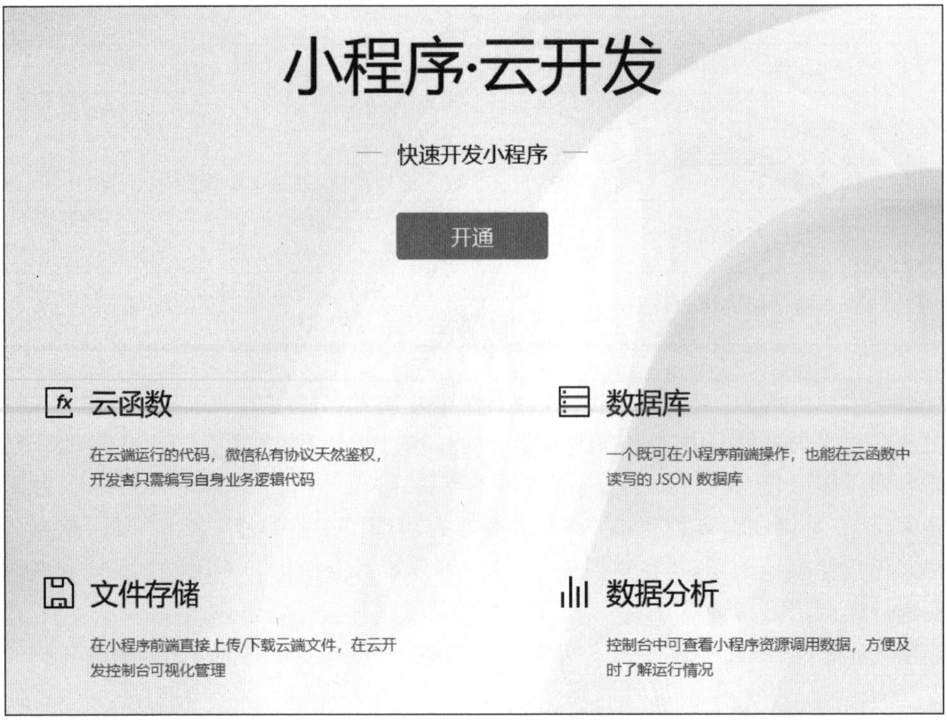

图 10.2　单击“开通”按钮

图 10.3　创建云环境界面

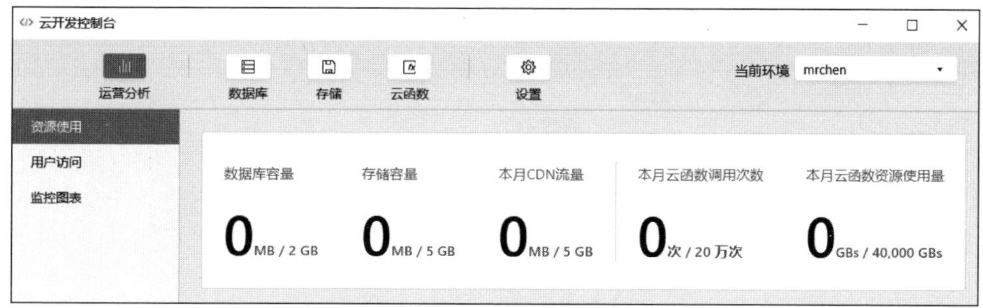

图 10.4　初始的云开发控制台

以上配额参数外，小程序云开发资源还包括以下系统参数限制。

云函数(单次运行)运行内存：256MB；

云函数数量：50 个；

数据库流量：单次出包大小为 16MB；

数据库单集合索引限制：20 个。

以上均是一个环境的配额，不是所有环境的限制总额。如需申请上调，开发者可以以“申请调整小程序云开发调用资源上限”为主题，发送邮件至 miniprogram@tencent.com 申请调整，并在正文中注明小程序账号 AppID、需要调整的环境名称、需要调整的资源上限(仅限资源配额中所列内容且非系统参数限制)、小程序服务类目(可在小程序基本设置中查询)、资源调整原因以及产品计划上线时间。

当看到如图 10.4 所示界面的时候，说明云开发功能开通成功了。

10.1.2　云开发控制台使用

云开发控制台提供了资源环境管理、云数据库管理、云存储管理、云函数管理和运营分析等功能。

1. 资源环境管理

在实际开发中，建议每一个正式环境都搭配一个测试环境，所有功能先在测试环境测试完毕后再进入正式环境。以初始可创建的两个环境为例，建议一个创建为 test 测试环境，一个创建为 release 正式环境。在如图 10.5 所示的云开发控制台右上角的“当前环境”栏目可以创建新的资源环境，单击“创建新环境”按钮之后即进入图 10.3 环境配置界面创建新的环境。

图 10.5　当前环境

2. 云数据库管理

云开发提供了一个 JSON 数据库。顾名思义，数据库中的每条记录都是一个 JSON 格式的对象。一个数据库可以有多个集合(相当于关系型数据库中的表)，集合可看作一个 JSON 数组，数组中的每个对象就是一条记录，记录的格式是 JSON 对象。云数据库中存在从高到低的三个层次：集合、记录和字段，它们依次对应关系型数据库中的表、记录和字段的概念，如图 10.6 所示，依次添加集合、记录和字段。

"创建集合"界面如图 10.7 所示。"添加记录"界面如图 10.8 所示。"添加字段"界面如图 10.9 所示。

图 10.6　云数据库的 3 个层次

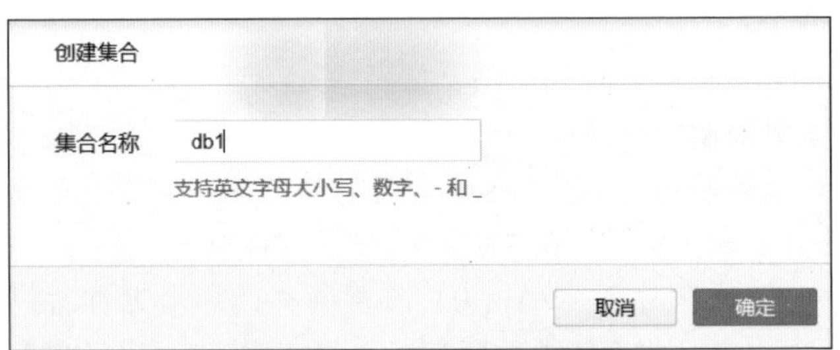

图 10.7　"创建集合"界面

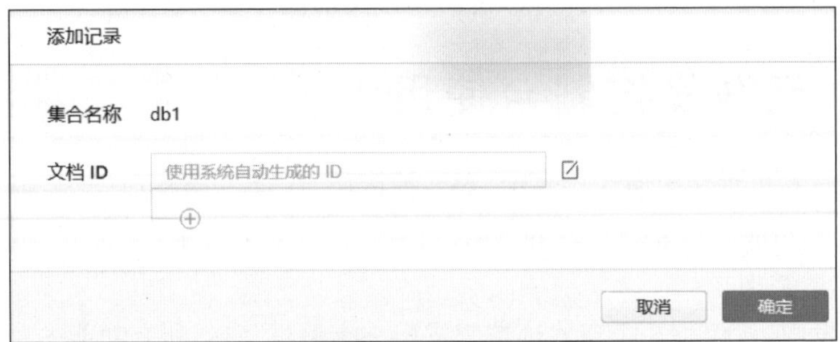

图 10.8　"添加记录"界面

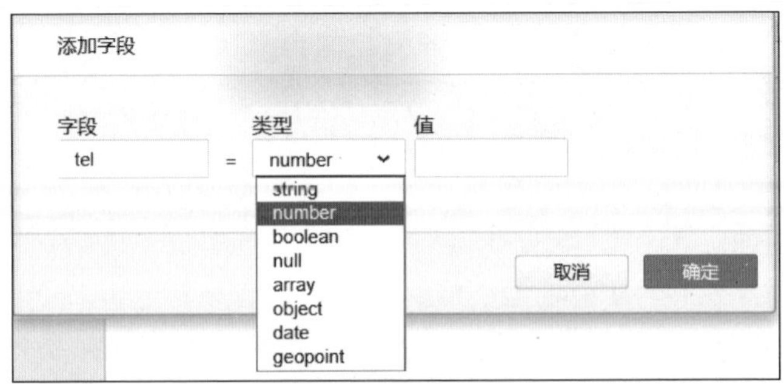

图 10.9　"添加字段"界面

每条记录都有一个“_id”字段用以唯一标识一条记录、一个“_openid”字段用以标识记录的创建者,即小程序的用户。需要注意的是在云控制台创建的记录没有“_openid”字段,而通过小程序的代码操作创建的记录会被自动添加“_openid”字段,也就是说不管是通过云开发控制台还是通过前端代码,它们创建的记录的“_openid”字段都不需要用户插入。开发者可以自定义“_id”,但不可以自定义和修改“_openid”。“_openid”是在文档创建时由系统根据小程序用户默认创建的。

数据库 API 分为小程序端 API 和服务端 API 两部分。小程序端 API 拥有严格的调用权限控制,开发者可在小程序内直接调用 API 进行非敏感数据的操作。对于有更高安全要求的数据,可在云函数内通过服务端 API 进行操作。云函数的环境是与客户端完全隔离的,在云函数上可以私密且安全地操作数据库。

3. 云存储管理

云开发免费为开发者提供了一块存储空间,提供了上传文件到云端、带权限管理的云端下载能力,开发者可以在小程序端和云函数端通过 API 使用云存储功能。如图 10.10 所示,单击“新建文件夹”按钮可以在云端创建文件夹,单击“上传文件”按钮可以和百度云盘一样将本地文件上传到云端。

图 10.10　云存储管理

图 10.10 中的 File ID:“cloud://mrchen201901.6d72-mrchen201901/img/垃圾邮件2.jpg”即为云端文件的引用地址。

4. 云函数管理

云函数是一段运行在云端的代码,无须管理服务器,在开发工具内编写、一键上传部署即可运行的后端代码。小程序内提供了专门用于云函数调用的 API。开发者可以在云函数内使用 wx-server-sdk 提供的 getWXContext 函数获取每次调用的上下文(appid、openid等),无须维护复杂的鉴权机制,即可获取天然可信任的用户登录态(openid)。

单击“新建云函数”,如图 10.11 所示,即可弹出“创建云函数”界面,如图 10.12 所示,填写“云函数名称”后单击“确定”按钮即完成了云函数的创建。

5. 运营分析

运营分析下面有资源使用、用户访问和监控图表三大功能,可以对小程序的统计数据做简单的分析。

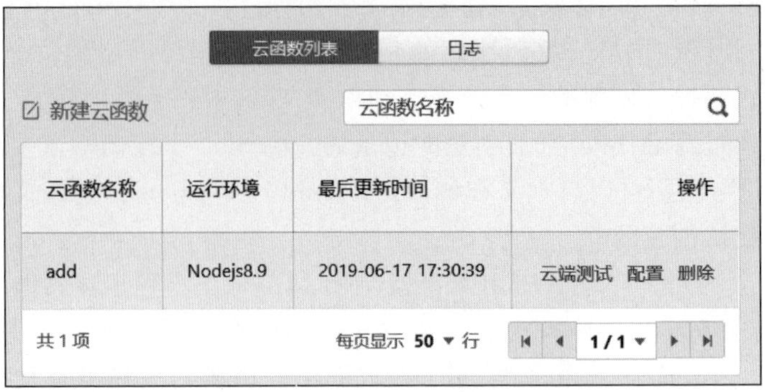

图 10.11 云函数管理界面

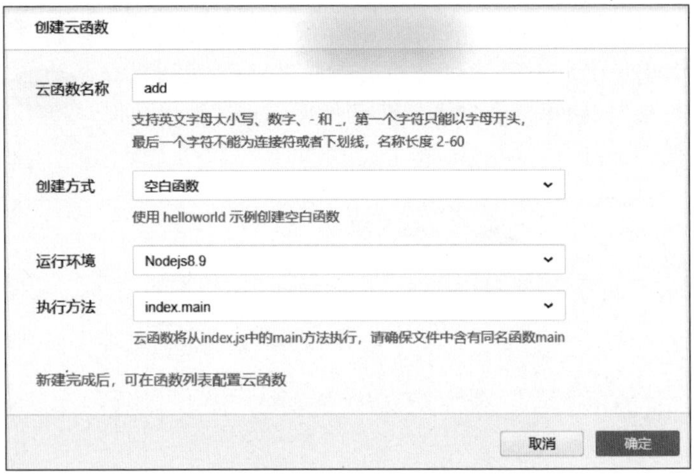

图 10.12 创建云函数

10.1.3 第一个云开发小程序

例 10-1 通过 10.1.2 节已经成功开通了云开发功能，本例创建第一个云开发项目。新建小程序项目如图 10.13 所示，在 AppID 栏目填写真实 AppID，后端服务选项选择“小程序·云开发”单选按钮，单击“新建”按钮，系统即自动创建了一个云开发项目 demo，项目结构图如图 10.14 所示。

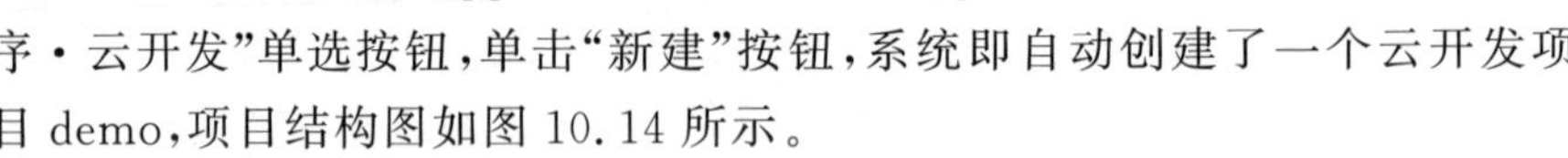

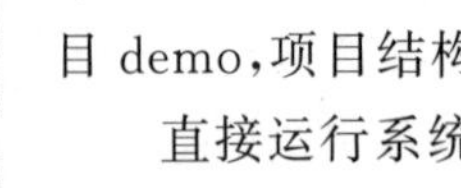

视频讲解

直接运行系统自动生成的 demo，会弹出如图 10.15 所示窗口提示云函数调用失败，这是因为在云端还没有云函数。

右击云函数 login 对应的 login 文件夹，弹出如图 10.16 所示界面，选择“创建并部署：云端安装依赖(不上传 node_modules)”菜单，云函数即会上传部署至云端。上传部署成功的话，再看项目结构的时候会发现 login 和 openapi 两个函数在图 10.14 中的文件夹图标已经变成了图 10.17 的云朵形状。打开云开发控制台，如图 10.18 所示，可以发现云函数中已经有了 login 和 openapi 两个云函数，这说明这两个函数已经成功上传并部署至云端，这时再运行 demo 程序则能正常运行。

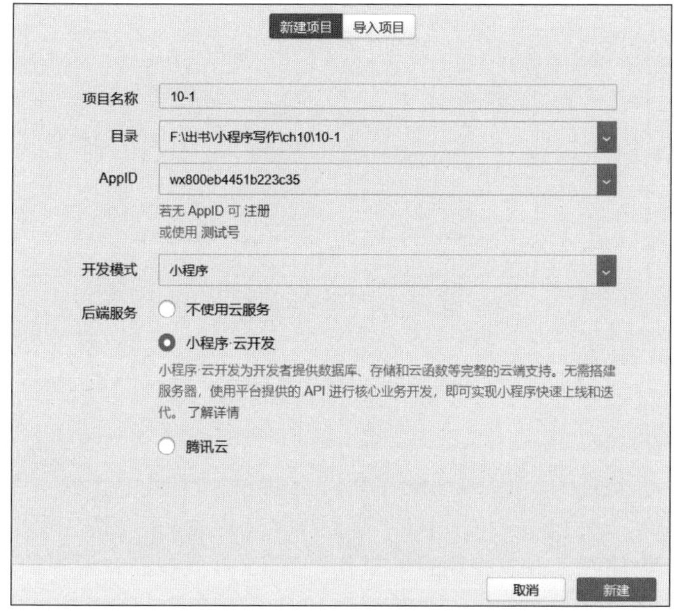

图 10.13　新建云开发项目

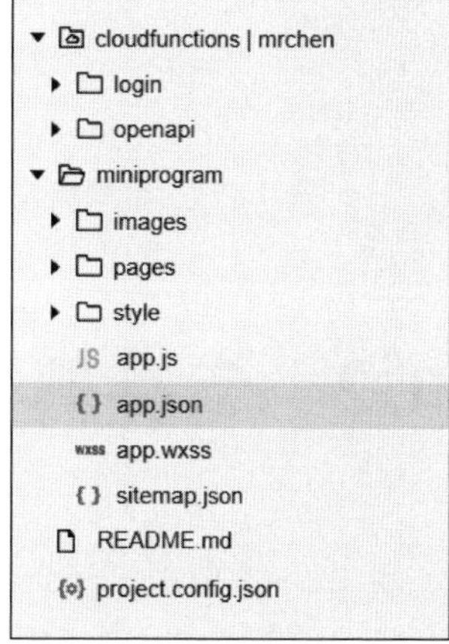

图 10.14　云开发项目目录结构

图 10.15　云函数调用失败

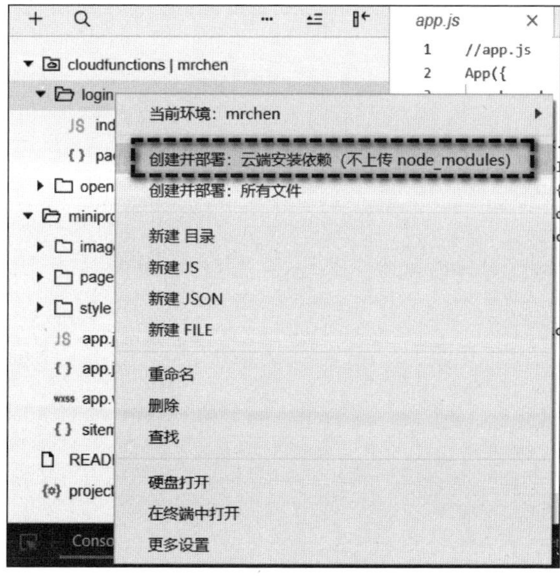

图 10.16　创建并部署

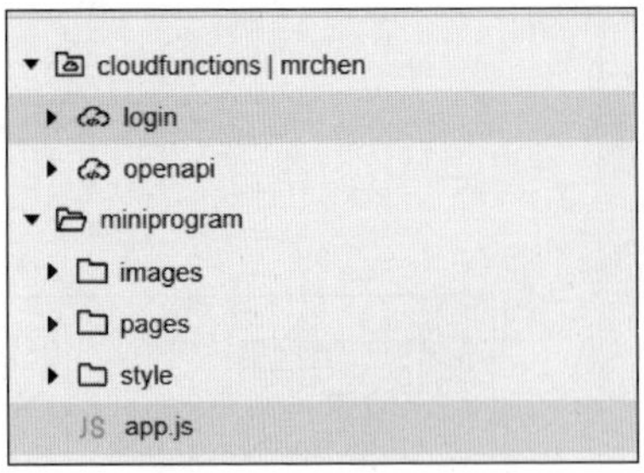

图 10.17　部署成功

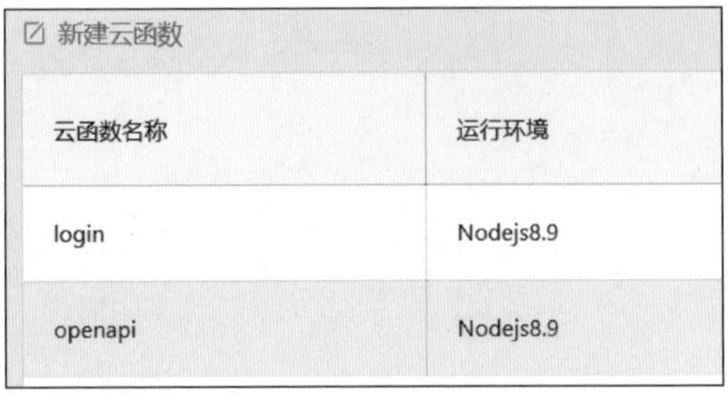

图 10.18　云开发控制台中的云函数

10.2　云存储

在 10.1.2 节中已经介绍过云存储在云开发控制台的操作，但是云开发控制台更多的是对项目中的初始文件的操作管理，例如项目的 Logo 图片可以通过云开发控制台提前上传到云端。项目在执行的过程中也会涉及文件的操作，例如用户上传图片的操作，这时就需要用到云开发存储 API。

小程序云开发提供了一系列存储操作 API，有 uploadFile() 上传文件接口、downloadFile()下载文件接口、deleteFile()删除文件接口和 getTempFileURL()换取临时链接接口。

wx.cloud.uploadFile()接口参数如表 10.2 所示。如果采用 Callback 风格(详见 10.3.2 节)，调用回调函数 success、fail、complete 中的任一个，则会返回一个 UploadTask 对象(封装返回信息的对象)，通过 UploadTask 对象可监听上传事件。

表 10.2　wx.cloud.uploadFile()接口属性表

字　段	数据类型	必填	说　明
cloudPath	String	Y	云存储路径，命名限制见文件名命名限制
filePath	String	Y	要上传文件资源的路径
header	Object	N	HTTP 请求 Header，Header 中不能设置 Referer
config	Object	N	配置

例 10-2　使用 wx.cloud.uploadFile()接口上传图片至云端，程序运行效果如图 10.19 所示。

视频讲解

pages/uploadFile/uploadFile.wxml 的代码如下：

```
<!-- 上传图片 -->
<view class="uploader">
  <view class="uploader-text" bindtap="doUpload">
    <text>上传图片</text>
  </view>
  <view class="uploader-container" wx:if="{{imgUrl}}">
```

```
        <image class = "uploader - image" src = "{{imgUrl}}" mode = "aspectFit" bindtap =
"previewImg"></image>
      </view>
    </view>
```

图 10.19 成功上传图片至云端并返回结果

pages/uploadFile/uploadFile.js 的代码如下：

```
    //上传图片
  doUpload: function() {
    //选择图片
    wx.chooseImage({
      count: 1,
      sizeType: ['compressed'],
      sourceType: ['album', 'camera'],
      success: function(res) {
        console.log(res)
        wx.showLoading({
          title: '上传中',
        })
        const filePath = res.tempFilePaths[0]
        var timestamp = (new Date()).valueOf();
        wx.cloud.uploadFile({
          cloudPath: "img/" + timestamp + ".jpg",          //上传至云端的路径
          filePath: filePath,                              //小程序临时文件路径
          success: res => {
            console.log('[上传文件] 成功：', res)
            app.globalData.fileID = res.fileID
            app.globalData.cloudPath = cloudPath
            app.globalData.imagePath = filePath
            wx.navigateTo({
              url: '../storageConsole/storageConsole'
            })
          },
          fail: e => {
            console.error('[上传文件] 失败：', e)
            wx.showToast({
              icon: 'none',
              title: '上传失败',
            })
```

```
            },
            complete: () => {
              wx.hideLoading()
            }
          })
        },
        fail: e => {
          console.error(e)
        }
      })
    }
```

【代码讲解】 本例先调用了 wx. chooseImage()接口选择一幅图片，然后调用 wx. cloud. uploadFile()接口上传图片至云端。cloudPath 字段是上传文件在云端的文件名字，为了不重复，本例采用了当前时间戳来命名云端文件名。filePath 字段是本地文件的路径，它的值取 wx. chooseImage()接口的回调参数 res. tempFilePaths[0]。

wx. cloud. downloadFile()接口从云存储空间下载文件的示例代码如下：

```
wx.cloud.downloadFile({
  fileID: '',                                    //文件 ID
  success: res => {
    //返回临时文件路径
    console.log(res.tempFilePath)
  },
  fail: console.error
})
```

wx. cloud. deleteFile()接口删除云端文件的示例代码如下：

```
wx.cloud.deleteFile({
  fileList: ['a7xzcb'],
  success: res => {
    //handle success
    console.log(res.fileList)
  },
  fail: console.error
})
```

10.3 云函数

云函数是部署在云端的函数，它和小程序本地的函数存在很大的区别，云函数应用涉及云端云函数定义和本地引用云端云函数的 API 接口两个问题。

10.3.1 云函数 API 和云函数创建

1. 小程序云函数 API 接口

小程序云函数 API 接口是指小程序调用云端函数的接口，即和第 6 章 wx. request()类似的接口。小程序提供了 wx.cloud.callFunction()接口作为云函数 API 接口，它的属性如表 10.3 所示。

表 10.3　wx.cloud.callFunction()接口属性表

参　数	数据类型	必填	说　明
name	String	是	云函数名
data	Object	否	传递给云函数的参数
config	Object	否	局部覆写 wx.cloud.init 中定义的全局配置
success	Function	否	返回云函数调用的返回结果
fail	Function	否	接口调用失败的回调函数
complete	Function	否	接口调用结束的回调函数(调用成功、失败都会执行)

示例代码：

```
wx.cloud.callFunction({
 name: 'add',                                      //云函数名称
  data: {
  a: 1,                                            //传给云函数的参数
  b: 2,
 },
 success: function(res) {
    //成功返回时的操作
 },
 fail:  function(res) {
//失败返回时的操作
}
})
```

2. 云端函数的创建

创建云函数的方法是右击项目中的 cloudfunctions 文件夹，如图 10.20 所示，选择“新建 Node.js 云函数”选项，开发者工具会自动生成一个文件夹，如图 10.21 所示，在文本框中输入的文件夹的名字即是云函数名。

图 10.20　创建云函数

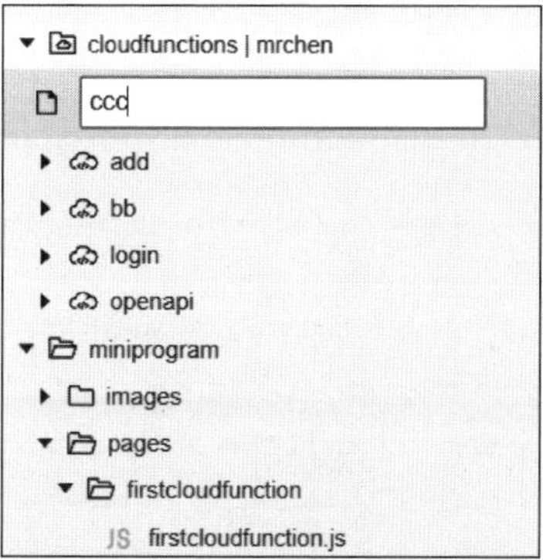

图 10.21　输入云函数名

输入文件名(云函数名)之后,按回车键结束名字的输入。以回车操作结束则 index.js 文件会自动生成代码,如图 10.22 所示。

```
▼ cloudfunctions | mr...
  ▶ add
  ▶ bb
  ▼ ccc
      index.js
      package.json
  ▶ login
  ▶ openapi
▼ miniprogram
  ▶ images
  ▼ pages
    ▼ firstcloudfunction
```

```
// 云函数入口文件
const cloud = require('wx-server-sdk')

cloud.init()

// 云函数入口函数
exports.main = async (event, context) => {
  const wxContext = cloud.getWXContext()

  return {
    event,
    openid: wxContext.OPENID,
    appid: wxContext.APPID,
    unionid: wxContext.UNIONID,
  }
}
```

图 10.22　系统自动创建的云函数

在创建好云函数之后,即可根据自己的需要在 main 函数中编写自己的代码。使用回车结束云函数的创建还有一个好处是可以不用像图 10.16 那样上传和部署云函数,系统会自动上传和部署函数至云端,即云函数对应的文件夹图标会自动变成云朵模样。

视频讲解

例 10-3　本例分别采用云函数和本地函数实现加法操作和减法操作,请对比区别,程序运行效果如图 10.23 和图 10.24 所示。

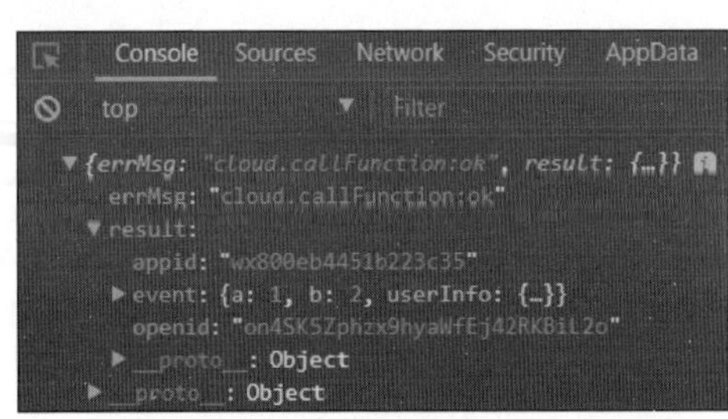

图 10.23　云函数在控制台返回的结果

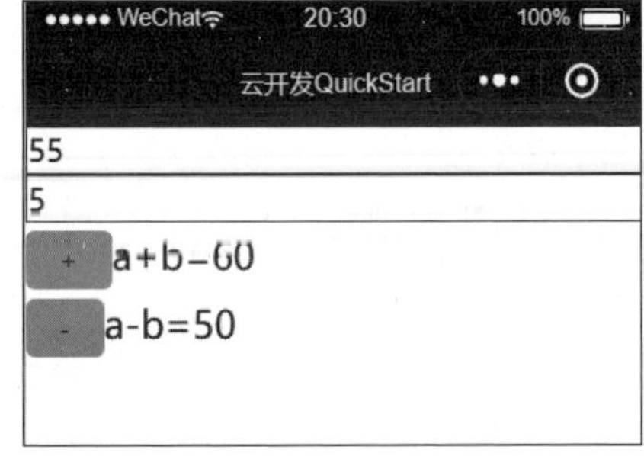

图 10.24　小程序端运行结果

pages/firstcloudfunction/firstcloudfunction.wxml 的代码如下:

```
<view class="data"><input placeholder="请输入 a" focus="true" bindinput="binda"></input></view>
<view class="data"><input placeholder="请输入 b" focus="true" bindinput="bindb"></input></view>
<view class="partition"></view>
<view class="arithmetic"><view bindtap='add'><button size="mini" class='bt'>+</button></view><view>{{add}}</view></view>
<view class="arithmetic"><view bindtap='sub'><button size="mini" class='bt'>-</button></view><view>{{sub}}</view></view>
```

pages/firstcloudfunction/firstcloudfunction.js 的代码如下：

```
const app = getApp()
Page({
  data: {
    a:1,
    b:1,
  add:'',
  sub:''
  },
binda:function(e){
 this.setData({
    a: e.detail.value
 })
    console.log(e.detail.value)
  },
  bindb: function(e) {
    this.setData({
      b: e.detail.value
    })
    console.log(e.detail.value)
  },
  add:function(e){
    var that = this;
    wx.cloud.callFunction({
      //云函数名称
      name: 'add',
      //传给云函数的参数
      data: {
        a: that.data.a,
        b: that.data.b,
      },
      success: function (res) {
        console.log(res.result.add)
        var c = res.result.add
        that.setData({
          add: "a + b = " + c
        })
      },
      fail: console.error
    })
  },
  sub: function(e) {
    var c = Number(this.data.a) - Number(this.data.b)
    console.log(c)
    this.setData({
```

```
      sub: "a - b = " + c
    })
  }
})
```

云函数 add/index.js 的代码如下：

```
exports.main = async (event, context) => {
return{add:Number(event.a) + Number(event.b)}
}
```

【代码讲解】 binda 和 bindb 两个点击函数从 firstcloudfunction.wxml 文件中获取用户输入的两个操作数 a 和 b，减法函数 sub 是普通的本地 JavaScript 函数，而加法函数 add 则采用了 wx.cloud.callFunction()接口调用云函数 add。本例重点是云函数的创建和部署。

10.3.2 Callback 风格和 Promise 风格

云开发的 API 同时支持 Callback 风格和 Promise 风格。在传入 API 的 Object 参数中，如果有 success、fail、complete 字段，则认为是采用 Callback 风格，API 方法调用不返回 Promise；如果 success、fail、complete 这 3 个字段都不存在，则认为是采用 Promise 风格，API 方法调用返回一个 Promise，Promise 风格的 resolve 回调如同 Callback 风格的 success 回调，Promise 风格的 reject 回调如同 Callback 风格的 fail 回调，需注意的是 Promise 风格的 resolve 回调使用的关键字是 then，reject 回调使用的关键字是 catch。下面是 wx.cloud.uploadFile 接口的两种风格写法。

Callback 风格：

```
wx.cloud.uploadFile({
  cloudPath: 'example.png',
  filePath: '',                                    //文件路径
  success: res => {
    //get resource ID
    console.log(res.fileID)
  },
  fail: err => {
    //handle error
  }
})
```

Promise 风格：

```
wx.cloud.uploadFile({
  cloudPath: 'example.png',
  filePath: '',                                    //文件路径
}).then(res => {
  //get resource ID
```

```
  console.log(res.fileID)
}).catch(error => {
  //handle error
})
```

考虑到云开发的云函数涉及云端云函数的定义和小程序端云函数的调用，而云端和小程序端都可以支持 Callback 风格和 Promise 风格，所以云函数和小程序存在如表 10.4 所示的风格对应关系。

表 10.4 云函数与小程序的 4 种风格对应关系

名 称	云 函 数	小 程 序
对应关系 1	Callback 风格结果返回形式	Callback 风格调用
对应关系 2	Callback 风格结果返回形式	Promise 风格调用
对应关系 3	Promise 风格结果返回形式	Callback 风格调用
对应关系 4	Promise 风格结果返回形式	Promise 风格调用

例 10-4 本例在例 10-3 的基础上，设计了云函数与小程序的 4 种风格关系，介绍 Callback 风格和 Promise 风格，程序运行结果如图 10.25 和图 10.26 所示。

视频讲解

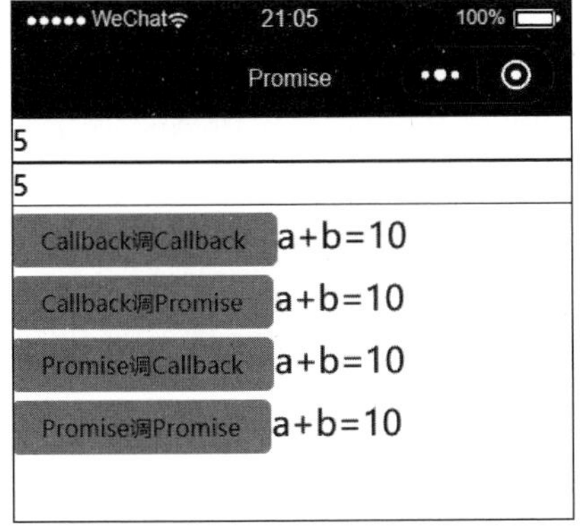

图 10.25 小程序端运行结果

图 10.26 控制台返回的结果

pages/firstcloudfunction/firstcloudfunction.wxml 的代码如下：

```
<view class="data"><input placeholder="请输入 a" focus="true" bindinput="binda">
</input></view>
<view class="data"><input placeholder="请输入 b" focus="true" bindinput="bindb">
</input></view>
<view class="partition"></view>
<view class="arithmetic"><view bindtap='Callback1'><button size="mini" class='bt'>
Callback 调 Callback</button></view><view>{{Callback1}}</view></view>
<view class="arithmetic"><view  bindtap='Callback2'><button size="mini" class='bt'>
Callback 调 Promise</button></view><view>{{Callback2}}</view></view>
<view class="arithmetic"><view  bindtap='Promise1'><button size="mini" class='bt'>
```

```
Promise调Callback</button></view><view>{{Promise1}}</view></view>
<view class="arithmetic"><view  bindtap='Promise2'><button size="mini" class='bt'>
Promise调Promise</button></view><view>{{Promise2}}</view></view>
```

pages/firstcloudfunction/firstcloudfunction.js 的代码如下：

```
const app = getApp()
Page({
  data: {
    a:1,
    b:1,
    Callback1:'',
    Callback2:'',
    Promise1:'',
    Promise2:''
  },
binda:function(e){
 this.setData({
   a: e.detail.value
 })
    console.log("a 的值是" + e.detail.value)
  },
  bindb: function(e) {
    this.setData({
      b: e.detail.value
    })
    console.log("b 的值是" + e.detail.value)
  },
  Callback1:function(e){
    var that = this;
    wx.cloud.callFunction({
      //云函数名称
      name: 'add',
      //传给云函数的参数
      data: {
        a: that.data.a,
        b: that.data.b,
      },
      success: function(res) {
        console.log("Callback 风格 add 云函数返回结果是" + res.result.add)
        var c = res.result.add
        that.setData({
          Callback1: "a+b=" + c
        })
      },
      fail: console.error
    })
  },
  Callback2: function(e) {
    var that = this;
    wx.cloud.callFunction({
```

```
    //云函数名称
    name: 'Padd',
    //传给云函数的参数
    data: {
      a: that.data.a,
      b: that.data.b,
    },
    success: function(res) {
      console.log("Promise风格padd云函数返回结果:" + res.result)
      var c = res.result
      that.setData({
        Callback2: "a + b = " + c
      })
    },
    fail: console.error
  })
},
Promise1: function(e) {
  var that = this;
  wx.cloud.callFunction({
    //云函数名称
    name: 'add',
    //传给云函数的参数
    data: {
      a: that.data.a,
      b: that.data.b,
    }
  }).then(res => {
    console.log("Callback风格add云函数返回结果:" + res.result.add)
      var c = res.result.add
      that.setData({
        Promise1: "a + b = " + c
      })
  }).catch(error => {
    console.log("Callback风格add云函数出错")
  })
},
Promise2: function(e) {
  var that = this;
  wx.cloud.callFunction({
    //云函数名称
    name: 'Padd',
    //传给云函数的参数
    data: {
      a: that.data.a,
      b: that.data.b,
    }
  }).then(res => {
    console.log("Promise风格Padd云函数返回结果:" + res.result)
    var c = res.result
```

```
        that.setData({
          Promise2: "a+b=" + c
        })
      }).catch(error => {
        console.log("Promise 风格 Padd 云函数出错")
      })
    }
})
```

云函数 add/index.js 的代码如下：

```
exports.main = async(event, context) => {
return{add:Number(event.a) + Number(event.b)}
}
```

云函数 Padd/index.js 的代码如下：

```
//云函数入口文件
const cloud = require('wx-server-sdk')
cloud.init()
//云函数入口函数
exports.main = async(event, context) => {
  return new Promise((resolve,reject) =>{
    setTimeout(() => {
      resolve(Number(event.a) + Number(event.b))
    }, 1000)
  })
}
```

【代码讲解】 本例根据表 10.4 设计了 4 个函数：Callback1 函数以 Callback 风格调用形式调用了 Callback 风格结果返回的 add 云函数；Callback2 函数以 Callback 风格调用形式调用了 Promise 风格结果返回的云函数 Padd；Promise1 函数以 Promise 风格调用形式调用了 Callback 风格结果返回的 add 云函数；Promise2 函数以 Promise 风格调用形式调用了 Promise 风格结果返回的 Padd 云函数。请注意 4 种对应关系之间的区别，并选择自己熟悉和喜欢的方式作为日后云函数的编写风格。

通常，小程序需要在云函数中处理一些异步操作，在异步操作完成后再将结果返回给调用方。所以例 10-4 中的 Padd 云函数在采用 Promise 返回结果的同时，还采用了 setTimeout()方法来实现异步操作，注意此处设置的等待时请小于 3000 毫秒。

10.3.3 npm 和 wx-server-sdk

1. npm

云函数的运行环境是 Node.js，在 7.1 节 WebSocket 内容中使用了 Node.js 环境，新版的 Node.js 已自带 npm，安装 Node.js 时会一起安装，npm 的作用就是对 Node.js 依赖的包进行管理，也可以把 npm 理解为用来安装、卸载 Node.js 需要的组件的一个管理器。

在云函数中如果需要引入第三方依赖来帮助更快地开发，开发者需要使用 npm 安装第三方依赖。例如需要使用 Node.js 提供的原生 HTTP 接口在云函数中发起网络请求，这时需要使用 npm 来安装第三方依赖 wx-server-sdk。

开发者在本地开发好云函数后选择云端安装依赖(不上传 node_modules 文件夹)或全量上传(同时上传 node_modules 文件夹)操作来上传和部署云函数至云端，而后就可以通过 API 接口来调用云函数，说明腾讯云已经默认为开发者的云端安装好了 Node.js 环境。

2. wx-server-sdk

wx-server-sdk 是微信小程序提供的第三方依赖包，云函数中使用 wx-server-sdk 需在对应云函数目录下安装 wx-server-sdk 依赖。在创建云函数之后选择"上传并部署：所有文件"时，开发者工具会提示本地没有安装 wx-server-sdk 依赖，如图 10.27 所示。请注意云函数的运行环境是 Node.js，因此在本地安装依赖包时务必保证已安装 Node.js，同时 node 和 npm 都已经在环境变量中设置好。

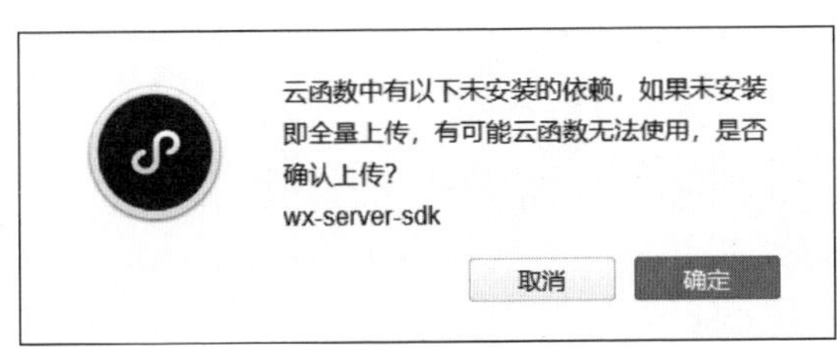

图 10.27　开发者工具本地提示没有安装 wx-server-sdk

下面介绍 wx-server-sdk 依赖的安装方法。

(1) 新建一个云函数，函数名为 sdk，右击该云函数，选择"在终端中打开"选项，如图 10.28 所示。

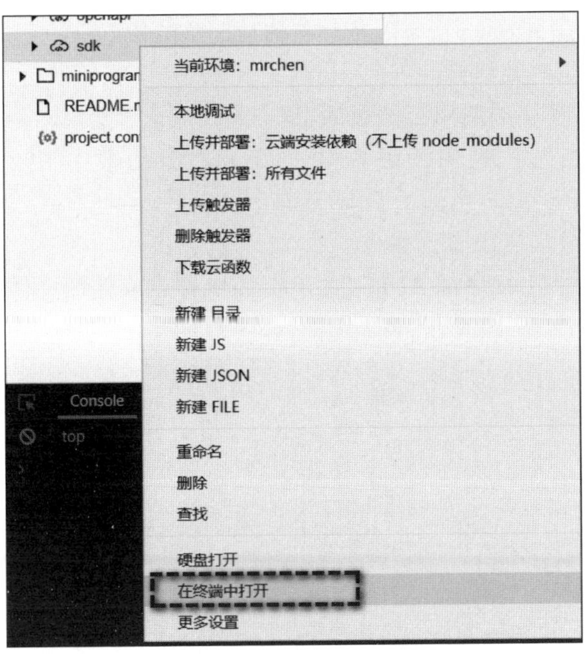

图 10.28　在终端中打开

（2）在弹出的cmd窗口中输入“npm install --save wx-server-sdk@latest”，如图10.29所示，等待大概20秒钟，wx-server-sdk依赖包即安装完成。

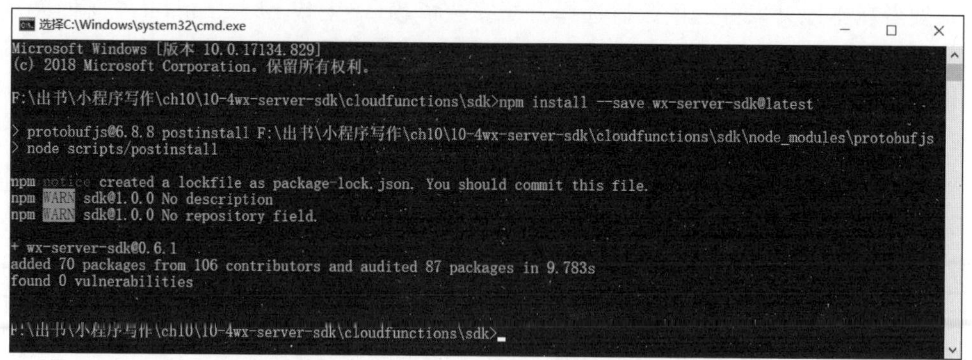

图10.29 cmd安装wx-server-sdk

（3）当安装完成之后，打开项目下的cloudfunctions文件夹下的sdk文件夹，如图10.30所示，文件夹下多了一个node_modules文件夹和package-lock.json文件，这说明wx-server-sdk依赖包安装成功。

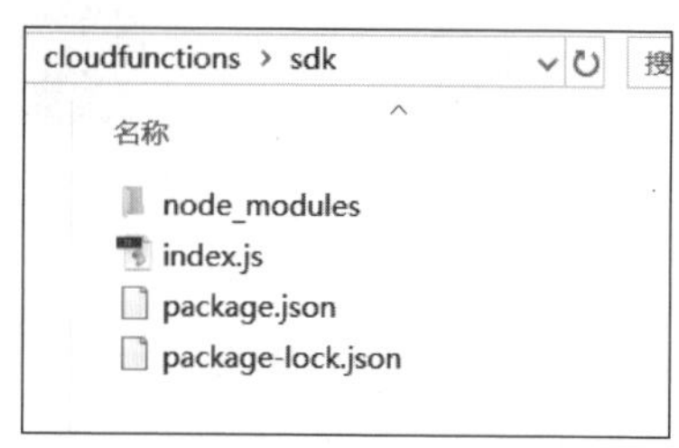

图10.30 sdk文件夹

3. 什么时候需要用到wx-server-sdk

大部分接触云开发的初学者不清楚什么时候需要安装wx-server-sdk依赖包，导致在创建任何云函数时都安装它，而wx-server-sdk依赖包的大小大约为9MB，如果每一个云函数都对其进行安装的话，云函数上传和部署至云端之后，有限的云空间就会不够用。现对需要使用wx-server-sdk依赖包的情况总结如下。

（1）在云函数中调用其他云函数。

一般情况下是小程序端调用云函数的，这时候不需要使用wx-server-sdk依赖包。假设开发者要在云函数A中调用云函数B，则需要在云函数A中安装wx-server-sdk依赖包。

（2）需要使用服务端API的云函数。

云开发的API分为小程序端API和服务端API，以云存储的文件上传功能为例：

小程序端文件上传API示例代码：

```
wx.cloud.uploadFile({
  cloudPath: 'example.png',
  filePath: '',                                   //文件路径
  success: res => {
    //get resource ID
    console.log(res.fileID)
  },
  fail: err => {
    //handle error
  }
})
```

服务器文件上传 API 示例代码：

```
const cloud = require('wx-server-sdk')
const fs = require('fs')
const path = require('path')
exports.main = async(event, context) => {
  const fileStream = fs.createReadStream(path.join(_dirname, 'demo.jpg'))
  return await cloud.uploadFile({
    cloudPath: 'demo.jpg',
    fileContent: fileStream,
  })
}
```

在云函数中使用服务端 API 接口时，必须先安装 wx-server-sdk 依赖包。

需要特别说明的是云函数属于管理端，在云函数中运行的代码拥有不受限的数据库读写权限和云文件读写权限。如果想不受限制地对数据库和云文件进行操作，需要在云函数中使用服务端 API 接口。例如在小程序端用户只能对自己创建的数据进行修改和删除操作，如果想让普通用户可以对所有的数据进行修改和删除操作，可以把相关代码写在云函数中，这样就突破了权限的限制问题。

当云函数中使用了 wx-server-sdk 依赖包，云函数按照图 10.28～图 10.30 所示的步骤部署至云端时，需选择"上传并部署：所有文件"选项，而不是"上传并部署：云端安装依赖(不上传 node_modules)"选项。

视频讲解

例 10-5　本例设计一个云函数调用另一个云函数，根据本节中关于 wx-server-sdk 应用场景的知识点，主调云函数需要安装 wx-server-sdk 依赖包。程序运行结果如图 10.31 和图 10.32 所示。

图 10.31　小程序端运行结果

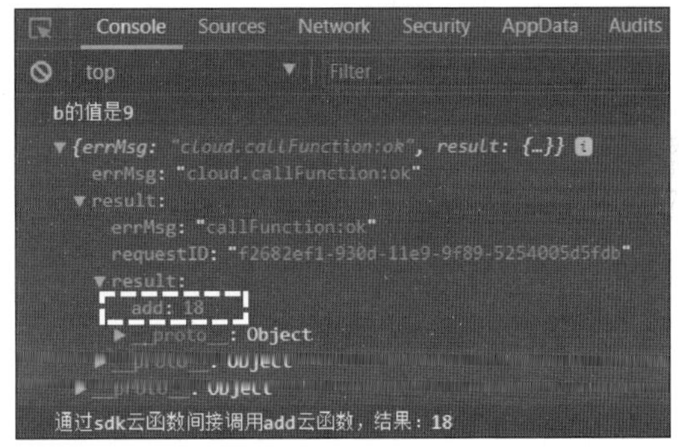

图 10.32　控制台中返回结果

pages/sdk/sdk.wxml 的代码如下：

```
<view class = "data"><input placeholder = "请输入 a" focus = "true" bindinput = "binda">
</input></view>
<view class = "data"><input placeholder = "请输入 b" focus = "true" bindinput = "bindb">
</input></view>
<view class = "partition"></view>
<view class = "arithmetic"><view bindtap = 'sdk'><button size = "mini" class = 'bt'> sdk 调用
```

```
add 加法</button></view><view>{{sdk}}</view></view>
```

pages/sdk/sdk.js 的代码如下：

```
const app = getApp()
Page({
  data: {
    a:1,b:1,sdk:''
  },
binda:function(e){
 this.setData({
 a: e.detail.value
 })
   console.log("a 的值是" + e.detail.value)
  },
  bindb: function(e) {
    this.setData({
      b: e.detail.value
    })
    console.log("b 的值是" + e.detail.value)
  },
  sdk:function(e){
    var that = this;
    wx.cloud.callFunction({
      //云函数名称
      name: 'sdk',
      //传给云函数的参数
      data: {
        a: that.data.a,
        b: that.data.b,
      },
      success: function(res) {
        console.log(res)
        console.log("通过 sdk 云函数间接调用 add 云函数,结果: " + res.result.result.add)
        var c = res.result.result.add
        that.setData({
          sdk: "a + b = " + c
        })
      },
      fail: console.error
    })
  }
})
```

云函数 sdk/index.js 的代码如下：

```
const cloud = require('wx-server-sdk')
cloud.init()
exports.main = async(event, context) => {
  return await cloud.callFunction({
```

```
    name:"add",
    data:{
      a: event.a,
      b: event.b
    }
  })
}
```

云函数 add/index.js 的代码如下：

```
exports.main = async(event, context) => {
  return { add: Number(event.a) + Number(event.b) }
}
```

【代码讲解】 本例在小程序端 sdk.js 中的本地函数 sdk 调用云函数 sdk，云函数 sdk 又调用了另外一个云函数 add，在云函数 sdk 中需要安装 wx-server-sdk 依赖包。本例需要特别注意的是如图 10.32 所示的返回结果中 add 所在的层次，云函数 sdk 调用云函数 add 的返回结果为 res.result.add，而本地小程序调用云函数 sdk 的返回结果为 res.result.result.add。

10.4 云数据库

在 10.1.2 节中介绍了使用云开发控制台管理云数据库，本节从代码操作的角度讲解云数据库的操作，包含云数据库的增删改查操作。

10.4.1 数据类型和权限控制

1. 云数据库中的数据类型

云开发数据库提供以下几种数据类型。

- String：字符串；
- Number：数字；
- Object：对象；
- Array：数组；
- Bool：布尔值；
- GeoPoint：地理位置点；
- Date：时间；
- Null：占位符。

下面对几个需要额外说明的字段做以下补充说明。

1）Date

Date 类型用于表示时间，精确到毫秒，在小程序端可用 JavaScript 内置的 Date 对象创建。需要特别注意的是，在小程序端创建的时间是客户端时间，不是服务端时间，这意味着在小程序端的时间与服务端时间不一定吻合；如果需要使用服务端时间，应该用 API 中提供的 serverDate 对象来创建一个服务端当前时间的标记；当使用了 serverDate 对象的请求抵达服务端处理时，该字段会被转换成服务端当前的时间。

2）GeoPoint

GeoPoint 类型用于表示地理位置点，用经度和纬度唯一标记一个点，这是一个特殊的数据存储类型。注意，如果需要对类型为地理位置的字段进行查找，一定要建立地理位置索引。具体的地理位置 API 可参考 Geo API 文档。

3）Null

Null 相当于一个占位符，表示一个字段存在但是值为空。

2. 云数据库中的权限控制

对云数据库的权限控制分为小程序端和管理端，管理端包括云函数端和控制台。小程序端对数据库的操作是受权限控制限制的，而管理端对数据库的操作则不受权限控制限制。

云开发数据库开放以下几种权限配置按照权限级别从宽到紧排列。

- 仅创建者可写，所有人可读：数据只有创建者可写、所有人可读，如文章；
- 仅创建者可读写：数据只有创建者可读写，其他用户不可读写，如私密相册；
- 仅管理端可写，所有人可读：该数据只有管理端可写，所有人可读，如商品信息；
- 仅管理端可读写：该数据只有管理端可读写，如后台用的不暴露的数据。

每个集合可以拥有一种权限配置，权限配置的规则是作用在集合的每个记录上的。出于易用性和安全性的考虑，云开发为云数据库做了小程序深度整合，在小程序中创建的每个数据库记录都会带有该记录创建者（即小程序用户）的信息，以_openid 字段将用户的 openid 保存在每个相应用户创建的记录中。因此，权限控制也相应围绕着一个用户是否应该拥有权限操作其他用户创建的数据展开。

在云开发控制台可以对数据库进行权限设置，如图 10.33 所示。

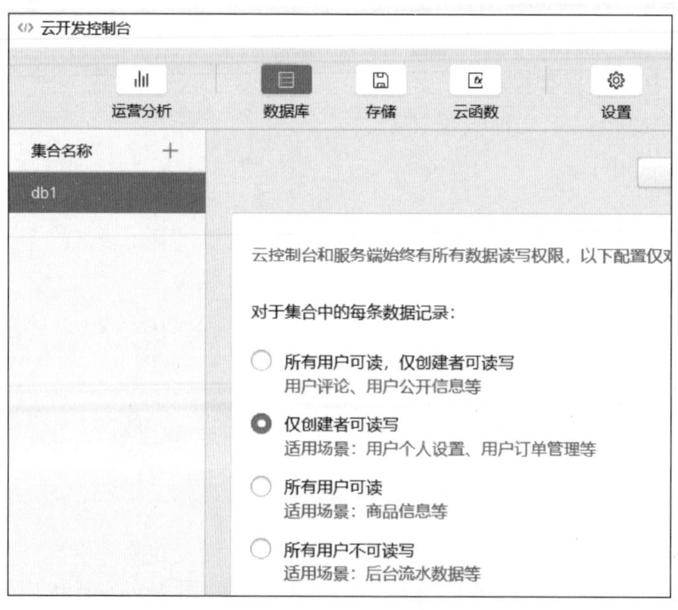

图 10.33　云开发控制台中设置数据库权限

从图 10.33 可以看出数据库权限是设置在集合这个层次，云开发数据库的集合对应于关系型数据库中的表的概念。

小程序端对数据库的操作权限如表 10.5 所示。

表 10.5　小程序端对数据库操作权限表

模　　式	小程序端读自己创建的数据	小程序端写自己创建的数据	小程序端读他人创建的数据	小程序端写他人创建的数据	管理端读写任意数据
仅创建者可写，所有人可读	√	√	√	×	√
仅创建者可读写	√	√	×	×	√
仅管理端可写，所有人可读	√	×	√	×	√
仅管理端可读写（该数据只有管理端可读写）	×	×	×	×	√

10.4.2　查询数据

在开始使用数据库 API 进行增删改查操作之前，需要先获取数据库的引用。以下调用获取默认环境的数据库的引用：

```
const db = wx.cloud.database()
```

要操作一个集合，需先获取它的引用。在获取了数据库的引用后，就可以通过数据库引用上的 collection 方法获取一个集合的引用了，例如获取待办事项清单集合：

```
const collection = db.collection('db1')
```

（1）通过 collection.doc 获取某一条记录。

如果需要对具体的某一条进行更新和删除操作，假设有一个待办事项的 id 为 record20190801，那么可以通过 collection.doc 方法获取它的引用：

```
const document = db.collection('db1').doc('record20190801')
```

（2）通过 collection.get 获取所有记录的数据。

通过 collection.get 获取集合中的所有数据，或获取根据查询条件筛选后的集合数据，其成功回调函数 success 的结果及 Promise resolve 的结果 Result 是一个数组对象，对应多条记录。

（3）通过 document.get 获取某一条记录的数据。

通过 document.get 获取具体某一条记录的数据，或获取根据查询条件筛选后的记录数据，其 success 回调的结果及 Promise resolve 的结果 Result 是一个对象，对应一条记录。

例 10-6 通过云开发控制台创建集合 db1，然后添加多条记录，记录的字段分别为 name（string）、password（string）、age（number）、sex（string）、tel（string）。本例通过 collection.get 获取集合中的所有数据，通过 document.get 获取具体某一条记录的数据，获取的云数据库的数据如图 10.34 所示。

视频讲解

图 10.34 get 操作运行结果

pages/get/get.wxml 的代码如下：

```
<view class = "doc" bindtap = "docget"> doc.get 获取一条记录</view>
   <view class = "list - item" bindtap = "testCgi">
     <text class = "request - text">{{docget.name}} -- {{docget.password}}
       -- {{docget.age}} -- {{docget.sex}} -- {{docget.tel}}</text>
   </view>
 <view class = "doc" bindtap = "Collectionget"> Collection.get 获取多条记录</view>
 <view wx:for = "{{Collectionget}}">
 <view class = "list - item" bindtap = "testCgi">
     <text class = "request - text">{{item.name}} -- {{item.password}}
       -- {{item.age}} -- {{item.sex}} -- {{item.tel}}</text>
  </view>
 </view>
```

pages/get/get.js 的代码如下：

```
const app = getApp()
Page({
  data: {
    docget:{},
    Collectionget:[]
  },
```

```
    docget:function(e){
      var that = this
      const db = wx.cloud.database()
  db.collection('db1').doc('b9eb782f-9818-45df-ab22-8f0f56fa05fc').get({
        success: function(res) {
          console.log(res.data)
          that.setData({
            docget: res.data
          })
        }
      })
    },
  Collectionget:function(e){
    var that = this
    const db = wx.cloud.database()
    db.collection('db1').get({
      success: function(res) {
        console.log(res.data)
        that.setData({
          Collectionget: res.data
        })
      }
    })
  }
})
```

【代码讲解】 docget 函数通过 document.get()接口获取具体某一条记录的数据。Collectionget 函数通过 collection.get()接口获取集合中的所有数据，注意两个方法获取的数据都在 res.data 中。

(4) 通过 collection.count 获取集合记录数量。

通过 collection.count 统计集合记录数或统计查询语句对应的结果记录数，注意这与集合权限设置有关，一个用户仅能统计其有读权限的记录数。success 回调的结果及 Promise resolve 的结果 Result 是一个 number 类型的对象 total。

(5) 通过 collection.where 条件查询。

通过 collection.where 指定筛选条件，方法签名如下：

```
function where(rule: object): Query
```

该方法接受一个必填对象参数 rule，用于定义筛选条件，示例代码如下：

```
const db = wx.cloud.database()
db.collection('db1').where({
  age: 30                                    //查询年龄为 30 岁的记录
}).get({
  success: console.log,
  fail: console.error
})
```

(6) 通过 collection.orderBy 排序查询。

通过 collection.orderBy 指定查询排序条件,方法签名如下:

```
function orderBy(fieldName: string, order: string): collection | query
```

该方法接受一个必填字符串参数 fieldName 用于定义需要排序的字段,一个字符串参数 order 用于定义排序顺序,只能取 asc 或 desc。

如果需要对嵌套字段排序,需要用"点表示法"连接嵌套字段,例如 style.color 表示字段 style 里的嵌套字段 color。如果要按多个字段排序,多次调用 orderBy 即可,多字段排序时会按照 orderBy 调用顺序先后对多个字段排序。

按一个字段排序示例代码:

```
const db = wx.cloud.database()
db.collection('db1').orderBy('age', 'asc')                    //按 age 升序排序
.get().then(console.log).catch(console.error)
```

按多个字段排序示例代码:

```
const db = wx.cloud.database()
db.collection('db1')
.orderBy('age', 'desc')                                       //按 age 降序
.orderBy('tel', 'asc')                                        //按 tel 升序
.get().then(console.log).catch(console.error)
```

例 10-7 本例和例 10-6 使用同一个云数据库集合 db1。本例使用 where 条件查询和 orderby 排序查询对云数据库的数据进行查询操作。程序运行结果如图 10.35 所示。

视频讲解

图 10.35 where 和 orderby 操作

pages/where-orderby/where-orderby. wxml 的代码如下：

```
< view class = "doc" bindtap = "where"> where 查询男人记录</view >
 < view wx:for = "{{where}}">
    < view class = "list - item" bindtap = "testCgi">
      < text class = "request - text">{{item.name}} -- {{item.password}}
        -- {{item.age}} -- {{item.sex}} -- {{item.tel}}</text >
    </view >
</view >
 < view class = "doc" bindtap = "orderby"> orderby 按 age 升序排序</view >
 < view wx:for = "{{orderby}}">
 < view class = "list - item" bindtap = "testCgi">
      < text class = "request - text">{{item.name}} -- {{item.password}}
        -- {{item.age}} -- {{item.sex}} -- {{item.tel}}</text >
 </view >
 </view >
```

pages/where-orderby/where-orderby. js 的代码如下：

```
const app = getApp()
Page({
  data: {
    where:[],
    orderby:[]
  },
    where:function(e){
      var that = this
      const db = wx.cloud.database()
      db.collection('db1').where({
        sex:"man" }).get({
        success: function(res) {
          console.log(res.data)
          that.setData({
            where: res.data
          })
        }
      })
    },
  orderby:function(e){
    var that = this
    const db = wx.cloud.database()
    db.collection('db1').orderBy('age', 'asc') .get({
      success: function(res) {
        console.log(res.data)
        that.setData({
          orderby: res.data
        })
      }
    })
  }
})
```

【代码讲解】 请注意 where 和 orderby 操作应该在 collection 和 get 操作的中间，因为 collection.get 是获取集合中的所有记录，对所有记录排序和条件查询，where 和 orderby 操作应在 get 操作之前。另外需要注意的是，get 操作得到的数据是多条记录，应该赋值给一个数组。

(7) 通过 collection.limit 指定查询结果集数量上限。

通过 collection.limit 指定查询结果集数量上限，方法签名如下：

```
function limit(max: number): collection | query
```

示例代码如下：

```
const db = wx.cloud.database()
db.collection('todos').limit(10)
.get().then(console.log).catch(console.error)
```

(8) 通过 collection.skip 跳过若干条记录。

通过 collection.skip 指定查询返回结果时从指定序列后的结果开始返回，常用于分页，方法签名如下：

```
function skip(offset: number): collection | query
```

示例代码如下：

```
const db = wx.cloud.database()
db.collection('todos').skip(10)
.get().then(console.log).catch(console.error)
```

(9) 通过 collection.field 指定返回字段。

通过 collection.field/query.field/document.field 指定返回结果中记录需返回的字段，方法签名如下：

```
function field(definition: object): collection | query | document
```

该方法接受一个必填字段用于指定需返回的字段，示例代码如下：

```
const db = wx.cloud.database()
db.collection('todos').field({
  name: true, sex: true, tel: true
}).get().then(console.log).catch(console.error)
```

视频讲解

例 10-8 本例和例 10-6 使用同一个云数据库集合 db1，输入了 7 条数据，并在每一条数据的 name 字段以数字开头。本例使用 field、limit 和 skip 进行查询操作，程序运行结果如图 10.36 所示。

pages/field-limit-skip/field-limit-skip.wxml 的代码如下：

```
<view class = "doc" bindtap = "field"> field 只显示 name、age 和 tel 字段</view>
<view wx:for = "{{field}}">
<view class = "list - item" bindtap = "testCgi">
    <text class = "request - text">{{item.name}} -- {{item.age}}
      -- {{item.tel}}  </text>
```

```
  </view>
  </view>
<view class = "doc" bindtap = "limit"> limit 只要 3 条记录</view>
  <view wx:for = "{{limit}}">
    <view class = "list - item" bindtap = "testCgi">
      <text class = "request - text">{{item.name}} -- {{item.password}}
        -- {{item.age}} -- {{item.sex}} -- {{item.tel}} </text>
    </view>
</view>
  <view class = "doc" bindtap = "skip"> skip 跳 4 条记录</view>
  <view wx:for = "{{skip}}">
  <view class = "list - item" bindtap = "testCgi">
      <text class = "request - text">{{item.name}} -- {{item.password}}
        -- {{item.age}} -- {{item.sex}} -- {{item.tel}} </text>
  </view>
  </view>
```

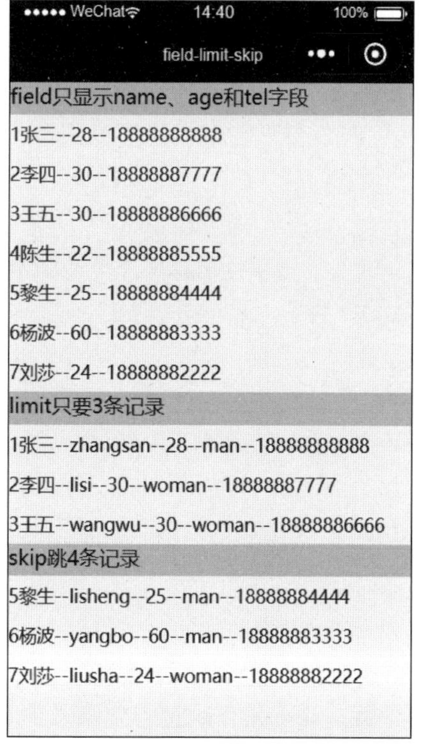

图 10.36 field、limit 和 skip 操作

pages/field-limit-skip/field-limit-skip.js 的代码如下：

```
Page({
  data: {
    limit:{},skip:[],field:[]
  },
  limit:function(e){
      var that = this
```

```
      const db = wx.cloud.database()
    db.collection('db1').limit(3).get({
        success: function(res) {
          console.log(res.data)
          that.setData({
            limit: res.data
          })
        }
      })
    },
  skip:function(e){
    var that = this
    const db = wx.cloud.database()
    db.collection('db1').skip(4).get({
      success: function(res) {
        console.log(res.data)
        that.setData({
          skip: res.data
        })
      }
    })
  },
  field: function(e) {
    var that = this
    const db = wx.cloud.database()
    db.collection('db1').field({
      name: true, age: true, tel: true
    }).get({
      success: function(res) {
        console.log(res.data)
        that.setData({
          field: res.data
        })
      }
    })
  }
})
```

【代码讲解】 field操作只获取指定的字段，本例中只显示name、age和tel 3个字段。limit操作限定获取记录的总条数。skip操作跳过指定条数的记录，常用于分页显示。这3种操作是对get操作的有效补充。

(10) 通过command查询指令。

原始的where条件查询只能查询"等于"的情况，例如以下语句是查询性别是man的所有记录：

```
db.collection('todos').where({
  sex: "man"
})
```

当需要查询"大于""小于""大于或等于""与"等特殊条件查询的时候，where语句已经不能胜任,这时需要使用 db. command 获取数据库查询指令。command (db. command)对象查询指令如表 10.6 所示。

表 10.6　command 查询指令

API	说　明	API	说　明
eq	等于指定值	gte	大于或等于指定值
neq	不等于指定值	in	在指定数组中
lt	小于指定值	nin	不在指定数组中
lte	小于或等于指定值	and	条件与,表示需同时满足另一个条件
gt	大于指定值	or	条件或,表示如果满足另一个条件也匹配

示例代码如下:

```
sex: _.eq("man")                        //sex 等于 man 的记录
age: _.lt(30)                           //age 小于 30 的记录
age: _.lte(30)                          //age 小于或等于 30 的记录
age: _.gt(30)                           //age 大于 30 的记录
age: _.gte(30)                          //age 大于或等于 30 的记录
age: _.in([20,30,40,50])                //age 是 20、30、40、50 的记录
age: _.nin([20,30,40,50])               //age 不是 20、30、40、50 的记录
age: _.gt(30).and(_.lt(50))             //age 大于 30 小于 50 的记录
age: _.gt(50).or(_.lt(30))              //age 大于 50 或者小于 30 的记录
```

视频讲解

例 10-9　本例和例 10-6 使用同一个云数据库集合 db1,使用 command. let 和 command. in 对数据库进行查询操作,程序运行结果如图 10.37 所示。

图 10.37　command 查询指令操作

pages/command/command.wxml 的代码如下：

```
<view class="doc" bindtap="lte"> lte 小于等于 30 岁的记录</view>
 <view wx:for="{{lte}}">
 <view class="list-item" bindtap="testCgi">
     <text class="request-text">{{item.name}} -- {{item.password}}—
       {{item.age}} -- {{item.sex}} -- {{item.tel}}</text>
 </view>
 </view>
<view class="doc" bindtap="in"> in 在 25 和 30 岁之间的记录</view>
 <view wx:for="{{in}}">
   <view class="list-item" bindtap="testCgi">
     <text class="request-text">{{item.name}} -- {{item.password}}—
       {{item.age}} -- {{item.sex}} -- {{item.tel}}       </text>
   </view>
</view>
```

pages/command/command.js 的代码如下：

```
Page({
  data: {
    lte:{}, in:[]
  },
  lte:function(e){
      var that = this
      const db = wx.cloud.database()
    const _ = db.command
    db.collection('db1').where({
      age: _.lte(30)
    }).get({
        success: function (res) {
          console.log(res.data)
          that.setData({
            lte: res.data
          })
        }
      })
    },
   in:function(e){
    var that = this
    const db = wx.cloud.database()
    const _ = db.command
    db.collection('db1').where({
      age: _.in([25, 30])
    }).get({
      success: function(res) {
        console.log(res.data)
        that.setData({
```

```
          in: res.data
        })
      }
    })
  }
})
```

【代码讲解】 in 操作需要特别注意,age: _.in([25, 30])是 age 等于 25 或者等于 30 的记录,不是 25～30 的记录。

10.4.3 插入数据

10.1.2 节中介绍了使用云开发控制台创建集合和记录,但对数据的操作更多是发生在程序端。小程序提供了 add 方法向集合中插入一条记录,方法签名如下:

```
function add(options: object): Promise < Result >
```

add 方法属性如表 10.7 所示。

表 10.7 add 方法属性表

字段名	类型	必填	说　明
data	Object	是	新增记录的定义
success	Function	否	成功回调,回调传入的参数 Result 包含查询的结果,Result 定义见下方
fail	Function	否	失败回调
complete	Function	否	调用结束的回调函数(调用成功、失败都会执行)

add 方法 success 回调的结果及 Promise resolve 的结果 Result 是一个 String 类型的_id,即新增的记录的 ID。

视频讲解

例 10-10 本例和例 10-6 使用同一个云数据库集合 db1,使用 add 方法向集合中插入记录,程序运行结果如图 10.38 所示。

图 10.38 使用 add 插入数据前后

pages/add/add.wxml 的代码如下：

```
<view class="doc" bindtap="lte">插入之前大于等于 30 岁的记录</view>
 <view wx:for="{{lte}}">
 <view class="list-item" bindtap="testCgi">
     <text class="request-text">{{item.name}} -- {{item.password}} -- {{item.age}}
       -- {{item.sex}} -- {{item.tel}}</text>
 </view>
 </view>
<view class="doc" bindtap="add">add 插入之后大于等于 30 岁的记录</view>
 <view wx:for="{{add}}">
   <view class="list-item" bindtap="testCgi">
     <text class="request-text">{{item.name}} -- {{item.password}} --
       {{item.age}} -- {{item.sex}} -- {{item.tel}}      </text>
   </view>
</view>
```

pages/add/add.js 的代码如下：

```
Page({
  data: {
    lte:{},add:[]
  },
  lte:function(e){
      var that = this
      const db = wx.cloud.database()
    const _ = db.command
    db.collection('db1').where({
      age: _.gte(30)
    }).get({
        success: function(res) {
          console.log(res.data)
          that.setData({
            lte: res.data
          })
        }
      })
    },
  add:function(e){
    var that = this
    const db = wx.cloud.database()
    const _ = db.command
    db.collection('db1').add({
      data: {                                   //插入的值
        age: 60, name: "韩老师", password:"han", sex:"man", tel:"18888880002"
      },
```

```
        success: function(res) {                //插入成功之后查询
          console.log(res)
          db.collection('db1').where({
            age: _.gte(30)
          }).get({
            success: function(res) {
              console.log(res.data)
              that.setData({
                add: res.data
              })
            }
          })
        }
      })
    }
  })
```

【代码讲解】 add 操作会返回一个"_id",该值是成功插入的记录的 id 值,程序可以通过检查是否返回了"_id"来判断是否成功插入,若没有返回"_id"则说明没有插入成功。

10.4.4 更新数据

小程序提供了两个 API 对云数据库进行记录的更新操作:update 方法更新一条或多条记录的局部字段;set 方法替换更新一条记录。

1. update 更新

使用 update 方法可以局部更新一个记录或一个集合中的记录,局部更新意味着只有指定的字段会得到更新,其他字段不受影响。示例代码如下:

```
db.collection('db1').doc('id20190901').update({
  data: {
      age: 30
  },
  success: function(res) {     }
})
```

2. set 更新

如果需要替换更新一条记录,可以在记录上使用 set 方法,替换更新意味着用传入的对象替换指定的记录。

```
db.collection('db1').set({
    data: {                                       //插入的值
    age: 60, name: "韩老师", password:"han", sex:"man", tel:"18888880002"
    },
  success: function(res) {
        }
})
```

例 10-11 本例和例 10-6 使用同一个云数据库集合 db1,使用 update 和 set 方法对数据进行更新操作,程序运行结果如图 10.39 所示。

视频讲解

图 10.39 update 和 set 操作

pages/update-set/update-set.wxml 的代码如下:

```
< view class = "doc" bindtap = "update1">韩老师记录 update 之前</view >
    < view class = "list - item" bindtap = "testCgi">
      < text class = "request - text">{{update1.name}} -- {{update1.password}}
        -- {{update1.age}} -- {{update1.sex}} -- {{update1.tel}}        </text >
</view >
< view class = "doc" bindtap = "update2">韩老师记录 update 之后</view >
    < view class = "list - item" bindtap = "testCgi">
      < text class = "request - text">{{update2.name}} -- {{update2.password}}
      -- {{update2.age}} -- {{update2.sex}} -- {{update2.tel}}        </text >
</view >
< view class = "doc" bindtap = "set1">李老师记录 set 之前</view >
    < view class = "list - item" bindtap = "testCgi">
      < text class = "request - text">{{set1.name}} -- {{set1.password}}
      {{set1.age}} -- {{set1.sex}} -- {{set1.tel}}        </text >
</view >
< view class = "doc" bindtap = "set2">李老师记录 set 之后</view >
    < view class = "list - item" bindtap = "testCgi">
      < text class = "request - text">{{set2.name}} -- {{set2.password}}—
      {{set2.age}} -- {{set2.sex}} -- {{set2.tel}}        </text >
</view >
```

pages/update-set/update-set.js 的代码如下:

```
Page({
  data: {update1:{},update2:{}, set1:{}, set2:{} },
  update1:function(e){
    var that = this;
    const db  =  wx.cloud.database()
    db.collection('db1').doc('94b1e1fc5d0df4ff05fc607934a3aba0').get({
      success: function (res) {
        console.log(res.data)
        that.setData({
          update1: res.data
```

```
      })
    }
  })
},
update2: function(e) {
  var that = this;
  const db = wx.cloud.database()
  db.collection('db1').doc('94b1e1fc5d0df4ff05fc607934a3aba0')
  .update({
    data: {age:55},
    success: function(res) {
      console.log(res.data)
      db.collection('db1').doc('94b1e1fc5d0df4ff05fc607934a3aba0').get({
        success: function(res) {
          console.log(res.data)
          that.setData({
            update2: res.data
          })
        }
      })
    }
  })
},
set1: function(e) {
  var that = this;
  const db = wx.cloud.database()
  db.collection('db1').doc('08560c9e5d0e2153061a3a3837585483').get({
    success: function(res) {
      console.log(res.data)
      that.setData({
        set1: res.data
      })
    }
  })
},
set2: function(e) {
  var that = this;
  const db = wx.cloud.database()
  db.collection('db1').doc('08560c9e5d0e2153061a3a3837585483')
    .set({
    data: {
      age: 60,name:"李教授",password:"li",sex:"woman",tel:"18888881111"
    },
    success: function(res) {
      console.log(res.data)
      db.collection('db1').doc('08560c9e5d0e2153061a3a3837585483')
      .get({success: function (res) {
          console.log(res.data)
          that.setData({
            set2: res.data
          })
        }
      })
    }
```

```
    })
  }
})
```

【代码讲解】 注意在 update 和 set 操作之后不可以接着就进行 get 操作，本例把 get 操作写在了 update 操作的 success 方法中。

10.4.5 删除数据

小程序提供了 remove 方法来删除一条记录，示例代码如下：

```
db.collection('db1').doc('id20190901').remove({
  success: function(res) {
    console.log(res.data)
  }
})
```

如果需要删除多条记录，需在 Server 端进行操作（云函数），并通过 where 语句选取多条记录进行删除，有权限删除的记录才会被删除。

例 10-12 本例和例 10-6 使用同一个云数据库集合 db1，使用小程序端和云函数同时对数据进行删除操作，由于云数据库操作权限问题，小程序端的删除操作失败，而云函数端的删除操作成功。程序运行结果如图 10.40 所示。

视频讲解

```
Console  Sources  Network  Security  AppData
top  Filter
▼{stats: {…}, errMsg: "document.remove:ok"}
   errMsg: "document.remove:ok"
  ▶stats: {removed: 0}
  ▶__proto__: Object
成功删除0条记录
▼{errMsg: "cloud.callFunction:ok", result: {…}}
   errMsg: "cloud.callFunction:ok"
  ▼result:
     errMsg: "collection.remove:ok"
    ▶stats: {removed: 1}
    ▶__proto__: Object
  ▶__proto__: Object
成功删除1条记录
```

图 10.40　删除云数据库数据

pages/remove/remove.wxml 的代码如下：

```
<view class = "doc" bindtap = "remove1">小程序端删除赵老师记录</view>
    <view class = "list - item" bindtap = "testCgi">
      <text class = "request - text">{{remove1}}</text>
</view>
<view class = "doc" bindtap = "remove2">服务器端删除赵老师记录</view>
    <view class = "list - item" bindtap = "testCgi">
      <text class = "request - text">{{remove2}}</text>
</view>
```

pages/remove/remove.js 的代码如下：

```
Page({
```

```
  remove1:function(e){
    var that = this
    const db = wx.cloud.database()
    db.collection('db1').doc('e7bc15af - 5983 - 4c50 - 9083 - dd465ebcfd2b')
  .remove({success: function (res) {
        console.log(res)
        console.log("成功删除" + res.stats.removed + "条记录")
      }
    })
  },
  remove2: function (e) {
    wx.cloud.callFunction({
      name: 'remove',                          //云函数名称
      data: {                                  //传给云函数的参数
        name: "赵老师"
      },
      success: function (res) {
        console.log(res)
        console.log("成功删除" + res.result.stats.removed + "条记录")
      },
      fail: console.error
    })
  }
})
```

云函数 remove/index.js 的代码如下：

```
const cloud = require('wx - server - sdk')
cloud.init()
const db = cloud.database()
const _ = db.command
exports.main = async (event, context) => {
    return await db.collection('db1').where({
      name:event.name
    }).remove({
    })
}
```

【代码讲解】 小程序端只能对用户在小程序端插入的数据进行删除操作，所以小程序端对数据的删除没有成功；云函数有最高的数据库操作权限，所以云函数端对数据的删除成功了。读者通过本例应该更好地认识到小程序端和云函数端对数据库操作的区别。

10.5 实训项目——基于云数据库的许愿墙

视频讲解

本实训项目以云开发的云数据库为基础，制作一个简易的许愿墙。顾名思义，"云数据库"就是把本项目中的愿望的数据全部存储在云端。

首先在云开发控制台新建一个集合 wishwall，然后添加几条记录，每一条记录的 3 个字段分别是 title(string)、date(string)和 address(string)，分

别表示一个愿望的名称、日期和地点。云数据库设计如图 10.41 所示。

图 10.41　云数据库设计

本项目包括一个 wishwall(愿望墙)页面、add(增加愿望)页面和 details(愿望详情)页面。

(1) add 页面,把页面 title、date 和 address 插入云数据库"wishwall"作为一条新的记录,插入成功跳回 wishwall 页面。add 页面执行效果如图 10.42 所示。

(2) wishwall 页面,自动加载云数据库中所有的愿望并显示出来,当点击某一条具体的愿望时,则进入 details 页面,此时需要传递愿望 id 值给 details 页面;当点击"增加愿望"区域时,则进入 add 页面。wishwall 页面效果如图 10.43 所示。

(3) details 页面,根据 wishwall 页面传递过来的 id 查询该条记录的 title、date 和 address 值并显示出来。details 页面执行效果如图 10.44 所示。

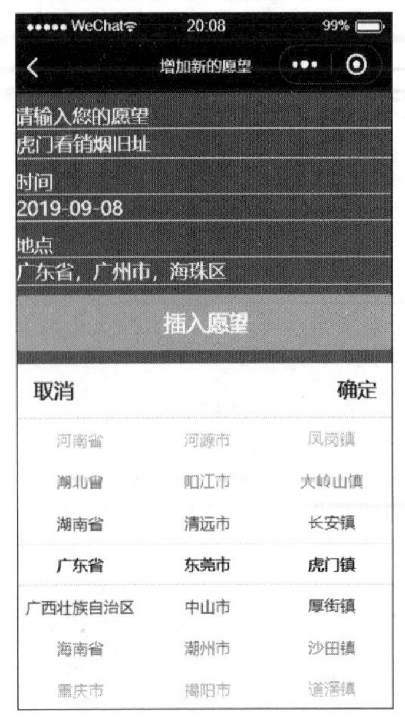

图 10.42　add 页面

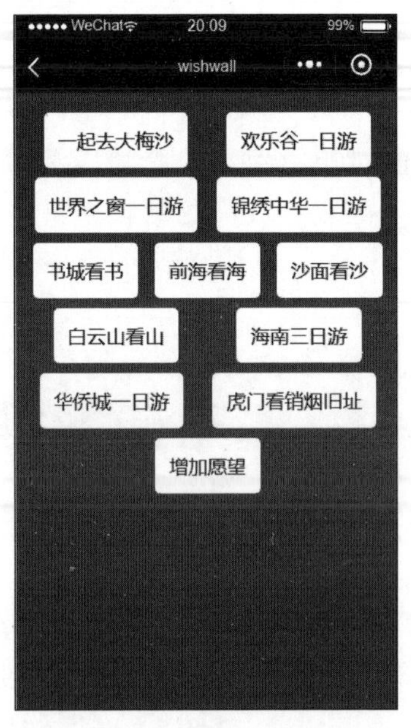

图 10.43　wishwall 页面

图 10.44　details 页面

pages/wishwall/wishwall.wxml 的代码如下：

```
<view class="zong">
    <view class="yang1" wx:for="{{wishs}}"
    id="{{item._id}}" bindtap='details'>
    {{item.title}}
    </view>
    <view class="yang1" bindtap='add'>
    增加愿望
    </view>
</view>
```

pages/wishwall/wishwall.js 的代码如下：

```
const app = getApp()
Page({
  data: {
    wishs: []
  },
  onLoad: function(e) {
    var that = this
    const db = wx.cloud.database()
    db.collection('wishwall').get({
      success: function(res) {
        console.log(res.data)
        that.setData({
          wishs: res.data
        })
      }
    })
  },
  details:function(e){
    console.log(e.target.id)                //点击了那条愿望
    wx.navigateTo({
      url: "../details/details?id=" + e.target.id
    })
  },
  add:function(e){
    wx.navigateTo({
      url: '../add/add',
    })
  }
})
```

pages/wishwall/wishwall.wxss 的代码如下：

```
page {
   display: flex; flex-direction: column;
  justify-content: flex-start; background-color: #005F8C;
}
.zong{
  display: flex; flex-direction:row;
```

```
  flex-wrap: wrap; padding: 20rpx;
  align-items: center; justify-content: space-around;
}
.yang1{
padding: 30rpx; background-color: #ffffff;
margin-top: 20rpx; border-radius:10rpx;
}
.yang2{
padding: 30rpx; background-color: #f1b0e6;
margin-top: 20rpx; border-radius:10rpx; width: 100rpx;
}
```

【代码讲解】 wishwall.js 的 onLoad()函数自动执行对云数据库的查询操作，获取云数据库中所有的愿望数据，赋值给 wishs，并通过数据绑定的方式在 wishwall.wxml 中进行渲染显示。

pages/add/add.wxml 的代码如下：

```
<view class='title'>
<view>请输入您的愿望</view>
<view><input class='in' auto-focus bindinput="title"></input></view>
</view>
<view class='title'>
<view>时间</view>
<picker mode="date" value='{{date}}' start="2019-08-01"
end="2020-08-08" bindchange='date'>
<view class='in'>{{date}}</view>
</picker></view>
<view class='title'>
<view>地点</view>
 <picker mode="region" bindchange="bindRegionChange"
value="{{region}}" custom-item="{{customItem}}">
<view class="in">
     {{region[0]}},{{region[1]}},{{region[2]}}
   </view>
</picker></view>
<view class="title"></view>
<button type="primary" bindtap="add" > 插入愿望</button>
```

pages/add/add.js 的代码如下：

```
Page({
  data: {
    title:'',
region: ['广东省', '广州市', '海珠区'],
date: "2019-08-08",address:"广东省广州市海珠区"
  },
  title:function(e){
  this.setData({
    title: e.detail.value
  })
  },
```

```
  date: function(e) {
    console.log(e)
    this.setData({
      date:e.detail.value
    })
  },
  bindRegionChange: function(e) {
    console.log('携带值为', e.detail.value[0] +
    e.detail.value[1] + e.detail.value[2])
    this.setData({
      region: e.detail.value,
      address: e.detail.value[0] + e.detail.value[1] + e.detail.value[2]
    })
  },
  add:function(e){
    var that = this;
    const db = wx.cloud.database()
    db.collection('wishwall').add({
      data: {
      title:that.data.title,
        date: that.data.date,
        address: that.data.address
      },
      success: function(res) {
        console.log(res)
      }
    })
    wx.navigateTo({
      url: '../wishwall/wishwall',
    })
  }
})
```

pages/add/add.wxss 的代码如下：

```
.in{
  border: 1px solid #ffffff;
}
.title{
  margin-top: 20rpx; margin-bottom: 20rpx; color:#ffffff;
}
page{
  background-color: #005F8C;
}
```

【代码讲解】 add.js 获取 add.wxml 由用户填入表单的数据，并执行对云数据库的插入操作，插入成功之后再跳转回 wishwall.wxml 页面。

pages/details/details.wxml 的代码如下：

```
<view class="title">{{item.title}}</view>
<view class='date'>{{item.date}}</view>
<view class="address">{{item.address}}</view>
```

pages/details/details.js 的代码如下：

```
Page({
  data: {
    item: {
      title: "白云山看山",date: "2019-08-08",address: "广东省广州市白云区"}
  },
  onLoad: function(options) {
    console.log("传过来的数据是")
    console.log(options.id)
    var id = options.id
    var that = this;
    const db = wx.cloud.database()
    db.collection('wishwall').where({
      _id:id
    }).get({
      success: function (res) {
        console.log(res.data)
        that.setData({
          item: res.data[0]
        })
      }
    })
  }
})
```

pages/details/details.wxss 的代码如下：

```
.title{
  margin-top: 100rpx;font-size: 2.5cm; color: #ffffff; text-align:center;
}
page{
  background-color: #005F8C;
}
.date{
  margin-top: 50rpx; font-size: 1.5em; color: #ffffff; text-align:center;
}
.address{
   margin-top: 30rpx; font-size: 1em; color: #ffffff; text-align:center;
}
```

【代码讲解】 details.js 的 onLoad()函数根据 wishwall.js 传递过来的 id 值，查询云数据库中被用户点击的那条愿望信息，获取的记录值赋值给 item，并通过数据绑定的方式在 detail.wxml 中进行渲染显示。

第11章 数码产品类电商小程序项目

通过前面十章的学习,相信读者已经具备了开发完整小程序项目的能力。本章将以电商类小程序项目为例,系统地完成一个项目的开发全过程。通过本章的学习,读者将提升商业项目的设计与实操能力。

本章主要目标

- 综合应用所学知识创建完整的电商小程序项目;
- 熟练掌握与实现页面中的交互;
- 了解项目开发中的流程步骤。

11.1 需求分析

本项目一共需要6个页面,即数码商城首页、商品分类页、商品详情页、购物车页面、支付页和个人主页。

(1) 数码商城首页:数码商城首页应包含顶部的轮播图、中间商品列表部分和底部tabBar菜单,其中,轮播图可以使用swiper实现,中间商品列表部分利用wx:for来渲染,底部tabBar菜单在app.json文件中配置。当用户点击某款商品时,可以跳转到该款商品的详情页中。底部tabBar菜单包含“首页”“分类”“购物车”“我的”4个切换菜单。

(2) 商品分类页:当用户点击左边的分类区域后,在右侧按分类显示对应的商品。当用户点击某款商品时,可以跳转到该款商品的详情页中。

(3) 商品详情页:通过商城首页和分类页的点击事件跳转而来,根据商城首页和分类页传递过来的{{index}}字段来调用正确的商品信息,用户在此页面可将商品添加进购物车。

(4) 购物车页面:用户可以进行商品数量的修改、删除某件商品、清空购物车等操作,点击“结算”按钮进入支付页面。

（5）支付页：显示购买的商品数量、设置配送时间、备注以及付款总额。

（6）个人主页：查看个人信息。

11.2 页面设计与实现

本节完成商城首页的设计、商品分类页的设计以及商品详情页的设计。

11.2.1 全局文件的设计与实现

创建一个名为 digitalMall 的小程序项目，在 app.json 文件中注册项目需要的页面、底部 tabBar 菜单以及项目标题的配置。本项目的商品数据均放在本地 app.js 文件中，有兴趣的读者可以按第 6 章的知识点把商品数据存放在服务器或者按第 10 章的知识点把商品数据存放在云数据库中。

app.json 文件代码如下：

视频讲解

```
{
  "pages": [
    "pages/goods - index/goods - index",
    "pages/goods - detail/goods - detail",
    "pages/sort/sort",
    "pages/cart/cart",
    "pages/pay/pay",
    "pages/my/my"
  ],
  "window": {
    "navigationBarBackgroundColor": "#B3B3B3"
  },
  "tabBar": {
    "color": "#4D4D4D",
    "selectedColor": "#FF0000",
    "borderStyle": "black",
    "list": [
      {
        "selectedIconPath": "icon/index0.png",
        "iconPath": "icon/index.png",
        "pagePath": "pages/goods - index/goods - index",
        "text": "首页"
      },
      {
        "selectedIconPath": "icon/sort0.png",
        "iconPath": "icon/sort.png",
        "pagePath": "pages/sort/sort",
        "text": "分类"
      },
      {
```

```
        "selectedIconPath": "icon/cart0.png",
        "iconPath": "icon/cart.png",
        "pagePath": "pages/cart/cart",
        "text": "购物车"
      },
      {
        "selectedIconPath": "icon/me0.png",
        "iconPath": "icon/me.png",
        "pagePath": "pages/my/my",
        "text": "我的"
      }
    ]
  },
  "sitemapLocation": "sitemap.json"
}
```

app.js 文件代码如下：

```
App({
  globalData: {
    data: [{
        id: 0,
        sort: 1,
        title: "(HUAWEI)P30",
        titleTwo: "超感光徕卡三摄|逆光智美自拍",
        price: "4288.00",
        image: "https://res.vmallres.com/pimages//product/6901443293513
        /800_800_1555464685019mp.png",
        imageone: "//img10.360buyimg.com/n7/jfs/t1/31698/11/11865/218814
        /5cb68870Ebf26e1bd/dbe080c29fb0aeff.jpg",
        imagetwo: "//img14.360buyimg.com/n7/jfs/t1/11352/31/13456/324178
        /5c98c88dE9419c2ca/4be2efca1d9e2b38.jpg",
        imagethree: "//img12.360buyimg.com/n7/jfs/t1/30693/17/7599/332089
        /5c98cc1fE43eafa3c/3b3515c7537efeaf.jpg",
        imgdetail: [{
          image: "//img20.360buyimg.com/vc/jfs/t1/37127/23/887/799049
        /5cadd35dE9ec3cc24/766b05c2232b3b23.jpg",
          mode: "widthFix"
        }, ],
        imgdeparameter: [{
          image: "//img20.360buyimg.com/vc/jfs/t1/59649/26/1793/4847556
        /5d0208b5E39a5ab59/300c87f3e3915620.jpg",
          mode: "widthFix"
        }, ]
      },
      {
        id: 1,
        sort: 1,
        title: "(HUAWEI)荣耀 V20",
        titleTwo: "魅眼视屏 4800 万深感相机",
        price: "3699.00",
```

```
    image: "//img12.360buyimg.com/n7/jfs/t25954/134/1930444050/488286
    /31587d0d/5bbf1fc9N3ced3749.jpg",
    imageone: "//img14.360buyimg.com/n7/jfs/t25420/83/1918969044/436308
    /64a55b49/5bbf1fbfN96f4ba7e.jpg",
    imagetwo: "//img14.360buyimg.com/n7/jfs/t26017/221/2094289784
    /437463/b9ee4f1f/5bc47489Nebd5e8b0.jpg",
    imagethree: "//img10.360buyimg.com/n7/jfs/t27568/359/1014140699
    /436308/64a55b49/5bbf1f3cN3363b2b2.jpg",
    imgdetail: [{
      image: "//img20.360buyimg.com/vc/jfs/t1/2989/1/11532/3806781
    /5bcfd0ceE9cb3575c/c7a5b12aada8552b.jpg",
      mode: "widthFix"
    }, ],
    imgdeparameter: [{
      image: "//img20.360buyimg.com/vc/jfs/t27433/167/1405654888
      /234469/b3aadbf1/5bc85d45N25be7ed6.jpg",
      mode: "widthFix"
    }, {
      image: "//img20.360buyimg.com/vc/jfs/t26110/279/1327305885
      /330746/ae320b81/5bc59e47Ne94d3efe.jpg",
      mode: "widthFix"
    }]
  },
  {
    id: 2,
    sort: 2,
    title: "IE80S 入耳式监听耳机",
    titleTwo: "森海塞尔 HiFi 音乐耳机",
    price: "2399.00",
    image: "https://images.wincheers.net/UpLoad/Web/ProductImg
    /2018-08-13/NEW_XM/IE80S.jpg",
    imageone: "//img11.360buyimg.com/n7/jfs/t1/51871/24/2765/90692
    /5d09ea26Eca12e23a/3183d39bae509977.jpg",
    imagetwo: "//img13.360buyimg.com/n7/jfs/t1/37032/4/10505/56477
    /5d09df52E32104653/27cfa82d534591dd.jpg",
    imagethree: "//img13.360buyimg.com/n7/jfs/t1/37032/4/10505/56477
    /5d09df52E32104653/27cfa82d534591dd.jpg",
    imgdetail: [{
        image: "//img13.360buyimg.com/cms/jfs/t1/44079/31/8065/242506
    /5d1a281dEb085f2d0/7d4c6766d6741784.jpg",
        mode: "widthFix"
      },
      {
        image: "//img10.360buyimg.com/cms/jfs/t1/53747/12/4070/157955
      /5d1b2820Ea16a5fac/36d136d2104ff491.jpg",
        mode: "widthFix"
      },
      {
        image: "//img30.360buyimg.com/jgsq-productsoa/jfs/t1/28069/8
      /4582/207938/5c32c4b1Ec5c9026c/a8dc4b5e9a93c3be.jpg",
        mode: "widthFix"
```

```
        }
      ],
      imgdeparameter: [{
          image: "//img13.360buyimg.com/cms/jfs/t1/70502/7/3355/209359
        /5d1a29a0E438ae9e4/88e090a64f796dcd.jpg",
          mode: "widthFix"
        },
        {
          image: "//img11.360buyimg.com/cms/jfs/t1/62580/4/3467/246593
        /5d1a2a12E0196f6eb/fcd04b7d68691caf.jpg",
          mode: "widthFix"
        },
      ]
    },
    {
      id: 3,
      sort: 2,
      title: "IE60 入耳式 HiFi 耳机",
      titleTwo: "森海塞尔入耳式 HIFI 耳机",
      price: "799.00",
      image: "https://images.wincheers.net/UpLoad/Web/ProductImg
      /2017-08-14/NEW_XM/IE-60-4.png",
      imageone: "//img12.360buyimg.com/n5/jfs/t11584/3/384240986/69087
        /fe5d3b68/59eebd52N9d786155.jpg",
      imagetwo: "//img10.360buyimg.com/n5/jfs/t2080/174/1543078580/76751
      /ceebd204/5667e903Nc1688332.jpg",
      imagethree: "//img12.360buyimg.com/n7/jfs/t9853/237/2116508249
      /69087/fe5d3b68/59eebd40N23b9d715.jpg",
      imgdetail: [{
        image: "//img30.360buyimg.com/jgsq-productsoa/jfs/t1/56674/3
      /6339/105009/5d414287E74c8e487/1401f285551831db.jpg",
        mode: "widthFix"
      }, ]
    }
  ],
  imgdeparameter: [{
    image: "//img10.360buyimg.com/imgzone/jfs/t3136/220/2277407509
/120182/79fdaa99/57df5245N2c4750c3.jpg",
    mode: "widthFix"
  }, ]
 }
})
```

11.2.2　商城首页的设计与实现

首先设计商城首页的布局，然后从 app.js 中获取数据，通过 wx:for 列表渲染在页面中显示每款商品的信息，当用户点击某一商品时，页面跳转到该款商品的详情页。商城首页的运行效果如图 11.1 所示。

图 11.1 商城首页效果图

pages/goods-index/goods-index.wxml 文件代码如下：

视频讲解

```
<!-- 轮播图 -->
<swiper indicator-dots autoplay interval="3000">
  <swiper-item>
    <image src='/images/ban1.jpg'></image>
  </swiper-item>
  <swiper-item>
    <image src='/images/ban2.jpg'></image>
  </swiper-item>
  <swiper-item>
    <image src='/images/ban3.jpg'></image>
  </swiper-item>
</swiper>
<!-- 商品展示部分 -->
<view class="demo-box">
  <block wx:for="{{main_key}}" wx:for-item="item">
    <view class="goods-box" bindtap="btntodetail"  data-id="{{item.id}}">
      <image class="goods-pic" src="{{item.image}}"></image>
      <view class="goods-title">{{item.title}}</view>
      <view class="goods-titleTwo">{{item.titleTwo}}</view>
      <view class="row">
        <view class="goods-price">¥{{item.price}}</view>
```

```
        <text class = "goods - btn">看相似</text>
      </view>
    </view>
  </block>
</view>
```

pages/goods-index/goods-index. js 文件代码如下：

```
const App = getApp();
Page({
  //获取数据
  onLoad: function(options) {
    this.setData({
      main_key: App.globalData.data
    })
  },
  //点击商品列表跳转商品详情页
  btntodetail: function(e) {
    var listid = e.currentTarget.dataset.id
    console.log("你点击了第" + (listid + 1) + "个商品")
    wx.navigateTo({
      url: '../../pages/goods - detail/goods - detail?listid = ' + listid
    })
  }
})
```

pages/goods-index/goods-index. wxss 文件代码如下：

```
/* 轮播图样式 */
swiper {
  width: 100 %; height: 380rpx;
}
/* 轮播图中图片样式 */
swiper image {
  width: 100 %; height: 380rpx;
}
/* 商品展示外部样式 */
.demo - box {
  display: flex; flex - direction: row; flex - wrap: wrap; padding: 30rpx;
}
.goods - box {
  width: 49 %; margin - bottom: 20rpx;
}
/* 商品内容样式 */
.goods - pic {
  width: 220rpx; height: 250rpx;
  margin: 0 auto; display: block;
}
```

```
.row {
  display: flex; flex-direction: row; justify-content: space-around;
}
/*看相似样式*/
.goods-btn {
  border: 1px solid #e3e3e3; border-radius: 20rpx;
  margin-top: 20rpx; width: 100rpx;
  height: 40rpx; line-height: 40rpx;
  font-size: 24rpx; color: #aaa;
  letter-spacing: 2rpx; text-align: center;
}
/*商品标题样式*/
.goods-title {
  font-size: 30rpx; font-weight: 600; text-align: center;
}
.goods-titleTwo {
  font-size: 24rpx; margin-top: 10rpx; text-align: center;
}
/*商品价格样式*/
.goods-price {
  font-size: 30rpx; margin-top: 20rpx; color: #ee3b3b;
}
```

【代码讲解】 goods-index.js 文件的 onLoad()函数通过语句 main_key: App.globalData.data 获取 app.js 文件中的商品数据，再使用 wx:for="{{main_key}}"列表渲染把所有的商品在页面中显示出来。点击事件 btntodetail()实现首页到详情页的跳转，在 goods-index.js 文件中自定义该函数如下：

```
  btntodetail: function(e) {
    var listid = e.currentTarget.dataset.id
    console.log("你点击了第" + listid + "个商品")
    wx.navigateTo({
      url: '../../pages/goods-detail/goods-detail?listid=' + listid
    })
  }
})
```

由于在 goods-index.wxml 文件中设置了自定义数据 data-id="{{item.id}}"，goods-index.js 通过语句 var listid = e.currentTarget.dataset.id 获取用户具体点击的那件商品在商品数组中的下标，商品详情页通过 listid 准确显示被点击的商品的信息。

11.2.3 商品分类页的设计与实现

商城首页没有对商品进行分类处理，而是通过 wx:for 对商品数组进行全部循环渲染。商品分类页实现商品的分类功能，当用户点击“手机”分类，页面的右侧只显示所有的“手机”商品。图 11.2 是分类页的“手机产品”页面；图 11.3 是分类页的“时尚耳机”页面。

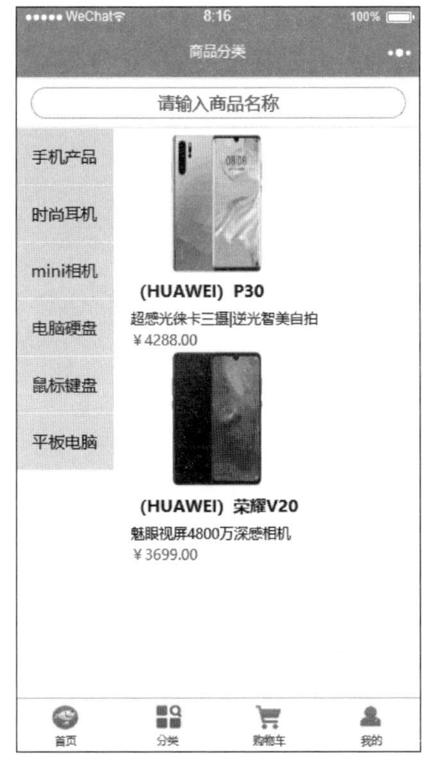

图 11.2　分类页之“手机产品”

图 11.3　分类页之“时尚耳机”

pages/sort/sort.wxml 文件代码如下：

视频讲解

```
<! -- 分隔线 -->
<view class = "hr"></view>
<! -- 搜索框 -->
<view class = "header">
  <input placeholder = '请输入商品名称'></input>
</view>
<view class = "hr"></view>
<view class = "content">
  <! -- 左侧内容 -->
  <view class = "left">
    <scroll - view scroll - y>
      <view id = "1" bindtap = "switchTab">手机产品</view>
      <view id = "2" bindtap = "switchTab">时尚耳机</view>
      <view id = "3" bindtap = "switchTab"> mini 相机</view>
      <view id = "4" bindtap = "switchTab">电脑硬盘</view>
      <view id = "5" bindtap = "switchTab">鼠标键盘</view>
      <view id = "6" bindtap = "switchTab">平板电脑</view>
    </scroll - view>
  </view>
  <! -- 右侧内容 -->
  <view class = "right">
    <block wx:for = "{{main_key}}" wx:for - item = "item" class = "left">
```

```
      <block wx:if = "{{item.sort == sort}}">
        <view bindtap = "btntodetail" data - id = "{{item.id}}">
          <image class = "right - image" src = "{{item.image}}"></image>
          <view class = "title - box">
            <view class = "goods - title">{{item.title}}</view>
            <text class = "goods - titleTwo">{{item.titleTwo}}</text>
            <view class = "goods - price">¥{{item.price}}</view>
          </view>
        </view>
      </block>
    </block>
  </view>
</view>
```

pages/sort/sort.js 文件代码如下：

```
const App = getApp();
Page({
  //数据初始化
  data: {
    flag: 0,
    list: ["手机产品", "时尚耳机", "mini 相机", "电脑硬盘", "鼠标键盘", "平板电脑"],
    sort: 1
  },
  //获得数据
  onLoad: function(options) {
    this.setData({
      main_key: App.globalData.data
    })
  },
  //点击商品列表跳转商品详情页
  btntodetail: function(e) {
    var listid = e.currentTarget.dataset.id
    console.log("你点击了第" + (listid + 1) + "个商品")
    wx.navigateTo({
      url: '../../pages/goods - detail/goods - detail?listid = ' + listid
    })
  },
  //Tab 栏切换
  switchTab: function(e) {
    console.log(e)
    this.setData({
      sort: e.currentTarget.id
    })
  },
})
```

pages/sort/sort.wxss 文件代码如下：

```
/* 外部样式 */
.content {
  display: flex; flex - direction: row;
```

```
}
/* 分隔线样式 */
.hr {
  border: 1px solid #ccc; opacity: 0.2;
}
/* 输入框样式 */
input {
  margin: 15rpx 30rpx; border: 1px solid #ccc;
  border-radius: 50rpx; text-align: center;
  font-size: 32rpx;
}
/* 左边样式 */
.left {
  width: 25%; font-size: 30rpx;
}
/* 左边内容样式 */
.left view {
  text-align: center; height: 50px; line-height: 50px;
}
scroll-view {
  height: 90%;
}
.left view {
  border-bottom: 1px solid #f0ffff; background-color: #eee9e9;
}
/* 右边内容样式 */
.right {
  width: 75%; padding: 10rpx;
}
/* 商品图片样式 */
.right-image {
  width: 200rpx; height: 250rpx; margin-left: 80rpx;
}
/* 商品标题样式 */
.title-box {
  margin-left: 20rpx;
}
.goods-title {
  font-size: 30rpx; font-weight: bold;
}
.goods-titleTwo {
  font-size: 26rpx;
}
/* 商品价格样式 */
.goods-price {
  color: #ee3b3b; font-size: 26rpx;
}
```

【代码讲解】 sort.wxml实现商品的分类功能，页面的左侧是一个<scroll-view>组件用来显示所有的分类，右侧显示具体分类下的所有商品，左侧和右侧之间的交互是本文件的关键点。<scroll-view>组件的每一个分类都对应一个 id，sort.js 的 select 函数接收该 id 赋值给页面变量 sort。sort.wxml 在使用 wx:for 循环渲染所有商品的时候，<block wx:if="{{item.sort==sort}}">实现只渲染符合该分类的商品，从而实现了左侧与右侧的交互行为。

11.2.4 商品详情页的设计与实现

商品详情页根据商城首页和分类页传递过来的 listid 变量来定位具体的商品，data: App.globalData.data[id]中的 id 即是 listid，该语句从商品数组中根据 id 值取出正确的那件商品的信息。商品详情页的运行效果如图 11.4～图 11.6 所示，其中，图 11.4 是“商品详情”内容，图 11.5 是“产品参数”内容，图 11.6 是“加入购物车”的状态。

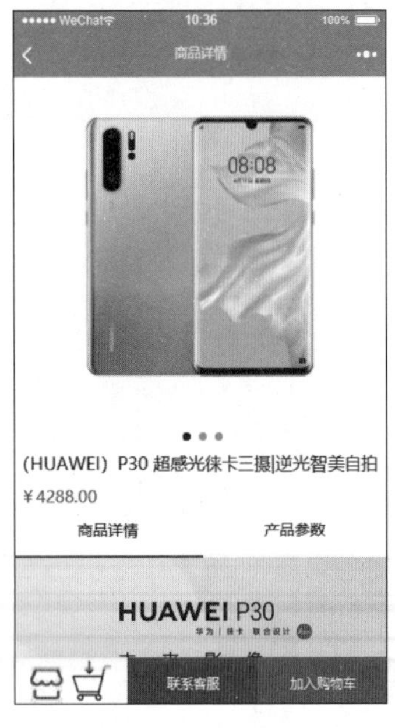

图 11.4 商品详情

图 11.5 产品参数

图 11.6 加入购物车

pages/goods-detail/goods-detail.wxml 文件代码如下：

视频讲解

```
<! -- 商品展示 -- >
< swiper indicator - dots autoplay interval = "3000">
  < swiper - item >
    < image src = "{{data.imageone}}" />
  </swiper - item >
  < swiper - item >
    < image src = "{{data.imagetwo}}" />
  </swiper - item >
  < swiper - item >
    < image src = "{{data.imagethree}}" />
  </swiper - item >
</swiper >
<! -- 商品标题和价格 -- >
```

```
<view class = 'box - demo'>
  <text class = 'title'>{{data.title}} {{data.titleTwo}}</text>
  <text class = 'price'>¥ {{data.price}}</text>
</view>
<! -- 商品详情展示 -->
<view>
  <view class = "tab {{HomeIndex == 0 ?'active':''}}" bindtap = "switchTab" data - index = "0">
商品详情</view>
  <view class = "tab {{HomeIndex == 1 ?'active':''}}" bindtap = "switchTab" data - index = "1">
产品参数
  </view>
  <view>
    <view wx: if = "{{HomeIndex  = =  0}}" wx: for = "{{data. imgdetail}}" wx: for - item =
"imgdetail">
      <image src = "{{imgdetail. image}}" mode = "{{imgdetail. mode}}" class = "fix - box">
</image>
    </view>
    <view wx:if = "{{HomeIndex  = =  1}}" wx: for = "{{data. imgdeparameter}}" wx: for - item =
"imgdeparameter">
      <image src = '{{imgdeparameter. image}}' mode = "{{imgdeparameter. mode}}" class = "fix -
box"></image>
    </view>
  </view>
</view>
<! -- 底部加入购物车 -->
<view class = "bottom">
  <view class = "left">
    <image src = "/images/jindian. png" style = ' height: 80rpx; width: 90rpx ' bindtap =
"backtoindex"></image>
    <image bindtap = "tocart" src = "/images/cart. png" style = "height: 90rpx; width: 90rpx">
</image>
    <text class = "carts - icon - num" wx:if = "{{hasCarts}}">{{num}}</text>
  </view>
  <view class = "right">
    <text class = "textone" bindtap = "addcart" data - id = "{{data. id}}" data - title = "{{data.
title}}" data - price = "{{data. price}}" data - image = "{{data. image}}">加入购物车</text>
  </view>
  <view class = "right">
    <text class = "texttwo">联系客服</text>
  </view>
</view>
```

pages/goods-detail/goods-detail. js 文件代码如下：

```
const App = getApp();
Page({
  data: {                                    //数据初始化
    HomeIndex: 0,
    hasCarts: false,
    num: 0
  },
```

```
//页面加载时获得 app.js 中的数据
onLoad: function(option) {
  var id = option.listid
  this.setData({
    data: App.globalData.data[id]
  })
},
//Tab 栏切换
switchTab: function(e) {
  var index = parseInt(e.currentTarget.dataset.index)
  this.setData({
    HomeIndex: index
  })
},
//返回商品首页
backtoindex: function() {
  wx.switchTab({
    url: "../../pages/goods-index/goods-index"
  })
},
//增加图标中的数量
addcart: function(e) {
  let num = this.data.num;
  num++;
  this.setData({
    num: num,
    hasCarts: true
  })
  //将商品信息放入缓存中
  var cartItems = wx.getStorageSync("cartItems") || []
  var exist = cartItems.find(function(el) {
    return el.id == e.target.dataset.id
  })
  //当购物车里已经存在该商品数量时加 1
  if (exist) {
    exist.value = parseInt(exist.value) + 1
  } else {
    cartItems.push({
      id: e.target.dataset.id,
      title: e.target.dataset.title,
      image: e.target.dataset.image,
      price: e.target.dataset.price,
      value: 1,
      selected: true
    })
  }
  //弹窗显示
  wx.showToast({
    title: "加入购物车",
    duration: 1000
  })
```

```
    //更新缓存数据
    wx.setStorageSync("cartItems", cartItems)
  },
  //跳转到购物车页面
  tocart: function() {
    wx.switchTab({
      url: "../../pages/cart/cart"
    })
  }
})
```

pages/goods-detail/goods-detail.wxss 文件代码如下：

```
/ * 轮播图样式 * /
swiper {
  width: 100 %; height: 720rpx;
}
/ * 轮播图内部图片样式 * /
swiper image {
  display: block; margin: 60rpx auto;
  width: 560rpx; height: 530rpx;
}
/ * 商品标题样式 * /
.title {
  font - size: 34rpx;
}
/ * 商品价格样式 * /
.price {
  color: #f00; font - size: 32rpx;
  padding - top: 20rpx; padding - left: 15rpx;
}
/ * 商品展示外部样式 * /
.box - demo {
  display: flex; flex - direction: column;
  width: 100 %; height: 100rpx;
}
/ * 切换内容样式 * /
.tab {
  display: inline - block; color: black;
  font - size: 30rpx; width: 50 %;
  height: 100rpx; line - height: 100rpx; mtext - align: center;
}
/ * 切换后样式 * /
.active {
  color: black; border - bottom: 5rpx solid black;
}
/ * 商品详情下方内容样式 * /
.fix - box {
  width: 100 %; height: auto; display: block;
}
/ * 底部样式 * /
```

```
.bottom {
  width: 100%; height: 90rpx; position: fixed;
  bottom: 0rpx; background: #fff;
}
/* 底部左边样式 */
.left {
  width: 25%; margin-left: 20rpx; float: left;
}
/* 底部右边样式 */
.right {
  width: 35%; height: 100%; float: right;
}
/* 底部文字样式 */
.textone {
  display: block; color: white;
  line-height: 100rpx; text-align: center;
  font-size: 26rpx; background: #f00;
}
.texttwo {
  display: block; color: white;
  line-height: 100rpx; text-align: center;
  font-size: 26rpx; background: green;
}
/* 加入购物车图标样式 */
.carts-icon-num {
  position: absolute; top: 5rpx;
  left: 160rpx; width: 40rpx;
  height: 40rpx; line-height: 40rpx;
  border-radius: 50%; background: #f00;
  color: #fff; font-size: 24rpx; text-align: center;
}
```

【代码讲解】 goods-detail.js 文件中的 onload()函数通过设置 var id = option.listid 来获取跳转过来的页面中用户浏览某款商品的 id,并且通过设置 data: App.globalData.data[id]来获取 app.js 文件中的商品数据。goods-detail.wxml 文件通过数据绑定的方式渲染 data 的数据即是商品的详情信息。

goods-detail.js 的点击事件 switchTab()用于实现页面内"商品详情"和"产品参数"的内容切换;点击事件 goPay()用于实现页面跳转到支付页面 pay;点击事件 addcart()用于实现购物车的更新,并把更新之后的购物车数据 cartItems 通过语句 wx.setStorageSync("cartItems", cartItems)设置到数据缓存区供支付页面和购物车页面使用。

11.3 购物车功能的设计与实现

购物车页面将数据缓存区的购物车数据 cartItems 显示出来,并提供一些与购物车相关的操作功能。购物车页面的运行效果如图 11.7～图 11.9 所示,其中,图 11.7 是全选购物车中的商品,图 11.8 是反选购物车中的商品,图 11.9 是清空购物车。

图 11.7　全选购物车

图 11.8　反选购物车

图 11.9　清空购物车

pages/cart/cart.wxml 文件代码如下：

视频讲解

```
<!-- 左侧图标 -->
<view wx:for="{{cartItems}}">
  <view data-id="{{item.id}}" class="icon-box" data-index="{{index}}">
    <view class='icon'>
      <icon wx:if="{{item.selected}}" type="success" color="#ff0000"
     size="22" bindtap="selectIcon" data-index="{{index}}" />
      <icon wx:else type="circle" bindtap="selectIcon" size="22" data-index="{{index}}" />
    </view>
    <view>
      <!-- 商品信息 -->
      <view class="left-image">
        <image class="addcart-image" src="{{item.image}}"></image>
      </view>
      <view class="left-detail">
        <text class="cart-title">{{item.title}}</text>
        <text class="cart-price">¥{{item.price}}</text>
        <text bindtap="reduce" class="input" data-index="{{index}}">-</text>
        <text class="input cart-amount">{{item.value}}</text>
        <text bindtap="add" class="input" data-index="{{index}}">+</text>
      </view>
      <!-- 删除图标 -->
      <view class="right">
        <image src="/images/delete.png" bindtap="delete"
       data-index="{{index}}"></image>
      </view>
    </view>
  </view>
</view>
```

```
</view>
<! -- 底部 -->
<view class = "cart - total">
  <text class = "total - text">合计: </text>
  <text class = "total - color">¥{{total}}元</text>
</view>
<view class = "total - bottom">
  <icon wx:if = "{{checkAll}}" class = "cart - icon" type = "success"
 color = "#ff0000" size = "22" bindtap = "select" data - index = "{{index}}" />
  <icon wx:else type = "circle" class = "cart - icon" size = "22" bindtap = "select"
 data - index = "{{index}}" />
  <text class = "checked - all">全选</text>
  <view class = "clear - cart">
    <text class = "pay" bindtap = "goPay">结算</text>
  </view>
  <view class = "clear - cart">
    <text class = "clear - text" bindtap = "clearcart" data - id = "{{item.id}}">清空购物
  车</text>
  </view>
</view>
```

pages/cart/cart.js 文件代码如下：

```
const App = getApp();
Page({
  data: {                                          //数据初始化
    cartItems: [],
    total: 0,
    checkAll: true
  },
  //获取缓存中的数据
  onShow: function() {
    var cartItems = wx.getStorageSync("cartItems")
    this.setData({
      cartList: false,
      cartItems: cartItems
    })
    this.getsumTotal()
  },
  //选中图标
  selectIcon: function(e) {
    var cartItems = this.data.cartItems            //获取购物车列表
    var index = e.currentTarget.dataset.index;     //获取当前点击事件的下标索引
    var selected = cartItems[index].selected;      //获取购物车里面的 value 值
    //未选中图标
    cartItems[index].selected = !selected;
    this.setData({
      cartItems: cartItems
    })
    this.getsumTotal();
    wx.setStorageSync("cartItems", cartItems)
```

```
},
//增加商品数量
add: function(e) {
  var cartItems = this.data.cartItems
  var index = e.currentTarget.dataset.index
  var value = cartItems[index].value
  value++
  cartItems[index].value = value
  this.setData({
    cartItems: cartItems
  });
  this.getsumTotal()
  wx.setStorageSync("cartItems", cartItems)     //存入缓存
},
//减少商品数量
reduce: function(e) {
  var cartItems = this.data.cartItems
  var index = e.currentTarget.dataset.index
  var value = cartItems[index].value
  if (value == 1) {
    value--;
    cartItems[index].value = 1
  } else {
    value--
    cartItems[index].value = value;
  }
  this.setData({
    cartItems: cartItems
  });
  this.getsumTotal()
  wx.setStorageSync("cartItems", cartItems)
},
//全选
select: function(e) {
  var checkAll = this.data.checkAll;
  checkAll = !checkAll
  var cartItems = this.data.cartItems
  for (var i = 0; i < cartItems.length; i++) {
    cartItems[i].selected = checkAll
  }
  this.setData({
    cartItems: cartItems,
    checkAll: checkAll
  })
  this.getsumTotal()
},
//删除商品列表
delete: function(e) {
  var cartItems = this.data.cartItems;
  var index = e.currentTarget.dataset.index;
  cartItems.splice(index, 1)
```

```
    this.setData({
      cartItems: cartItems
    });
    if (cartItems.length) {
      this.setData({
        cartList: false
      });
    }
    this.getsumTotal()
    wx.setStorageSync("cartItems", cartItems);
  },
  //清空购物车
  clearcart: function(e) {
    this.setData({
      cartItems: [],
      total: 0
    })
    wx.setStorageSync("cartItems", [])
  },
  //跳转到支付页面
  goPay: function(e) {
    wx.setStorageSync("cartItems", this.data.cartItems)
    wx.setStorageSync("total", this.data.total)
    wx.navigateTo({
      url: '../../pages/pay/pay'
    })
  },
  //合计商品总价
  getsumTotal: function() {
    var cost = 0;
    for (var i = 0; i < this.data.cartItems.length; i++) {
      if (this.data.cartItems[i].selected) {
        cost += this.data.cartItems[i].value * this.data.cartItems[i].price
      }
    }
    //更新数据
    this.setData({
      total: cost
    })
  },
})
```

pages/cart/cart.wxss 文件代码如下：

```
/* 图标外部样式 */
.icon-box {
  height: 250rpx;
}
/* 左边选中图标样式 */
.icon {
  float: left; margin: 88rpx 20rpx;
```

```
}
/* 购物车商品图片左浮动 */
.left-image {
  float: left;
}
/* 购物车商品图片样式 */
.addcart-image {
  width: 170rpx; height: 210rpx; float: left; padding: 10rpx;
}
/* 购物车商品信息左浮动 */
.left-detail {
  float: left; margin-top: 20rpx; line-height: 60rpx;
}
/* 购物车商品标题样式 */
.cart-title {
  margin-top: 24rpx; font-size: 32rpx;
}
/*  购物车商品价格样式 */
.cart-price {
  display: flex; color: #f00; font-size: 20rpx; ;
}
/* 输入框样式 */
.input {
  display: block; width: 65rpx;
  height: 70rpx; line-height: 70rpx;
  text-align: center; float: left;
}
/* 购买商品数量样式 */
.cart-amount {
  width: rpx; background: #f4f4f4; font-size: 30rpx;
}
/* 删除图标样式 */
.right image {
  width: 52rpx; height: 52rpx;
  float: right; margin-right: 34rpx; margin-top: 34rpx;
}
/* 底部样式 */
.total-bottom {
  position: fixed; width: 100%;
  height: 86rpx; bottom: 0; background: white;
}
/* 底部选中图标样式 */
.cart-icon {
  margin: 15rpx 22rpx; float: left;
}
/* 全选样式 */
.checked-all {
```

```
  line-height: 90rpx; font-size: 36rpx;
}
/*合计样式*/
.cart-total {
  width: 300rpx; height: 100%;
  line-height: 86rpx; float: right; text-align: center;
}
/*清空购物车和结算图标样式*/
.clear-cart {
  float: right; width: 230rpx;
}
/*清空购物车文字样式*/
.clear-text {
  display: block; line-height: 87rpx; text-align: center;
  font-size: 36rpx; color: white; background: #f00;
}
/*结算文字样式*/
.pay {
  display: block; line-height: 87rpx;
  text-align: center; font-size: 36rpx;
  color: white; background: #ff8c00;
}
/*合计价格样式*/
.total-color {
  color: #f00; font-size: 36rpx;
}
/*合计内容样式*/
.total-text {
  font-size: 36rpx;
}
```

【代码讲解】 cart.js 文件的 onload()函数通过语句 var cartItems = wx.getStorageSync("cartItems")从数据缓存区中获取购物车中的商品数据。

cart.js 文件中的 select()函数用于实现单件商品的选与不选；selectIcon()函数用于实现单件商品的选与不选；add()函数用于实现用户点击“+”按钮时商品数量增加；reduce()函数用于实现用户点击“-”按钮时商品数量减少；delete()函数用于实现用户点击删除图标时将当前的商品从购物车中删除；clearcart()函数用于清空购物车；getsumTotal()函数用于计算购物车商品的总价；goPay()函数用于实现当前页面跳转到支付页面。

11.4 支付页面的设计

在购物车页面中，当用户点击“结算”按钮后将跳转到支付页面。支付页面的运行效果如图 11.10 和图 11.11 所示。

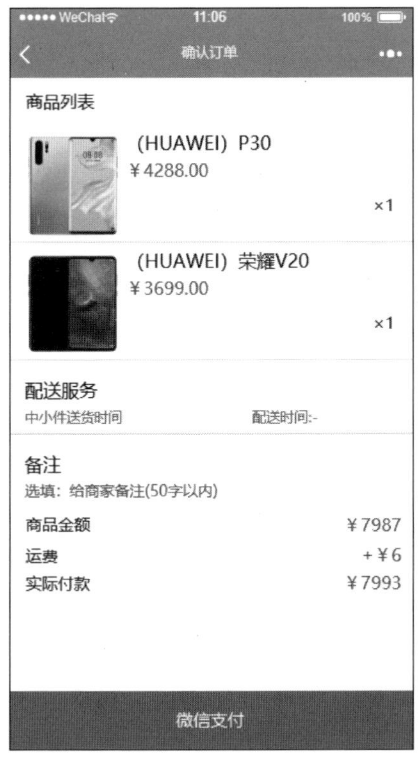

图 11.10　支付页面

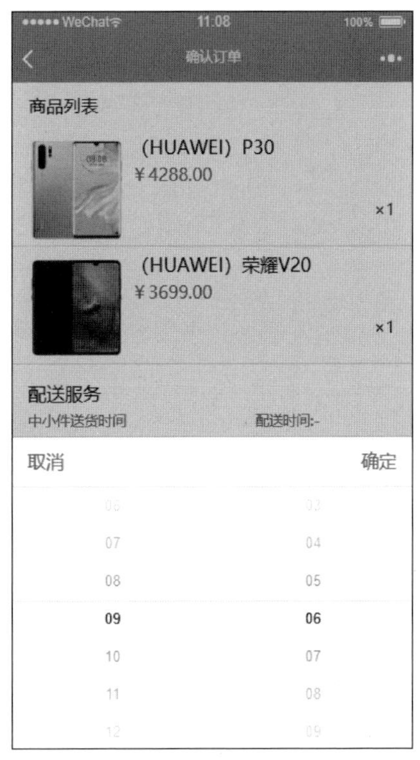

图 11.11　支付页面选择配送时间

pages/pay/pay.wxml 文件代码如下：

视频讲解

```
<view class="list-text">
  <text>商品列表</text>
</view>
<view wx:if="{{cartItems!=[]}}">
  <view class="list-goods" wx:for="{{cartItems}}">
    <!-- 商品展示 -->
    <image class="pic-goods" src="{{item.image}}"></image>
    <!-- 商品名称 -->
    <text class="title-goods">{{item.title}}</text>
    <!-- 价格图标和商品价格 -->
    <text class="price-goods">¥{{item.price}}</text>
    <!-- 删除图标 -->
    <text class="amount-goods">×{{item.value}}</text>
  </view>
</view>
<view class="deliver">
  <view>
    <view class="sevice">配送服务</view>
    <view class="time">中小件送货时间</view>
  </view>
  <view class="time-select">
    <picker mode="time" bindchange="timeChange1">
```

```
      <text>配送时间:{{time1}}</text>
    </picker>
    <picker mode="time" bindchange="timeChange2">
      <text>-{{time2}}</text>
    </picker>
  </view>
</view>
<view class="hr"></view>
<view class="remark">
  <view>备注</view>
  <input placeholder="选填: 给商家备注(50字以内)" auto-focus></input>
</view>
<view class="cost-one">
  <text class="left">商品金额</text>
  <text class="right">¥{{total}}</text>
</view>
<view class="cost-two">
  <text class="left">运费</text>
  <text class="right">+ ¥6</text>
</view>
<view class="cost-three">
  <text class="left">实际付款</text>
  <text class="right">¥{{total+6}}</text>
</view>
<view class="bottom">
  <text>微信支付</text>
</view>
```

pages/pay/pay.js 文件代码如下:

```
const App = getApp();
Page({
  onLoad: function(options) {                        //从缓存中获取购物车中的数据
    var cartItems = wx.getStorageSync("cartItems")
    var total = wx.getStorageSync("total")
    var payId = options.id
    var data = App.globalData.data
    this.setData({
      cartItems: cartItems,
      total: total
    })
  },
  timeChange1: function(e) {                         //选择配送时间
    var value1 = e.detail.value;
    this.setData({
      time1: value1
    });
  },
  timeChange2: function(e) {
    var value2 = e.detail.value;
    this.setData({
```

```
      time2: value2
    });
  },
})
```

pages/pay/pay.wxss 文件代码如下：

```
/* 商品列表 */
.list-text {
  height: 84rpx; line-height: 84rpx;
  font-size: 32rpx; padding-left: 32rpx;
}
/* 商品列表外部样式 */
.list-goods {
  position: relative;
  padding: 10rpx 10rpx 10rpx 220rpx;
  height: 190rpx;
  border-bottom: 1px solid #eee9e9;
}
/* 商品图片 */
.pic-goods {
  position: absolute; top: 20rpx;
  left: 20rpx; width: 200rpx; height: 180rpx;
}
/* 商品标题 */
.title-goods {
  display: block; font-size: 34rpx;
}
/* 商品价格 */
.price-goods {
  font-size: 32rpx; height: 100rpx; color: #f00;
}
/* 商品数量 */
.amount-goods {
  position: absolute; right: 10rpx;
  bottom: 20rpx; width: 90rpx;
  height: 90rpx; line-height: 90rpx;
  text-align: center; font-size: 30rpx;
}
/* 配送服务外部样式 */
.deliver {
  display: flex; flex-direction: row;
  height: 50rpx; padding: 30rpx;
}
/* 分隔线样式 */
.hr {
  margin-top: 20rpx; border: 1px solid #eee9e9;
}
/* 配送服务内容样式 */
.deliver .sevice {
  font-size: 34rpx;
```

```
}
/ * 中小件送货时间内容样式 * /
.deliver .time {
  margin - top: 10rpx; font - size: 26rpx; color: #919191;
}
/ * 配送时间外部样式 * /
.time - select {
  display: flex; flex - direction: row;
  margin - top: 42rpx; margin - left: 240rpx;
}
/ * 配送时间内容样式 * /
.time - select text {
  font - size: 26rpx; color: #919191;
}
/ * 备注外部样式 * /
.remark {
  font - size: 34rpx; height: 50rpx; padding: 30rpx;
}
/ * 输入框样式 * /
.remark input {
  font - size: 28rpx;
}
/ * 商品金额外部样式 * /
.cost - one {
  margin - top: 36rpx; height: 30rpx; line - height: 30rpx;
}
/ * 商品金额内容样式 * /
.cost - one .left {
  float: left; font - size: 30rpx; padding - left: 32rpx;
}
/ * 商品金额价格样式 * /
.cost - one .right {
  float: right; color: #f00; font - size: 32rpx; padding - right: 22rpx;
}
/ * 运费外部样式 * /
.cost - two {
  height: 80rpx; line - height: 80rpx;
}
/ * 运费内容样式 * /
.cost - two .left {
  float: left; font - size: 30rpx; padding - left: 32rpx;
}
/ * 运费价格样式 * /
.cost - two .right {
  float: right; color: #f00;
  font - size: 32rpx; padding - right: 22rpx;
}
/ * 实际付款外部样式 * /
.cost - three {
  height: 20rpx; line - height: 20rpx;
}
```

```
/* 实际付款内容样式 */
.cost - three .left {
  float: left; font - size: 30rpx; padding - left: 32rpx;
}
/* 实际付款价格样式 */
.cost - three .right {
  float: right; color: #f00;
  font - size: 32rpx; padding - right: 22rpx;
}
/* 底部 */
.bottom {
  width: 100%; height: 105rpx;
  line - height: 105rpx; background: #1aad19;
  position: fixed; bottom: 0;
  color: white; font - size: 32rpx; text - align: center;
}
```

支付页面的运行效果如图 11.10 所示，图 11.11 是支付页面选择配送时间的界面。

【代码讲解】 pay.js 文件的 onload() 函数使用语句 var cartItems = wx.getStorageSync("cartItems")获取数据缓存区的购物车商品数据；使用语句 var total = wx.getStorageSync("total")获取数据缓存中的总金额。pay.wxml 页面中使用 wx:for="{{cartItems}}"列表渲染显示购物车中的商品信息；使用<picker>组件选择送货时间；使用<input>组件添加订单的备注信息。

11.5　项目小结

通过数码产品类电商小程序实训项目主要复习了以下知识点和操作：

- 项目的创建步骤；
- 注册新页面；
- 从 app.js 文件中获取数据的方法；
- 页面布局和样式设计的基本方法；
- 使用双大括号{{}}实现 WXML 和 JS 文件的数据绑定操作；
- 使用 setData()重置动态数据的方法；
- wx:if 条件渲染和 wx:for 列表渲染的使用；
- 页面跳转中数据的传递；
- 数据缓存的使用。

第12章 基于云开发的新闻小程序项目

6.4 节介绍了基于 MySQL 数据库的新闻小程序，第 10 章介绍了云开发，本章将使用云开发改造 6.4 节的新闻小程序，同时也在 6.4 节的基础上增加了一些功能模块。通过本章的学习，读者可以使用云开发独立完成小程序项目的开发。

本章主要目标

- 综合应用所学知识开发完整新闻小程序项目；
- 熟练掌握云开发在项目中的应用；
- 了解项目开发中的软件工程方法学。

12.1 需求分析

本项目一共需要 5 个页面，即新闻列表页、新闻详情页、个人主页、新闻发布页和个人信息设置页面。

(1) 新闻列表页：新闻列表页应包含顶部栏目切换导航、中间文章列表部分和底部 tabBar 菜单，其中，顶部栏目切换导航可以使用 scroll-view 实现，中间文章列表部分可以使用 wx:for 来渲染，而底部 tabBar 菜单则需要在 app.json 文件中设置。顶部栏目切换导航包含"推荐""科技""财经""汽车""时尚"等目录，当用户点击某一栏目的时候，触发 JS 函数对云数据库进行对应栏目文章的查询操作，查询结果显示在中间文章列表部分。底部 tabBar 菜单包含"首页"和"我的"两个切换菜单。

(2) 新闻详情页：新闻详情页根据用户在首页点击的文章的 id 条件查询文章的 title、cTime、img 和 content 等内容。

(3) 个人主页：个人主页包含"关注""粉丝"和"7 天访问"等信息，还包含"我要爆料"和"系统设置"两个按钮。

(4) 新闻发布页：新闻发布页实现用户发布新闻的功能，即将数据插入云数据库中。

(5) 个人信息设置页面：个人信息设置页面对系统的"字体""自动播放"和"消息推送"等项目进行设置。

12.2　云存储的设计与实现

云开发包含云存储、云数据库和云函数 3 个核心功能，考虑到 JS 函数已经能够完成项目功能，本项目没有使用云函数，只使用了云存储和云数据库。

12.2.1　云存储在本项目中的意义

微信小程序代码容量被限制为不超过 2MB，限制大小是出于对小程序启动速度的考虑，希望用户在使用任何一款小程序时，都能获得一种"秒开"的体验。然而，2MB 的大小也限制了小程序功能的扩展，小程序业务的发展可能需要更大的体积。当开发者开发的小程序超过了 2MB 的大小，小程序将不可以真机调试，也不可以发布上线。

突破 2MB 的大小限制的方法有两种：第一种方法是采用小程序提供的"分包加载"技术，关于"分包加载"的技术实现，读者可以自行查询；考虑到小程序中最占空间的是图片，第二种方法则是采用"云存储"功能把项目中的图片存储在云端，从而降低小程序的容量。本例将使用"云存储"来存储项目中的新闻图片。

12.2.2　云存储的设计与实现

视频讲解

新建一个项目 ch12，如图 12.1 所示。在 AppID 下拉列表框中输入真实的开发者 AppID，"后端服务"选项选择"小程序 · 云开发"单选按钮，单击"新建"按钮完成项目的创建。进入项目之后单击"云开发"菜单进入云开发控制台，如图 12.2 所示。

新建项目　导入项目

项目名称　ch12

目录　F:\出书\小程序写作\ch12\ch12

AppID　wx800eb4451b223c35

若无 AppID 可 注册
或使用 测试号

开发模式　小程序

后端服务　不使用云服务

小程序·云开发

小程序·云开发为开发者提供数据库、存储和云函数等完整的云端支持。无需搭建服务器，使用平台提供的 API 进行核心业务开发，即可实现小程序快速上线和迭代。了解详情

腾讯云

取消　新建

图 12.1　新建云开发项目

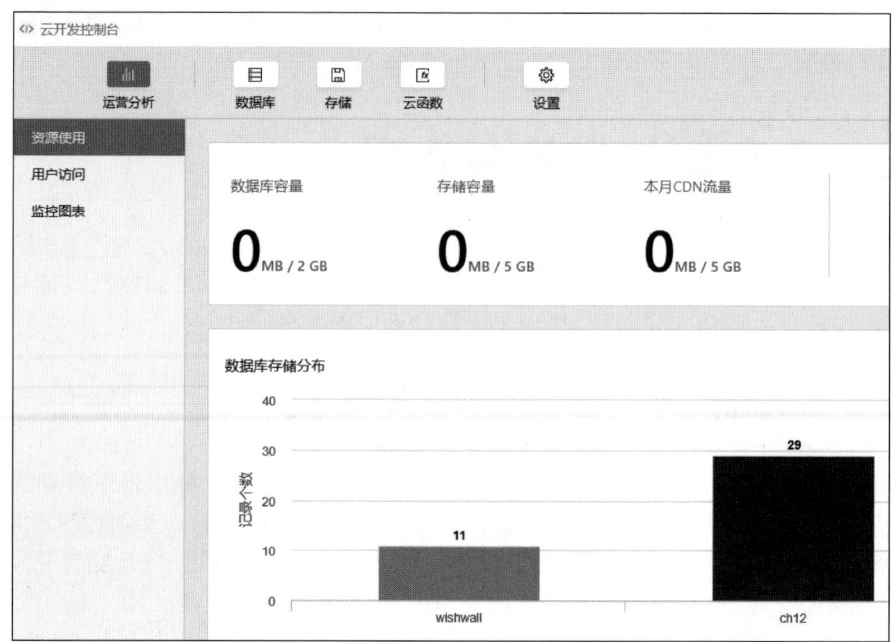

图 12.2　云开发控制台

在云开发控制台单击“存储”菜单，然后在出现的界面中单击“新建文件夹”按钮新建 img 文件夹，把本项目需要用到的新闻图片全部上传到 img 文件夹中，效果如图 12.3 所示。图片的 File ID 即是项目中需要引用的图片地址。

图 12.3　上传图片到云存储空间

单击“权限设置”按钮，选中“所有用户可读，仅创建者可读写”单选按钮，如图 12.4 所示。至此，项目的云存储部分操作完毕。

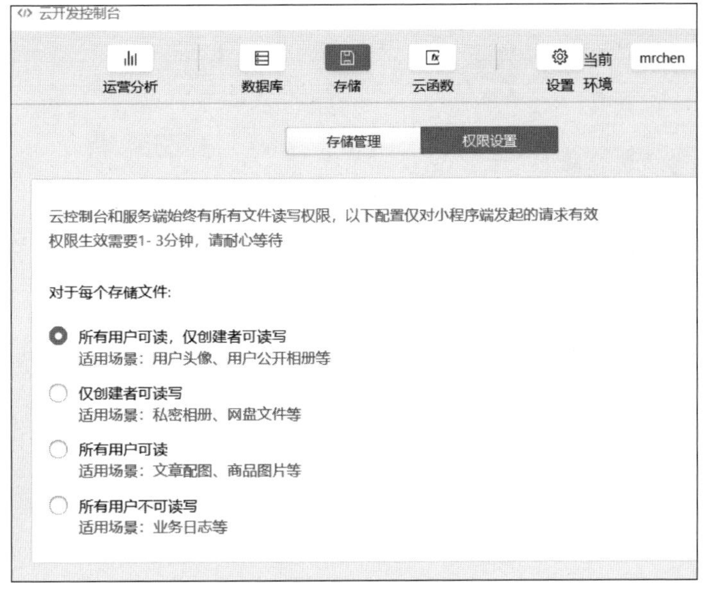

图 12.4　设置云存储权限

12.3　云数据库的设计与实现

视频讲解

本项目中新闻信息包含的字段有标题(title)、发布时间(cTime)、新闻图片(img)、新闻内容(content)和新闻类别(newsid)，其中，img 字段为 12.2 节中云存储产生的云端图片地址。因为 newsid 字段是小程序 WXML 传递给 JS 文件的 string 类型的 id，所以本例中 newsid 设置了 string 类型。为了操作方便，其他几个字段也设置为 string 类型。

单击云开发控制台的“数据库”菜单，新建集合 ch12，然后添加 50 条新闻记录，如图 12.5 所示。在添加记录时注意 newsid 的取值范围为 0～9，分别代表不同的新闻类别。

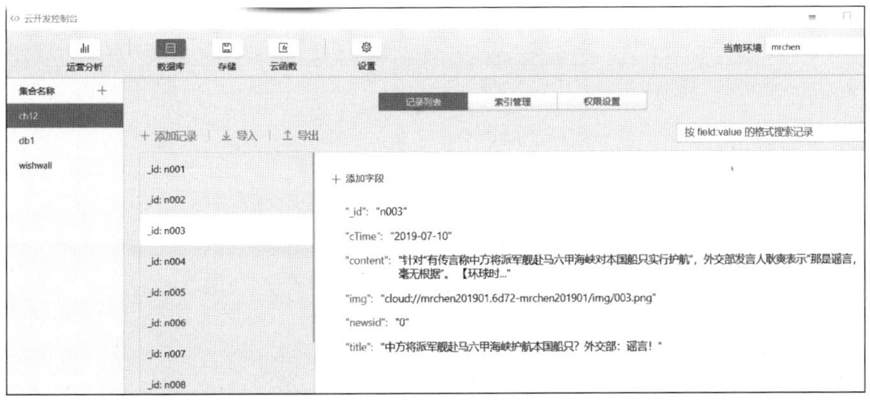

图 12.5　添加新闻记录

在添加好记录之后，把云数据库权限设置为“所有用户可读，仅创建者可读写”，如图 12.6 所示。

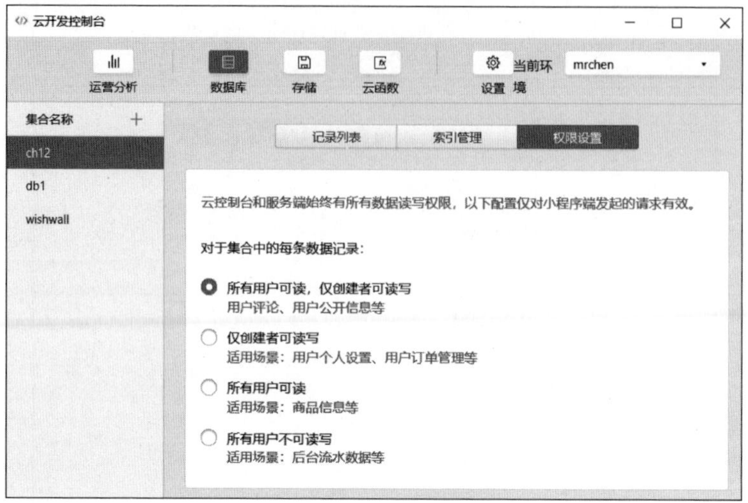

图 12.6　设置云数据库权限

12.4　小程序端的实现

12.4.1　项目效果图展示

本项目的项目结构如图 12.7 所示，项目中的 cloudfunctions|mrchen 文件夹是创建项目时自动生成的，img 文件夹下存放了除云存储之外的图片（app.json 文件识别不到云存储图片，只能使用本地图片，pages 文件夹下的 add、detail、list、me 和 setting 页面为项目内容页面。

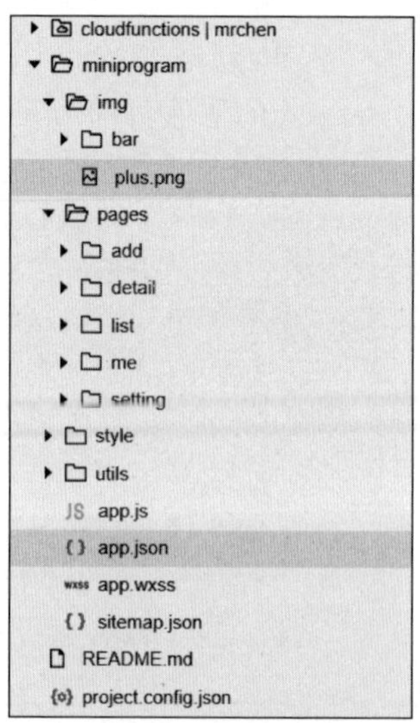

图 12.7　项目结构

list 页面效果如图 12.8 和图 12.9 所示；me 页面效果如图 12.10 所示；detail 页面效果如图 12.11 所示；add 页面效果如图 12.12 所示；setting 页面效果如图 12.13 所示。

图 12.8　list 页面“推荐”栏目

图 12.9　list 页面“科技”栏目

图 12.10　me 页面

图 12.11　detail 页面

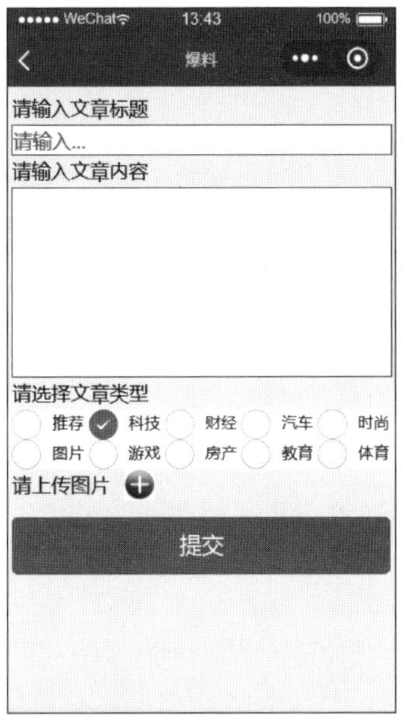

图 12.12　add 页面

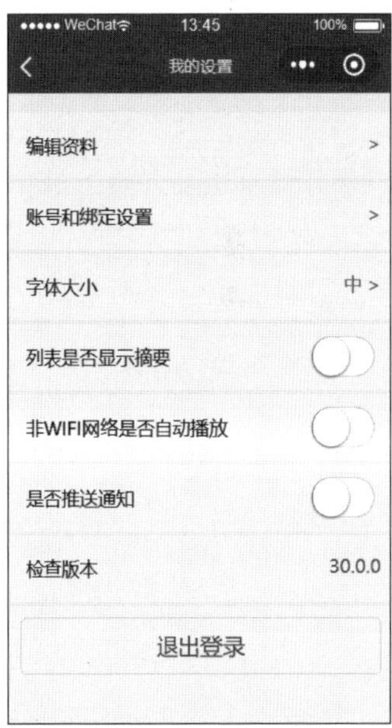

图 12.13　setting 页面

12.4.2 全局文件的实现

app.js 的代码如下：

```
App({
  onLaunch: function() {
      wx.cloud.init({
        traceUser: true
      })
    this.globalData = {}
  }
})
```

【代码讲解】

本项目因为是云开发项目，所以需要在 app.js 文件中使用 wx.cloud.init()进行初始化。

app.json 的代码如下：

视频讲解

```
{
  "pages": [
    "pages/list/list",
    "pages/setting/setting",
    "pages/add/add",
    "pages/detail/detail",
    "pages/me/me"
  ],
  "window": {
    "backgroundColor": "#F6F6F6",
    "backgroundTextStyle": "light",
    "navigationBarBackgroundColor": "#D53C3E",
    "navigationBarTitleText": "基于云开发的新闻小程序",
    "navigationBarTextStyle": "white"
  },
  "tabBar": {
    "selectedColor": "#D53C3E",
    "borderStyle": "black",
    "backgroundColor": "#F9F9F9",
    "list": [
      {
        "pagePath": "pages/list/list",
        "text": "首页",
        "iconPath": "/img/bar/index0.jpg",
        "selectedIconPath": "/img/bar/index1.jpg"
      },
      {
        "pagePath": "pages/me/me",
        "text": "我的",
        "iconPath": "/img/bar/aboutme0.jpg",
        "selectedIconPath": "/img/bar/aboutme1.jpg"
      }
    ]
```

```
  },
  "sitemapLocation": "sitemap.json"
}
```

【代码讲解】

app.json 文件通过 tabBar 的定义，控制 tabBar 菜单在 list 和 me 页面中显示。

12.4.3 其他页面的实现

pages/list/list.wxml 的代码如下：

视频讲解

```
<view class = "head">
  <view class = "head1">
    <scroll - view class = "scroll - view_class" scroll - x = "true">
      <view class = "scroll - view_class">
        <view>
          <view class = "{{flag == 0?'choose':'nochoose'}}"
         id = "0" bindtap = "select">推荐</view>
        </view>
        <view>
          <view class = "{{flag == 1?'choose':'nochoose'}}"
         id = "1" bindtap = "select">科技</view>
        </view>
        <view>
          <view class = "{{flag == 2?'choose':'nochoose'}}"
         id = "2" bindtap = "select">财经</view>
        </view>
        <view>
          <view class = "{{flag == 3?'choose':'nochoose'}}"
         id = "3" bindtap = "select">汽车</view>
        </view>
        <view>
          <view class = "{{flag == 4?'choose':'nochoose'}}"
         id = "4" bindtap = "select">时尚</view>
        </view>
        <view>
          <view class = "{{flag == 5?'choose':'nochoose'}}"
         id = "5" bindtap = "select">图片</view>
        </view>
        <view>
          <view class = "{{flag == 6?'choose':'nochoose'}}"
         id = "6" bindtap = "select">游戏</view>
        </view>
        <view>
          <view class = "{{flag == 7?'choose':'nochoose'}}"
         id = "7" bindtap = "select">房产</view>
        </view>
        <view>
          <view class = "{{flag == 8?'choose':'nochoose'}}"
         id = "8" bindtap = "select">教育</view>
        </view>
```

```
        <view>
          <view class="{{flag==9?'choose':'nochoose'}}"
         id="9" bindtap="select">体育</view>
        </view>
      </view>
    </scroll-view>
  </view>
</view>
<view class="body">
<!--文章列表模板 begin-->
<template name="itmes">
  <navigator url="../../pages/detail/detail?detail_id={{_id}}" hover-class=
  "navigator-hover">
    <view class="imgs"><image src="{{img}}" class="in-img" background-size=
  "cover" model="scaleToFill"></image></view>
    <view class="infos">
      <view class="title">{{title}}</view>
      <view class="date">{{cTime}}</view>
    </view>
  </navigator>
</template>
<!--文章列表模板 end-->
<!--循环输出列表 begin-->
<view wx:for="{{shuzu}}" class="list">
  <template is="itmes" data="{{...item}}" />
</view>
</view>
```

【代码讲解】

list.wxml 页面的顶部通过 scroll-view 组件实现多栏目切换；页面的中间部分是各个栏目的列表内容，对象 shuzu 是 JS 文件通过查询云数据库获取符合条件的文章列表内容。本例通过 wx:for 循环渲染输出文章的 img、title 和 cTime 字段。

pages/list/list.js 的代码如下：

```
Page({
  /*** 页面的初始数据 ***/
  data: {
    shuzu: [],
    detail_id:"n001"
  },
  /*** 生命周期函数--监听页面加载 ****/
  onLoad: function(options) {
    var that = this
    const db = wx.cloud.database()
    db.collection('ch12').where({
      newsid: "0"
    }).get({
      success: function(res) {
        console.log(res.data)
        that.setData({
          shuzu: res.data
        })
```

```
      }
    })
  },
  select:function(e)
  { console.log(e)
    var that = this
    var id = e.target.id
    const db = wx.cloud.database()
    db.collection('ch12').where({
      newsid:id
    }).get({
      success: function(res) {
        console.log(res.data)
        that.setData({
          shuzu: res.data
        })
      }
    })
  },
  /*** 用户点击右上角分享 */
  onShareAppMessage: function() {
  }
})
```

【代码讲解】

当用户没有在 list. wxml 选择新闻栏目时，需要自动加载默认推荐的新闻内容，这是通过 JS 文件的 onLoad 函数自动执行对云数据库 ch12 的条件查询实现的，查询 newsid 为 0 的记录。当用户点击了对应的栏目的时候，list. wxml 传递给 list. js 一个 id 值，则调用 select 函数查询对应的 newsid 为 id 的所有记录。

pages/list/list. wxss 的代码如下：

```
.head{
  background-color: #F6F5F3;  height: 36px;
  color: #000000;  display: flex;
  flex-direction: row;  align-items: center;
}
.head1{
  width: 85%; height: 36px;
}
.scroll-view_class{
  height: 40px; display: flex;
  flex-direction: row; margin-left:5px;
}
.choose{
  width: 40px; height: 40px;
  line-height: 40px; padding-left: 5px;
  padding-right: 5px; font-size: 14px; font-weight: bold;
}
.nochoose{
  width:40px; height: 40px;
```

```
  line-height: 40px; padding-left: 5px;
  padding-right: 5px; font-size: 14px;
}
.body{
    height: 100%; display: flex;
    flex-direction: column; padding: 20rpx;
}
navigator { overflow: hidden;}
.list {margin-bottom: 20rpx;height: 200rpx;position: relative;}
.imgs {float: left;}
.imgs image {display: block; width: 200rpx;height: 200rpx;}
.infos {float: left; width: 480rpx; height: 200rpx;padding: 20rpx 0 0 20rpx;}
.title {font-size: 16px; overflow:hidden;
text-overflow:ellipsis;
display: -webkit-box;
-webkit-box-orient:vertical;
-webkit-line-clamp:2;}
.date {font-size: 14px;color: #aaa; position: absolute;bottom: 0;}
.loadMore {text-align: center;margin: 30px;color: #aaa;font-size: 16px}
page{
  background-color: #fdfeff;
}
```

pages/detail/detail.wxml 的代码如下：

```
<!--detail.wxml-->
<view class="warp">
<view>
  <view class="title">{{detail_content.title}}</view>
  <view class="cTime">{{detail_content.cTime}}</view>
  <view class="img"><image src="{{detail_content.img}}" class="in-img" background-size="cover" model="scaleToFill"></image></view>
  <view class="content">{{detail_content.content}}</view>
  <view class="close" bindtap="closepage">返回</view>
</view></view>
```

【代码讲解】 detail.wxml 通过数据绑定的方式调用 detail.js 中的 detail_content 对象，并把 detail_content 对象的 title、cTime、img 和 content 字段在页面中显示出来。

pages/detail/detail.js 的代码如下：

视频讲解

```
var app = getApp()
Page({
  data: {
    id1: 1,
    detail_content: {}
  },
  onLoad: function(options) {
    console.log(options)
    var that = this
    var detail_id = options.detail_id;
    const db = wx.cloud.database()
```

```
    db.collection('ch12').doc(detail_id).get({
      success: function(res) {
        console.log(res.data)
        that.setData({
          detail_content: res.data
        })
      }
    })
  }
})
```

【代码讲解】

用户在 list. wxml 文件中点击了某一条新闻时，会传递一个 id 值给 detail. js，id 存储在 detail. js 的 onLoad 函数的 option 事件中，即 option. detail_id。

pages/detail/detail. wxss 的代码如下：

```
.warp {
    height: 100%; display: flex;
    flex-direction: column; padding: 20rpx;
    font-size: 16px;
}
.title {
    text-align: center; padding: 20rpx; font-size: 20px;
}
.cTime {
    color: #aaa;
}
.img {
    text-align: center; padding: 20rpx;
}
.img image {
    width: 120px; height: 120px;
}
.content {
    text-indent: 2em;
}
.close {
    text-align: center; margin: 30px;
    font-size: 20px; color: #aaa
}
page{
  background-color: #fdfeff;
}
```

pages/me/me. wxml 的代码如下：

```
<view class="all">
  <view class="top">
    <view class="top1">
      <view class="avatarUrl">
        <image src="{{userInfo.avatarUrl}}" style="width:70px;height:70px;"></image>
```

```
            </view>
            <view class = "nickName">
              {{userInfo.nickName}}
            </view>
        </view>
        <view class = "top2">
            <view class = "topitem">
              <view>1000</view>
              <view>关注</view>
            </view>
            <view class = "topitem">
              <view>500</view>
              <view>粉丝</view>
            </view>
            <view class = "topitem" style = "border:0px;">
              <view>3000</view>
              <view>7 天访客</view>
            </view>
        </view>
      </view>
      <view class = "part"></view>
      <view class = "mecontent">
        <view class = "mefuntion" bindtap = 'mefuntion'>我要爆料</view>
      </view>
       <view class = "part1"></view>
      <view class = "mecontent" bindtap = "setup">
        <view class = "mefuntion" bindtap = 'setup'>系统设置</view>
      </view>
    </view>
```

【代码讲解】 me.wxml 通过数据绑定的方式获取 me.js 中的 userInfo 对象的 avatarUrl 和 nickName 值。

pages/me/me.js 的代码如下：

视频讲解

```
var app = getApp();
Page({
  data:{
    userInfo:{}
  },
  onLoad:function(){
    var that = this
    wx.getUserInfo({
      success: function(res) {
       that.setData({
         userInfo:res.userInfo
       })
      }
    })
  },
  mefuntion:function(e){
  wx.navigateTo({
```

```
      url: '../add/add',
    })
  },
  setup:function(){
    wx.navigateTo({
      url: '../setting/setting',
    })
  }
})
```

【代码讲解】 me.js 在 onLoad 函数中调用 wx.getUserInfo()接口获取用户信息,文件中 mefuntion 和 setup 函数实现用户点击之后的跳转事件。

pages/me/me.wxss 的代码如下:

```
.top{
  width: 100%; height: 150px;
  background-color: #D53E37;
}
.top1{
  display: flex; flex-direction: row;
}
.top1 image{
  border-radius: 50%; margin-left: 10px;
}
.nickName{
  color: #ffffff; font-size: 15px;
  position: absolute; left: 100px; margin-top:30px;
}
.top2{
  margin-top:10px; display: flex; flex-direction: row;
}
.topitem{
  width:33%; text-align: center;
  font-size: 13px; color: #ffffff;
  line-height: 20px; border-right: 1px solid #cccccc;
}
.part{
  width:100%; height: 15px;
  background-color: #F4F5F6;
}
.mecontent{
  display: flex;
  flex-direction: row;
}
.part1{
  border: 1px solid #cccccc; opacity: 0.2;
}
.mefuntion{
  font-size: 1rem;
}
```

pages/add/add.wxml 的代码如下：

视频讲解

```
< view class = "container">
  <! -- < template is = "head" data = "{{title: 'editor'}}"/> -- >
  < view class = "page - body">
    < view class = 'wrapper'>
    < view class = "contentinfo">
 请输入文章标题
  </view >
    < view class = 'in'> < input bindinput = 'title' placeholder = "请输入..."
focus = "{{focus}}" /></view >
 < view class = "contentinfo">
 请输入文章内容
  </view >
 < view class = 'in'>< textarea bindinput = "content" maxlength = "300" /></view >
  < view class = "contentinfo">
 请选择文章类型
  </view >
  < view class = 'in'>
  < radio - group class = "radio - group" bindchange = "radioChange">
  < label class = "radio" wx:for = "{{items}}">
    < radio value = "{{item.name}}" checked = "{{item.checked}}"/>{{item.value}}
  </label >
</radio - group ></view >
< view >
< view class = 'pic'>
 < view class = "contentinfo">
 请上传图片
  </view >
< view bindtap = 'picfunction'>< image  class = "plus" src = '/img/plus.png'></image ></view >
</view ></view >
< button type = "primary" bindtap = 'submit'>提交</button >
  </view >
</view >
    </view >
```

【代码讲解】

用户在 add.wxml 页面输入爆料新闻的 title、content、img 和 newsid 字段并传递给 add.js，用于将数据插入云数据库中，其中，cTime 字段由 add.js 字段自动产生。

pages/add/add.js 的代码如下：

```
Page({
  data: {
    title: "", content:"",
    newsid: "", img: "", cTime: "",
    items: [
      { name: '0', value: '推荐' },{ name: '1', value: '科技', checked: 'true' },
      { name: '2', value: '财经' },{ name: '3', value: '汽车' },
      { name: '4', value: '时尚' },{ name: '5', value: '图片' },
      { name: '6', value: '游戏' }, { name: '7', value: '房产' },
```

```
    { name: '8', value: '教育' }, { name: '9', value: '体育' },
   ]
  },
  title:function(e){
  console.log(e.detail.value)
  this.setData({
    title: e.detail.value                          //待插入的文章 title 字段
  })
  },
  content: function(e) {
    console.log(e.detail.value)
    this.setData({
      content: e.detail.value                      //待插入的文章 content 字段
    })
  },
  radioChange: function(e) {
    console.log(e.detail.value)
    this.setData({
      newsid: e.detail.value                       //待插入的文章 newsid 字段
    })
  },
  picfunction:function() {
    var that = this
    wx.chooseImage({                               //选择图片
      count: 1,
      sizeType: ['compressed'],
      sourceType: ['album', 'camera'],
      success: function(res) {
        console.log(res)
        wx.showLoading({
          title: '上传中',
        })
        const filePath = res.tempFilePaths[0]
        var timestamp = (new Date()).valueOf();
        wx.cloud.uploadFile({
          cloudPath: "img/" + timestamp + ".jpg",//上传至云端的路径
          filePath: filePath,                      //小程序临时文件路径
          success: res => {
            console.log('[上传文件] 成功：', res)
            that.setData({
              img:res.fileID                       //待插入的文章 img 字段
            })
          },
          fail: e => {
            console.error('[上传文件] 失败：', e)
            wx.showToast({
              icon: 'none',
              title: '上传失败',
            })
          },
          complete: () => {
```

```
          wx.hideLoading()
        }
      })
    },
    fail: e => {
      console.error(e)
    }
  })
},
submit:function(e){
  var that = this
  var newDate = new Date();
  that.setData({
    cTime: newDate.getFullYear() + "-" + newDate.getMonth() + "-" + newDate.getDay()
  })
  const db = wx.cloud.database()
  db.collection('ch12').add({                    //插入新闻信息
    data: {
      title: that.data.title,
      content: that.data.content,
      newsid: that.data.newsid,
      img: that.data.img,
      cTime: that.data.cTime
    },
    success: function(res) {
      //res 是一个对象,其中有 _id 字段标记刚创建的记录的 id
      console.log("插入成功" + res)
    },
    fail: console.error
  })
}
})
```

【代码讲解】 add.js 中的 title、content 和 radioChange 函数负责获取用户在 add.wxml 输入的 title、content 和 newsid 字段的值；picfunction 函数调用 wx.chooseImage() 接口上传本地图片到云存储空间；submit 函数实现自动产生一个 cTime 字段，并把 title、content、newsid、img 和 cTime 等字段插入云数据库中。

pages/add/add.wxss 的代码如下：

```
.wrapper {
  padding: 5px;
}
.contentinfo{
  font-size: 1rem;
}
.iconfont {
  display: inline-block; padding: 8px 8px;
  width: 24px; height: 24px;
  cursor: pointer; font-size: 20px;
}
```

```
.toolbar {
  box-sizing: border-box; border-bottom: 0;
  font-family: 'Helvetica Neue', 'Helvetica', 'Arial', sans-serif;
}
.ql-container {
  box-sizing: border-box; padding: 12px 15px;
  width: 100%; min-height: 30vh; height: auto;
  background: #fff; margin-top: 20px;
  font-size: 16px; line-height: 1.5;
}
.ql-active {
  color: #06c;
}
.in{
  background-color: #fff;
}
input{
  border:1px solid #00F;
  font-size: 1rem; width: 100%;
}
textarea{
  border:1px solid #00F;
  font-size: 1rem; width: 100%;
}
image{
  height: 60rpx; width: 60rpx; margin-left: 30rpx;
}
.pic{
  display: flex; flex-direction: row;
}
```

pages/setting/setting.wxml 的代码如下：

视频讲解

```
<view class="head"></view>
<view class="element">
   <view class="content">编辑资料</view>
   <view class="right">
     <text>></text>
   </view>
</view>
<view class="part"></view>
<view class="element">
   <view class="content">账号和绑定设置</view>
   <view class="right">
     <text>></text>
   </view>
</view>
<view class="part"></view>
<view class="element">
   <view class="content">字体大小</view>
```

```
    <view class = "right">
      <text >中 ></text >
    </view >
  </view >
    <view class = "part"></view >
  <view class = "element">
    <view class = "content">列表是否显示摘要</view >
    <view class = "right">
      <switch type = "switch" />
    </view >
  </view >
  <view class = "part"></view >
  <view class = "element">
    <view class = "content">非 WiFi 网络是否自动播放</view >
    <view class = "right">
      <switch type = "switch" />
    </view >
  </view >
    <view class = "part"></view >
    <view class = "element">
    <view class = "content">是否推送通知</view >
    <view class = "right">
      <switch type = "switch" />
    </view >
  </view >
  <view class = "part"></view >
  <view class = "element">
    <view class = "content">检查版本</view >
    <view class = "right">
      <text > 30.0.0 </text >
    </view >
  </view >
   <view class = "part"></view >
   <button class = "btn">退出登录</button >
```

pages/setting/setting.js 的代码如下：

```
Page({
  onLoad:function(){
    wx.setNavigationBarTitle({
      title: '我的设置',
    })
  }
})
```

pages/setting/setting.wxss 的代码如下：

```
.head{
  width:100%; height: 15px;
  background-color: #F4F5F6;
}
.element{
```

```
  display: flex; flex-direction: row;
}
.content{
  padding-top:15px; padding-bottom: 15px;
  padding-left: 15px; font-size: 15px;
}
.right{
  font-size: 15px; position: absolute;
  right: 10px;  height: 50px;
  line-height: 50px; color: #888888;
}
.part{
  border: 1px solid #cccccc; opacity: 0.2;
}
.btn{
  margin: 10px; color: #D53E37;
}
```

12.5　项目小结

通过本新闻小程序项目的实训主要介绍了以下知识点和操作：

- 云开发小程序项目的创建步骤；
- 云存储的操作和权限设置；
- 云数据库的操作和权限设置；
- app.json 文件中 tabBar 菜单和导航栏标题的设置方法；
- 页面布局和样式设计的基本方法；
- 使用双大括号{{}}实现 WXML 和 JS 文件的数据绑定操作：
- 使用 setData()重置动态数据的方法；
- wx:for 条件渲染和 template 模板的使用；
- 函数的定义；
- 云数据库的普通查询、条件查询和插入操作；
- 云存储的代码操作；
- wx.getUserInfo()和 wx.chooseImage()等接口的使用；
- textarea、input、radio-group 和 button 等组件的使用；
- 页面导航跳转用法；
- WXML 和 JS 之间，WXML 和 WXML 之间的数据传递；
- 云开发控制台的使用，项目预览和真机调试的步骤。

图书资源支持

感谢您一直以来对清华版图书的支持和爱护。为了配合本书的使用，本书提供配套的资源，有需求的读者请扫描下方的"书圈"微信公众号二维码，在图书专区下载，也可以拨打电话或发送电子邮件咨询。

如果您在使用本书的过程中遇到了什么问题，或者有相关图书出版计划，也请您发邮件告诉我们，以便我们更好地为您服务。

我们的联系方式：

地　　址：北京市海淀区双清路学研大厦 A 座 701

邮　　编：100084

电　　话：010-83470236　010-83470237

资源下载：http://www.tup.com.cn

客服邮箱：tupjsj@vip.163.com

QQ：2301891038（请写明您的单位和姓名）

资源下载、样书申请

书圈

扫一扫，获取最新目录

课程直播

用微信扫一扫右边的二维码，即可关注清华大学出版社公众号"书圈"。